Fortschritte der Chemie organischer Naturstoffe

Progress in the Chemistry of Organic Natural Products

43

Founded by L. Zechmeister
Edited by W. Herz, H. Grisebach, G. W. Kirby

Authors:
J. L. Ingham, A. Koskinen,
M. Lounasmaa

Springer-Verlag
Wien New York 1983

Dr. W. HERZ, Professor of Chemistry, Department of Chemistry,
The Florida State University, Tallahassee, Florida, U.S.A.

Prof. Dr. H. GRISEBACH, Biologisches Institut II, Lehrstuhl für Biochemie der Pflanzen,
Albert-Ludwigs-Universität, Freiburg i. Br., Federal Republic of Germany

G. W. KIRBY, Sc. D., Regius Professor of Chemistry, Chemistry Department,
The University, Glasgow, Scotland

ISSN 0071-7886

ISBN-13: 978-3-7091-8705-0   e-ISBN-13: 978-3-7091-8703-6
DOI: 10.1007/978-3-7091-8703-6

Contents

List of Contributors

INGHAM, Dr. J. L., Department of Botany, Plant Science Laboratories, University of Reading, Whiteknights, Reading RG6 2AS, England.

KOSKINEN, Dr. A., Laboratory for Organic Chemistry, Department of Chemistry, Technical University of Helsinki, SF-02150 Espoo 15, Finland.

LOUNASMAA, Prof. Dr. M., Laboratory for Organic Chemistry, Department of Chemistry, Technical University of Helsinki, SF-02150 Espoo 15, Finland.

Naturally Occurring Isoflavonoids
(1855—1981)

By J. L. INGHAM, Department of Botany, University of Reading, Whiteknights, Reading, U.K.

Contents

I. Introduction

This review is mainly concerned with the structures, sources and biological properties of those natural products containing the 1,2-diphenyl-propane ring system (see Chart 1, and compare with the isomeric 1,3-diphenylpropane skeleton common to flavonoid compounds), and which collectively are known as *isoflavonoids*. Also briefly considered, however, are the glucosidic deoxybenzoin derivative, onospin (**630**), as well as several 2-arylbenzofurans (Table 2, Section R) of leguminous origin which lack a carbon atom corresponding to that marked (Chart 1) with an asterisk (*). These latter compounds and onospin are included because whilst they cannot be regarded as isoflavonoids, it is by no means uncommon to find that in the family Leguminosae they co-occur with (and in certain instances may originate from) isoflavonoids having similar aromatic substitution.

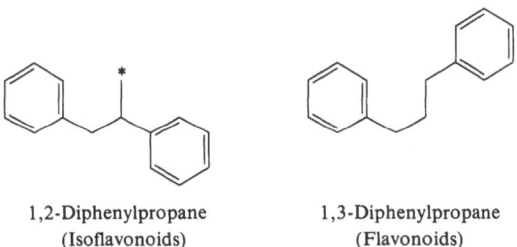

1,2-Diphenylpropane   1,3-Diphenylpropane
(Isoflavonoids)   (Flavonoids)

Chart 1. Basic structural units of isoflavonoid and flavonoid compounds

The branched C_6-C_3-C_6 skeleton illustrated in Chart 1 is characterised by its chemical flexibility, the many distinct isoflavonoid groups considered in Table 2 arising via comparatively simple stepwise oxidative and reductive modifications at one or more of several central, or near central, molecular sites. Importantly, the resulting aromatic/O-heterocyclic ring systems (all of which typically have an oxygen function at the position corresponding to C-7 of isoflavones; Table 2, Section A) seem in general to be highly amenable to further O- and C-substitution. Thus, despite their comparatively restricted distribution within the plant kingdom (see Section II), the isoflavonoids as a whole display a quite remarkable array of structures, well over 500 fully identified aglycones and glycosides having now been isolated from higher plants. All these compounds, and those produced by microbial modification of either natural or synthetic isoflavonoids, can be conveniently divided into two broad categories termed "simple" and "complex". Compounds of the latter type possess cyclic and/or acyclic isoprenoid (mainly C_5, but sometimes C_{10} or C_{15}) or isoprenoid-derived (furan/dihydrofuran) attachments in addition to any other substituents (e. g. OH/OCH_3 groups) which may be present, whereas those of the former do not. Most of the isoflavonoid groups and subgroups listed in Table 2 are, in fact, readily divisible on this basis.

Isoflavones are without doubt the most widespread and commonly encountered isoflavonoid compounds, their discovery dating from the mid-19th century when HLASIWETZ (344) reported the extraction of formononetin [5]-7-O-glucoside (ononin, 521) and its corresponding deoxybenzoin (onospin, 630) from roots of the thorny restharrow (Ononis spinosa; Leguminosae). Ononin is widely regarded as the very first isoflavonoid to be discovered in Nature but although HLASIWETZ (344) elegantly demonstrated its glucosidic nature and chemical relationship to onospin, he was unable to deduce the structure of either compound successfully, a task only accomplished in comparatively recent times (818, 842). Amongst other isoflavones known for many decades may be mentioned prunetin (84) and its 4'-O-glucoside, prunitrin (574), both obtained about 70 years ago by FINNEMORE (250) from the bark extract of an unidentified Prunus (Rosaceae) species, genistein (81), a weak mordant dye discovered at the beginning of this century in dyer's broom, Genista tinctoria (637, 638), and three glycosides, baptisin (538), pseudobaptisin (531) and iridin (609), all isolated, but only partially identified, prior to 1900 (198, 293, 294). Compounds (531) and (538) are found in roots of the shrubby legume, Baptisia tinctoria (293, 294), whilst iridin is a major constituent of commercial Florentine orris "root" (actually rhizomes of Iris florentina (Iridaceae) admixed in varying proportions with those of I. germanica and I. pallida) (36, 198), a commodity much prized — at least in former years — by the perfumery industry because of its highly desirable violet-like odour.

Apart from isoflavones, the early literature also contains a few scattered accounts of other compounds which, as the result of subsequent research, we now know to be isoflavonoids. For example, two optically active pterocarpans, homopterocarpin or baphinitone (280) and pterocarpin (284), were isolated in the latter half of the 19th century from heartwood extracts of Baphia nitida (280) and Pterocarpus santalinus (280 and 284) (18, 138). Shortly thereafter, GEOFFROY (277) successfully extracted rotenone (233), or nicouline as it was originally named, from Lonchocarpus (Robinia) *nicou*[1] although it was not until 1932 that the structure of this complex rotenoid was finally established following much patient effort by several quite independent research groups (*132, 503, 697*), formal priority for the identification now being generally attributed to LaForge and Haller (*503*). Whilst the formula assigned to rotenone has stood the test of time, that initially attributed (*548*) to pterocarpin (**284**) was revised in 1961 as the result of a ^{1}H NMR study which provided evidence to suggest that the methylenedioxy group occupied a linear (C-8/C-9) position instead of the angular (C-9/C-10) arrangement proposed by McGookin *et al.* (*548*). Even today, isoflavonoid compounds are still occasionally misidentified (see also Section II), examples of some recently revised structures including derrugenin (**122**) (*784*), cajanol (**208**) (*372*) and glyceollin I (**367**) (*126*).

The vast majority of isoflavonoids occur in Nature as "free" aglycones rather than combined with glucose (or some other mono- or di-saccharide) in the form of glycosides, this situation contrasting sharply with that observed amongst the ubiquitous flavonoid compounds where *O*-glycosylation is very frequent, and *C*-glycosylation by no means uncommon. Only within the simple 5-deoxy and 5-oxy isoflavone sub-groups can *O*-glycosides be regarded as prevalent (almost 100 examples have now been documented), and even here their occurrence tends to be rather erratic. However, it is probably true to say that in general isoflavone *O*-glycosides are most likely to be encountered in leaves, roots and flowers, especially those of species belonging to essentially herbaceous legume genera (e. g. *Baptisia, Lupinus Thermopsis* and *Trifolium*) indigenous to temperate regions of the world; seeds, bark and wood are alternative, but apparently much less important, sources. Outside the Leguminosae, at least eight isoflavone *O*-glycosides have been extracted from *Iris* and *Belamcanda* (Iridaceae) rhizomes, whilst the wood and bark of some *Prunus* (Rosaceae) species have furnished genistin (**551**) and prunitrin (**574**). Sissotrin (**566**), the 4'-*O*-methyl ether of genistin, occurs in *Cotoneaster* (another isoflavone-producing member of the Rosaceae), and iridin (**609**) hitherto known only from the Iridaceae was recently discovered in leaves of the Gymnosperm, *Juniperus macropoda* (Cupressaceae) (*722*).

Isoflavone *C*-glycosides are rare when compared with their *O*-glycosylated counterparts and only 12 examples have been reported. Interestingly, however, they are all variously found in species from the tropical/subtropical legume genera, *Pueraria* and *Dalbergia*, although two compounds occurring in *D. nitidula* (**563** and **579**) are likewise found in

---

[1] "Derrid", a substance obtained slightly earlier (1890 – 1893) by Greshoff from species of *Derris, Mundulea* and *Ormocarpum*, and "timboin" isolated in 1891 by Pfaff from a plant named as *Paullinia pinnata* (a member of the family Sapindaceae) but probably identical with *Lonchocarpus urucu*, are both thought to have contained impure rotenone (see *410*).

tissues of the temperate yellow lupin *(Lupinus luteus)*. *Dalbergia* in particular is rich in isoflavone C-glycosides and additionally has provided chemists with dalpanin (**617**), currently the sole C-glycosylated isofla-vanone. As can be seen from Table 3 (Sections B – E), a handful of other isoflavonoid (isoflavanone, rotenoid, pterocarpan and coumestan) O-glycosides have been described, mainly from north temperate Leguminosae. All are glucosides with the exception of amorphin (**619**) and amorphol (**622**) which contain the unusual disaccharide, vicianose (arabinosylglucose). Rotenoid glycosides have mostly been obtained from seeds — principally those of *Amorpha* species — whereas pterocarpan glycosides appear to be predominantly root constituents.

As with some flavonoids, a number of isoflavonoid glycosides possess an acyl substituent attached to the sugar (glucose in all cases) moiety. Isoflavone glycosides substituted in this fashion are 4',6''-di-O-acetylpuerarin (**519**) from *Pueraria tuberosa* (**67**), the 6''-O-acetyl de-rivatives of daidzin and genistin [compounds (**512**) and (**552**) respectively] found in seeds of the soybean *(Glycine max)* *(602, 603)*, and 6''-O-malonylononin (**527**) and 6''-O-malonylgenistin (**561**) which occur in leaves of the Yarloop strain of subterranean clover *(Trifolium subterraneum)*. Isoflavone (**527**) has also been found in foliage of the related fodder legume, red clover *(T. pratense)* *(49)* and may possibly occur in certain other *Trifolium* species *(263)*. Apart from compounds (**527** and **561**), the leaves of *T. subterraneum* and *T. pratense* contain appreciable amounts of a third acylated isoflavone, namely 6''-O-malonylsissotrin (**572**). This substance, first reported in 1969 following a detailed chemical examination of the constituents in *T. pratense* foliage, was initially formulated as sissotrin-5-malonate *(780)*, a structure later revised to (**572**) by BECK and KNOX *(49)*. There is evidence to suggest that the 6''-malonate, methyl esters (**528** and **573**) of (**527** and **572**) respectively, are also present in *T. subterraneum*, where they are probably accompanied by traces of the corresponding — but as yet only provisionally identified — genistin analogue (**562**) *(49)*. More recent additions to the steadily growing list of acylated isoflavonoids are the pterocarpan glucoside, 6'-O-acetyltrifolirhizin (**627**) *(473)*, and the iso-flavone aglycones corylinal (**10**) and glyzarin (**23**) where the acyl groups [formyl in the case of (**10**), and acetyl in the case of (**23**)] are attached directly to the aromatic ring system *(62, 310)*.

II. Distribution

Well over 90% of the fully characterised isoflavonoids (aglycones + glycosides) of plant origin are known to be produced (but not always exclusively) by species belonging to the large (about 480 genera; 12000 species) and taxonomically-advanced subfamily Papilionoideae of the Leguminosae. Elsewhere in the plant kingdom — even within the two other, admittedly much smaller and more primitive, leguminous subfamilies Caesalpinioideae (about 150 genera; 2800 species) and Mimosoideae (about 56 genera; 2800 species) — isoflavonoids are generally scarce although such compounds do appear to be locally abundant particularly in parts of the Iridaceae, Myristicaceae and Rosaceae. In contrast, however, investigation of Papilionoideae whether trees, shrubs or tiny short-lived herbs, and quite regardless of their geographic origin, has revealed the occurrence of isoflavonoids in a host of organs and tissues ranging from bark and non-living heartwood (a potentially rich source of isoflavonoid structures) to roots, seeds, hypocotyls, cotyledons, stems, leaves and even delicate flower petals.

Broadly speaking, plant isoflavonoids fall into two classes depending on their mode, or apparent mode, of formation. They may, for example, occur as components of normal, healthy tissues when the term "constitutive" (or more rarely "autonomous") isoflavonoid is applicable, but increasingly compounds are being discovered which accumulate *de novo* as a consequence of invasion and attempted (or actual) colonisation of a plant organ or tissue by potentially pathogenic micro-organisms, most often fungi and bacteria. Use of the term *de novo* implies (*770*) that the compounds concerned are formed directly from elementary building blocks such as acetate and shikimate, and not — as in the case of substances referred to as post-inhibitins (*358*) — by a simple hydrolytic or oxidative process utilising pre-formed plant constituents of relatively sophisticated molecular structure. Induced synthesis of enzymes catalyzing the necessary biosynthetic steps is considered to be an essential prerequisite for *de novo* production of isoflavonoid compounds.

Whilst these so-called "induced" isoflavonoids are frequently unique to microbially invaded tissues, it is not unknown for the same compounds (or perhaps their enantiomers) to occur constitutively in other legumes [e. g. medicarpin (**276**) and vestitol (**397**)] or, on rare occasions, in other parts of the same plant. Thus, maackiain (**279**) can be induced to form in leaves of *Sophora japonica* (*356*) whereas it occurs constitutively in the roots of this papilionoid tree (*469, 741*). In fact, inducible isoflavonoids are commonly encountered in the Papilionoideae having been isolated from the vast majority of species subjected, usually under strict laboratory conditions, to either microbial inoculation or treatment with appropriate non-living (abiotic) stimuli such as short wavelength (254 nm) UV light or one of countless phytotoxic chemicals. If isoflavonoids formed in this manner have antimicrobial properties they are usually called phytoalexins, a word derived from the Greek "phyton" and "alexein", and meaning literally "a plant warding-off compound" since their primary purpose seems to lie in limiting tissue colonisation by uncongenial micro-organisms. Phytoalexins are, in fact, a special subgroup of those compounds collectively named "stress metabolites"; this term is applicable to any substance (irrespective of its antimicrobial activity) which accumulates in the tissues of a plant following exposure to, or interaction with, harmful living or abiotic agents. The pathogen-induced coumestans (**453, 454, 456, 460, 463, 464**) found in *Medicago sativa* and *Trifolium repens* are stress metabolites but not phytoalexins because they apparently lack antifungal or antibacterial properties. A great many reviews dealing with phytoalexins in general or with isoflavonoid phytoalexins in particular have now been published (*32, 188, 306, 320, 357, 379, 380, 693, 770, 799*) and the reader is referred to these for more detailed biological information.

Over the past 20 years, phytoalexins have been obtained from many diverse plant groups (Gymnosperms and Angiosperms; monocotyledons and dicotyledons), the compounds associated with one family frequently differing profoundly from those of another in terms of both chemical structure and biosynthetic origin (*320*). It so happens, however, that species comprising the Papilionoideae typically accumulate one or two — sometimes several — isoflavonoid phytoalexins as an active defence response to microbial invasion; indeed, investigation of many hundreds of papilionate legumes (*356*) has clearly confirmed their overwhelming ability to produce isoflavonoid phytoalexins, the range of induced non-isoflavonoid fungitoxins being almost wholly limited to three benzofurans (compounds **500, 501** and **505**), seven furanoacetylenes found exclusively in species of *Vicia* and *Lens*, four stilbenes from *Arachis hypogaea*, and two unique ethylated chromones recently obtained from *Lathyrus odoratus* (*380, 700, 701, 703*).

All the isoflavonoid phytoalexins thus far described accumulate as aglycones rather than as glycosides, the most regularly encountered compounds being phenolic pterocarpans and isoflavans. These substances, which often occur together in legumes, have been recognised as phytoalexins in most tribes of the Papilionoideae although they are as yet poorly represented in, or absent from, groups such as the temperate Genisteae and Old World Desmodieae (*356*). In temperate Vicieae, the pterocarpan-producing genera *Pisum* and *Lathyrus* exist alongside *Vicia* and *Lens* where formation of non-isoflavonoid furanoacetylene phytoalexins is the norm (*701*). Even in *Lens* and *Vicia*, however, there is still a limited ability to produce fungitoxic pterocarpans; thus, medicarpin (**276**), unquestionably the most widespread of all isoflavonoid phytoalexins, has been detected (albeit in very small amounts) in broad bean (*Vicia faba*) tissues (*322, 323*) whilst the non-phenolic pterocarpan, variabilin (**354**) is found in both *Lens culinaris* (lentil) and *L. nigricans* (*700*).

In contrast to pterocarpans and isoflavans, recent surveys (*356*) have shown that isoflavone and isoflavanone phytoalexins are limited both in terms of their distribution within the Papilionoideae and in the number of compounds described. This is an interesting observation because, as will be apparent from Table 2, the isoflavone group contains by far the largest number of constitutive representatives. At present, isoflavone/isoflavanone phytoalexins are known to be comparatively plentiful only in tribes Genisteae and Desmodieae (where they presumably replace isoflavans and pterocarpans as defence compounds), and the pantropical Phaseoleae (most notably subtribes Cajaninae, Diocleinae and Clitoriinae) although elsewhere in this tribe, e. g. *Phaseolus* and *Lablab* (subtribe Phaseolinae) they are sometimes of secondary importance when compared with concurrently produced pterocarpans and isoflavans. Lastly, a few inducible coumestans have been described but apart from coumestrol (**452**), psoralidin (**472**) and possibly sojagol (**471**), they do not appear to function as phytoalexins (see above).

From studies carried out in these laboratories, it is becoming increasingly clear that tribes and genera of the Papilionoideae differ, often strikingly so, in terms of their phytoalexin chemistry. Of course, some chemical variation between distantly related and geographically separated legume groups is not unexpected since it is reasonable to assume that plants will evolve a defensive strategy peculiar to those potentially pathogenic micro-organisms found specifically in their environment. However, considerable variability in phytoalexin response is also evident amongst closely allied genera and tribes, and though such differences may not be absolute, they are frequently sufficient to be of some taxonomic value.

The utility of this phytoalexin approach to the study of systematic relationships within the Leguminosae is best illustrated by reference to four genera *(Medicago, Melilotus, Trifolium* and *Trigonella)* of the mainly Eurasian tribe Trifolieae. Here it has been shown that leaves of *Melilotus* species produce only medicarpin (**276**) following inoculation with the fungus *Helminthosporium carbonum* (*366*) whereas *Medicago* reveals a much more complex accumulation pattern (*371*) with (**276**) being regularly accompanied by isoflavan derivatives, principally vestitol (**397**) and sativan (**400**). The response of *Trifolium* is even more diverse, and in addition to compounds (**276, 397** and **400**), a range of other pterocarpans [e. g. maackiain, (**279**)] and isoflavans [e. g. isovestitol (**398**) and arvensan (**402**)] have been associated with this genus (*369*). Significantly, species of the fourth genus, *Trigonella* — noteworthy because of its intermediate position between *Melilotus* and *Medicago* — are divisible into three major groups on the basis of phytoalexin accumulation (*383, 390*). Two of them respectively link *Trigonella* to *Melilotus* on the one hand and to *Medicago* on the other, whilst the third [characterised by formation of maackiain, (**279**)] provides evidence for a connection to *Trifolium*. It is satisfying to note that the phytoalexin data support, to a very large degree, the results and conclusions of earlier morphological and anatomical investigations concerned particularly with the relationship between *Medicago* and *Trigonella* (*379*).

As well as the above mentioned work, phytoalexin induction has also been successfully employed to demonstrate generic similarities and differences within the Vicieae (*701*), and to provide supporting evidence for the view now held by plant taxonomists that *Cicer* should be excluded from this legume tribe (*382*). At the specific level, phytoalexin studies have clearly shown that the isoflavone-accumulator, *Neonotonia wightii* (formerly known as *Glycine wightii*) is chemically distinct from other *Glycine* species which typically produce 6a-hydroxypterocarpans in response to stress conditions (*391, 439, 527 – 529*), and that the grass-like *Lathyrus nissolia* is, by virtue of its ability to form the novel pterocarpans nissolin (**281**) and methylnissolin (**285**), correctly placed in an outlying section *(Nissolia)* of *Lathyrus* (*702*).

Few general statements can be made about the distribution of constitutive isoflavonoids in the subfamily Papilionoideae although there is a tendency for complex compounds — like some of their inducible counterparts — to be most prevalent in genera from the tropics and subtropics. This statement is particularly applicable to rotenoids (Table 2, Sections E and F) for apart from their appearance in the temperate North American taxon *Amorpha*, they are essentially restricted to genera (e. g. *Crotalaria, Derris, Lonchocarpus, Millettia, Mundulea, Pachyrrhizus, Piscidia* and *Tephrosia*) found in warmer regions of the world. Many rotenoids are toxic to insects (see Section III), and it is possible that their presence in seeds, roots and other tissues may protect plants which synthesise them from the predacious or burrowing insects which abound in tropical/subtropical areas. Another point worth mentioning is that most complex 6aH-pterocarpans and all the known 6a-hydroxy analogues (both constitutive and induced) have been obtained only from genera belonging to three papilionoid tribes, namely the Phaseoleae (in particular subtribes Erythrininae, Glycininae and Phaseolinae), and its close botanic relatives the Desmodieae *(Desmodium* and *Lespedeza)* and Psoraleeae *(Psoralea)*. The small African genus *Neorautanenia* (Phaseoleae, subtribe Phaseolinae) has proved to be extremely rich in complex isoflavonoids providing amongst others a quite remarkable collection of novel pterocarpan structures.

Unlike complex compounds, simple constitutive isoflavonoids largely transgress climatic boundaries appearing free or as glycosides (few complex isoflavonoids are glycosylated) in an immense variety of temperate, tropical and subtropical Leguminosae. Some simple isoflavonoids [e. g. daidzein (**3**), formononetin (**5**), genistein (**81**), medicarpin (**276**) and maackiain (**279**)] are very widely distributed but many, as Tables 2 and 3 will immediately reveal, are at best known from only a handful of species. None of these compounds can be considered as absolutely characteristic of one tribe or genus but some are locally common, for example 5-*O*-methylgenistein (**85**) in tribe Genisteae, and afrormosin (**28**) in genera *(Amphimas, Castanospermum, Cladrastis, Myrocarpus, Myroxylon* and *Pericopsis)* of the Sophoreae. Homopterocarpin (**280**) and pterocarpin (**284**) are both prevalent in genus *Pterocarpus,* whilst caviunin (**130**)/isocaviunin (**132**) and their glycosides appear exclusive to *Dalbergia.*

Lastly, examination of several genera *(Aldina, Cordyla, Mild-braedeodendron,* and *Swartzia)* comprising the primitive Swartzieae has revealed a considerable number of simple isoflavonoids as heartwood constituents, the presence of which upholds the current view that, as George Bentham believed, this rather controversial tribe is indeed of papilionoid stock. Many authorities have tended to assign Swartzieae to the legume subfamily Caesalpinioideae where isoflavonoids are almost unknown, a decision now effectively negated by both chemical and non-chemical data (*179*).

Although it is now beyond dispute that isoflavonoids are characteristic constituents of papilionaceous legumes, there is no evidence whatsoever to suggest that such compounds are correspondingly widespread in genera of the associated subfamilies Caesalpinioideae and Mimosoideae. Thus, only *Apuleia leiocarpa* (*98*) of the Caesalpinioideae, and *Albizia procera* (*210*) and *Prosopsis juliflora* (*747*) of the Mimosoideae have been reliably recognised as isoflavonoid producers. Additionally, however, some evidence has been obtained to suggest that rotenone and/or other rotenoid compounds may be present in the leaves of *Entada africana* (a member of the same mimosoid tribe [Mimoseae] as *Prosopsis)* but unfortunately this claim (*275*) lacks satisfactory experimental verification. In 1950, GAKHOKIDZE (*272*) reported the isolation of a substance (termed "olmelin") apparently identical with biochanin A (**83**) from pods of the caesalpinioid tree, *Gleditsia triacanthos* (honey locust) but considerable doubt now surrounds this identification. For example, "olmelin" crystallised as cherry-red plates, mp 287 – 291°, properties quite unlike those of authentic biochanin A which forms faintly yellow needles, mp 210° (*174*). It is just conceivable that "olmelin" may consist of biochanin A admixed with large quantities of a red contaminating pigment, and careful re-examination of the constituents in *G. triacanthos* pods might be a worthwhile exercise. It should be mentioned, however, that several attempts to detect isoflavonoids (including biochanin A) in fungus-inoculated hypocotyls of *G. triacanthos* have been completely unsuccessful (*356*).

Outside the Leguminosae, isoflavonoid formation is a rare phenomenon with compounds (often isoflavones, but occasionally rotenoids, pterocarpans and coumestans) occurring unpredictably in at least thirteen other Angiosperm (monocotyledon + dicotyledon) and two Gymnosperm families. Dicotyledon producers include such diverse groups as the Chenopodiaceae *(Beta)*, Compositae *(Eclipta, Wedelia* and *Wyethia)*, Menispermaceae *(Tinospora cordifolia; 551)*, Myristicaceae *(Osteophleum* and *Virola)*, Rosaceae *(Cotoneaster* and *Prunus)*, Scrophulariaceae *(Verbascum)* and Stemonaceae *(Stemona)*, but amongst the monocots, isoflavonoids are known with certainty only from the Iridaceae where, as aglycones or glycosides, they accumulate in bulbs/rhizomes/roots of several *Iris* species, and the blackberry lily, *Belamcanda (Pardanthus) chinensis.* The claim by SCARONE (see *410)* that Spanish moss *(Tillandsia usneoides;* Bromeliaceae) contains rotenone **(233)** has yet to be substantiated. Isoflavonoid records in the Gymnospermae are at present restricted to *Juniperus macropoda* (Cupressaceae) *(722, 723)* and *Podocarpus spicatus* (Podocarpaceae) *(113, 115)*.

Of the non-leguminous isoflavonoids, betavulgarin **(100)** produced by the fungus-infected leaves of sugar beet *(Beta vulgaris;* Chenopodiaceae) *(276)* and other *Beta* species *(692)* is notable in two respects. First, it is the sole isoflavonoid phytoalexin to be characterised outside the Papilionoideae, and secondly, like tlatlancuayin **(110)** in *Iresine celosoides* (Amaranthaceae), it has an extraordinary B-ring oxygenation pattern (2'-oxy; 4'-deoxy). Interestingly, though, isoflavones **(100** and **110)** are both associated with families belonging to the Order Centrospermae (Caryophyllales), a plant group also unusual in that nitrogen-containing betalain pigments mostly replace the customary flavonoidal anthocyanins found elsewhere in the Angiospermae. A somewhat similar 4'-deoxy-isoflavone, podospicatin **(115)**, occurs together with genistein **(81)** in the heartwood of *Podocarpus spicatus*, a taxoid conifer native to New Zealand *(113, 115)*.

The few reports of isoflavonoid production by micro-organisms *(274, 348)* are almost certainly spurious, merely reflecting modification of isoflavones present as contaminants in the growth medium rather than a genuine biosynthetic capability of the organism concerned. Similarly, isoflavone-like material detected in butterfly wings is presumably derived from appropriate food plants via feeding caterpillars *(246a)*.

In the years since FINNEMORE *(250)* determined the structure of prunetin **(84)**, a number of isoflavonoid compounds have been misidentified. Often, errors have simply involved the location of a substituent at an incorrect site on one or other of the aromatic rings, but from time to time investigators have assigned new structures and common names to compounds already reported in the literature. For example, some three decades ago NARASIMHACHARI and SESHADRI *(580)* described the extraction of prunetin **(84)** and two isoflavanones, padmakastein (considered to be 5,4'-dihydroxy-7-methoxyisoflavanone or dihydroprunetin) and its 4'-O-glucoside (padmakastin) from the bark of *Prunus puddum* (Rosaceae), the re-isolation of padmakastein (but not of padmakastin) being claimed in 1958 (*671*). Later, however, FARKAS and his colleagues (*241*) successfully synthesised both 5,4′-dihydroxy-7-methoxyisoflavanone (mp 146 – 147°) and the corresponding 4′-*O*-glucoside (mp 115 – 117°) and found that their melting points were very different from those of the *Prunus*-derived isoflavanones. In the absence of a direct comparison between natural and synthetic material these discrepancies are not easily resolved although it seems likely — as FARKAS *et al.* (*241*) suggest — that padmakastein (mp 236 – 240°; *580, 671*) is slightly impure prunetin (mp 240 – 241°; *579*), and that padmakastin (mp 225 – 227°; *580*) is merely prunitrin (**574**) (prunetin-4′-*O*-glucoside, mp 235 – 236°; *241*), itself a well-known constituent of *Prunus* bark (*250, 646*).

Another case of isoflavonoid misidentification involves four isoflavones which OKANO and BEPPU (*605*) apparently discovered in soybean (*Glycine max;* Leguminosae) seed meal, a recognised source of the glucosides daidzin (**511**) (daidzein-7-*O*-glucoside) and genistin (**551**) (genistein-7-*O*-glucoside) (*812*). These compounds were thought to be the aglycones, 5,4′-dihydroxy-8-methylisoflavone (named tatoin; mp 318°) and 5,7,4′-trihydroxy-8-methylisoflavone (methylgenistein, mp 298°), and the glucosides, isogenistin and methylisogenistin. Upon acid hydrolysis, isogenistin reportedly gave 5,7,2′-trihydroxyisoflavone[2] (isogenistein, mp 302°) whilst methylisogenistin seemed to yield 5,7,2′-trihydroxy-8-methylisoflavone (methylisogenistein, mp 301 – 302°) (*605*). BHANDARI *et al.* (*57*) also described the extraction of tatoin from soybean seed germ, and a synthetic route to methylgenistein (mp of synthetic product 296 – 300°) was devised at about the same time (*746*) although its validity was subsequently questioned by WHALLEY (*819*). In fact, the results of other syntheses have shown convincingly that the melting points of 5,4′-dihydroxy-8-methylisoflavone (mp 180°; *820*), 5,7,4′-trihydroxy-8-methylisoflavone (mp 252°; *819*), 5,7,2′-trihydroxyisoflavone (dimorphic crystals, mp 187° and 222 – 223°; *37*), and 5,7,2′-trimethoxy-8-methylisoflavone[3] (mp 185°; *819: cf.* methylisogenistein trimethyl ether, mp 152 – 153°; *605*) are significantly different from those of the compounds studied by OKANO and BEPPU (*605*). Attempted reisolation of the soybean isoflavones proved unsuccessful (the only compounds detected were daidzein and genistein; *11*), and it is now virtually certain that *Glycine*-derived methylgenistein, isogenistein and

---

[2] It is interesting to note that the 7-*O*-glucoside of 5,7,2′-trihydroxyisoflavone (**565**) was recently isolated from root bark of *Cajanus cajan* (*59*), a close taxonomic relative of *G. max*.

[3] All efforts to produce 5,7,2′-trihydroxy-8-methylisoflavone by demethylation of the corresponding trimethyl ether resulted in failure (*819*).

methylisogenistein are actually genistein (note its comparable mp of 300 – 301°; *88*) samples varying in their degree of purity, and that tatoin is principally daidzein (mp 325°; *88*) contaminated by traces of genistein (*11, 37*).

Finally, two further examples are worthy of mention. First, prunusetin (an extractive from the bark of *Prunus puddum*) was initially formulated as 5-methoxy-7,4′-dihydroxyisoflavone (*139*) but later found to be prunetin (*454, 579, 581*). Secondly, MERZ and SCHMIDT (*555*) claimed that in addition to dehydrodeguelin (**265**), the seeds of *Tephrosia vogelii* contained two hitherto unrecognised rotenoid compounds for which the names iso-deguelin and allotephrosin were proposed. In reality, these are identical with deguelin (**232**) and tephrosin (**251**) respectively (*81*), both having been discovered some years earlier in leaves of *T. vogelii* (*131, 168*).

III. Biological Activity and Uses

Apart from their preponderance in Leguminosae, the isoflavonoids are also remarkable in that they display an exceedingly diverse range of biological properties, some of which are particularly characteristic of certain compounds. Thus rotenone and many related rotenoid derivatives are highly effective fish poisons, an attribute not generally associated with other types of isoflavonoid although exceptions [e. g. luteone (**145**); (*269*)] have occasionally been noted. In fact, rotenoid-containing plants belonging to genera such as *Derris, Lonchocarpus, Piscidia* and *Tephrosia* have long been used as a fishing aid by tribal groups in parts of South America, tropical Africa and Asia (principally Indonesia and Malaysia). Here, roots or other plant parts are first crushed to release their active (rotenoid) constituents and are then simply thrown into ponds, pools and slow-running streams. The liberated rotenoids (particularly rotenone) rapidly stupefy any fish in the vicinity which float to the surface from where they can easily be collected. As the oral toxicity of rotenone to most warm-blooded animals is very low[4], any traces in fish for human consumption present a negligible health hazard.

Several 12aH- (e. g. rotenone) and 12a-hydroxyrotenoids [e. g. teph-rosin, (**251**)] are potent insecticides whereas dehydrorotenoids are generally

---

[4] According to the "Merck Index" (9th edition), the oral LD_{50} of rotenone to rats is 133 mg/kg body weight. This contrasts sharply with its intraperitoneal LD_{50} value of only 5 mg/kg.

inactive or only weakly active. Tests undertaken over many years suggest that rotenone — the chief constituent of commercially available *Derris* (tuba root) and *Lonchocarpus* (variously termed barbasco, cubé, haiari, nekoe and timbo) species — is by far the most insecticidal member of the group, its major physiological effect [and that of other rotenoids, e. g. toxicarol (**236**) and sumatrol (**237**)] being to depress cellular respiration by inhibiting a portion of the mitochondrial electron transport system (*267a, 516, 520*). Although rotenone-based insecticides have now been replaced for many purposes by wholly synthetic substitutes, *Derris* preparations (dusts or sprays) are still recommended for use on cane fruits against raspberry beetle (a serious pest of raspberries, loganberries and blackberries), and on some vegetable (mainly *Brassica*) crops to control blackfly, greenfly and caterpillars in general. An advantage of *Derris* formulations is that because of their low mammalian toxicity it is possible to use them safely at or near harvest. The relative insensitivity of mammals to rotenone when compared with fish and many insects presumably reflects either poor absorption into the bloodstream or more effective enzymic detoxification of the molecule *in vivo* since respiration in isolated animal systems (mouse liver slices/rat liver mitochondria) is also highly sensitive to the harmful effects of this complex isoflavonoid (*268, 516*).

A considerable number of pterocarpans and isoflavans as well as certain isoflavones and isoflavanones — especially those with isoprenoid substituents — are extremely toxic to saprophytic and plant pathogenic fungi (*187, 256, 269, 640, 642, 648, 688*) causing inhibition of spore germination, germ tube elongation or hyphal growth (*338, 756*). Effects on hyphae exposed to isoflavonoids such as kievitone (**213**) and phaseollin (**309**) include rapid cessation of cytoplasmic streaming, development of granules and vacuoles in the cytoplasm, shrinkage of the protoplast from hyphal tips (the growing point of the fungus) and occasionally swelling and bursting of apical hyphal cells (*32, 759, 798*). Amongst associated physiological changes may be mentioned uncontrolled leakage of metabolites and electrolytes from the mycelium resulting in a dry weight loss, and marked effects on both exogenous and endogenous respiration (*32, 798*). All these abnormalities imply that exposure of fungi to selected isoflavonoids brings about severe (and often irreversible) damage to membrane systems. As mentioned earlier (Section II), a great many Leguminosae, and a few *Beta* (Chenopodiaceae) species, typically produce isoflavonoid phytoalexins as a defence against fungal invasion. When considered in the light of their fungitoxic properties, the rate at which phytoalexins accumulate in affected tissues is entirely consistent with the now widely held belief that these inducible compounds play a major role in preventing successful fungal colonisation of both leguminous and non-leguminous species. Isoflavonoids found constitutively in heartwood, leaves, roots, seeds and other plant parts may also inhibit the growth and development of attacking micro-organisms (129, 321) and, in some cases at least, appear to act as effective deterrents to feeding insect larvae (707, 707a, 775).

Apart from their action on fungi, it has been observed that a few isoflavonoids are toxic to yeast cells (187, 419) whilst others are anti-bacterial (187, 287, 419, 530, 622, 816, 836) or nematistatic (421, 691). Recent studies have provided evidence to suggest that Gram-positive bacteria are somewhat more sensitive to isoflavonoids than are Gram-negative organisms (287). Knowledge that certain zoopathogens are vulnerable to the effects of isoflavonoids (269, 292, 419, 561) suggests that in future years these compounds (or synthetic analogues with considerably enhanced activity) may perhaps be employed as chemotherapeutic agents. The use of isoflavonoid phytoalexins as plant protectant fungicides has also been investigated but with disappointing results (686). Even at an applied spray concentration of 100 µg/ml, none of the compounds tested [kievitone (213), medicarpin (276), maackiain (279), phaseollin (309), vestitol (397), sativan (400) and phaseollinisoflavan (418)] gave satisfactory control of fungus diseases such as brown rust of wheat (caused by *Puccinia recondita*), downy mildew of vine *(Plasmopara viticola)*, peanut leaf spot *(Cercospora arachidicola)* and tomato leaf blight *(Phytophthora infestans)* whereas two commercially available fungicides, benomyl (a systemic protectant) and the dithiocarbamate mancozeb (a non-systemic protectant), proved to be highly effective against all four fungal pathogens at applied levels of 50 µg/ml. Spray trials with synthetic analogues of the 2-arylbenzofuran phytoalexin, vignafuran (501), have, however, given slightly more encouraging results, particularly against the broad bean rust fungus, *Uromyces viciae-fabae* (137).

As well as effects on micro-organisms, at least four pterocarpans — medicarpin (276), phaseollin (309), pisatin (356) and glyceollin I (367) — are demonstrably toxic to isolated animal systems causing rapid haemolysis (i.e. swelling and bursting) of human, bovine or avian red blood cells (erythrocytes) at concentrations of about 25 [compounds (276, 309 and 367)] and 300 µg/ml [compound (356)] respectively (606, 796, 798). When exposed to lower pisatin concentrations (100 µg/ml), human erythrocytes are deformed to give either crenated spheres (echinocytes) or cup-shaped structures termed stomatocytes (606). Pisatin has also been shown to uncouple oxidative phosphorylation by isolated rat liver mitochondria (606). Some isoflavonoids [pisatin, phaseollin, phaseollidin (312) and phaseollinisoflavan (418)] are lethal at very low levels to water snails and/or brine shrimps (33, 799).

Even plant cells themselves are not immune from the harmful properties of isoflavonoid compounds. Thus, most of the cells derived from a French bean *(Phaseolus vulgaris)* suspension culture were killed within 1 hour of immersion in a solution (30 μg/ml) of the phytoalexin phaseollin, marked inhibition of respiration being evident after an exposure time of only 2 minutes *(757)*. Similarly, wilting observed in pea *(Pisum sativum)* leaves heavily infected with the powdery mildew fungus, *Erysiphe pisi,* was attributed to the injurious effects of pisatin (produced as a result of the plant-fungus interaction) on the plasma membrane of pea cells *(744)*. Pisatin is also known to inhibit the growth of pea callus cultures *(28)*. In addition, the pterocarpan glucoside, trifolirhizin **(626)** and a range of isoflavone aglycones [e. g. daidzein **(3)**, formononetin **(5)**, genistein **(81)** and biochanin A **(83)**] and glucosides [e. g. daidzin **(511)**, ononin **(521)** and sissotrin **(566)**] have been found to reduce germination of red clover *(Trifolium pratense)* seeds, and to variously inhibit the hypocotyl and radicle/root growth of this and other legumes *(141, 504)*. The anti-auxin and mild anti-gibberellin activity associated with some isoflavonoids *(256, 504, 647)* has led to the suggestion that in plant species such as the yellow lupin *(Lupinus luteus)* they may perhaps function as natural growth regulators *(504)*. In contrast, however, sayanedine **(27)**, a simple phenolic isoflavone from pods of *Pisum sativum,* has kinetin-like activity stimulating root growth of rice seedlings, and cell division in tobacco pith callus *(398)*.

Certain plants commonly used for pasture grazing contain isoflavonoids that mimic the activity of animal oestrogens, regular ingestion of these substances for periods of 6 months to several years leading to the development of severe infertility problems in the female *(517)*. According to SHUTT *(748)* it has been estimated that each year a million Australian ewes fail to lamb because of this condition which, for obvious reasons, is often referred to as "clover disease". The legume most responsible for "clover disease" is *Trifolium subterraneum* (subterranean clover) since the leaves of certain varieties still widely employed in agriculture (e. g. the Dwalganup strain) may have a particularly high oestrogenic content, but other species implicated to a lesser degree include *Medicago sativa* (alfalfa or lucerne), *T. pratense* (red clover) and *T. repens* (white clover).

Several compounds have now been identified as either legume oestrogens or pro-oestrogens (i. e. compounds which may be changed into more active substances in the body of the animal), the most important of these being formononetin **(5)**, genistein **(81)**, biochanin A **(83)** and coumestrol **(452)**. Daidzein **(3)**, pratensein **(94)** and 9-*O*-methylcoumestrol **(453)** are very minor forage oestrogens. Of the first four isoflavonoids, genistein and biochanin A are only weakly active in ruminants whereas formononetin and coumestrol are considerably more potent, although much less so than either natural oestradiol or the synthetic oestrogen, diethylstilboestrol, to which they bear a structural resemblance (Chart 2).

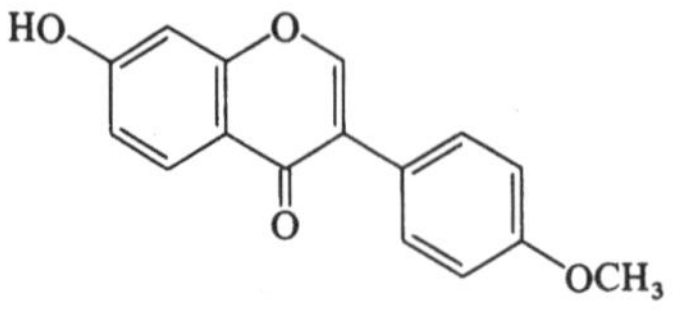

Formononetin

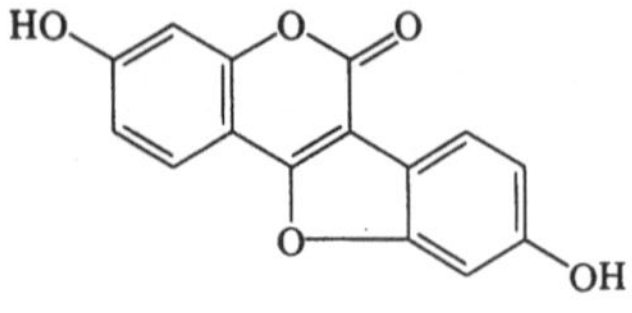

Coumestrol

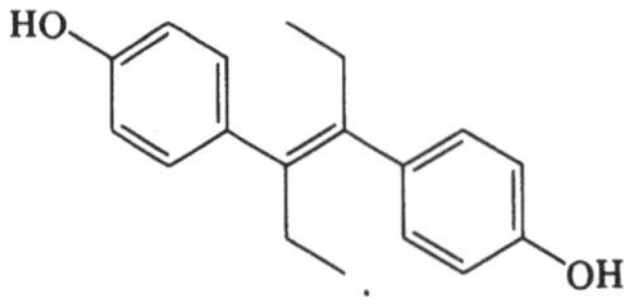

Diethylstilboestrol

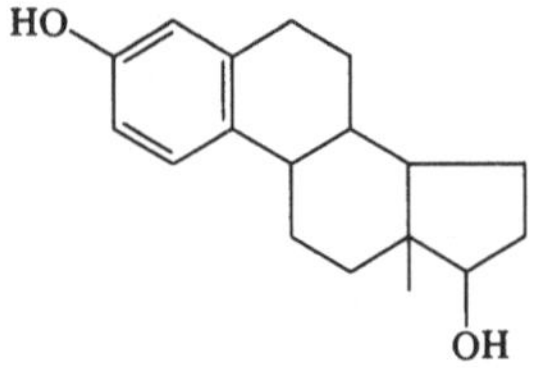

Oestradiol

Chart 2. Structures of formononetin (**5**) and coumestrol (**452**)
compared with diethylstilboestrol (a synthetic oestrogen) and oestradiol (an ovarian
oestrogen)

The profound difference in activity between 5-deoxy-(**5**) and 5-oxy-(**81, 83**) isoflavones is unquestionably related to their mode of degradation in the rumen of sheep and similar animals. Here, micro-organisms convert both genistein and biochanin A (after appropriate C-4′ demethylation) to inactive phenols such as *p*-ethylphenol (*89, 748*), a process which becomes increasingly efficient with time. In contrast, formononetin is first demethylated to daidzein and then reduced to yield the oestrogenic isoflavan equol (**394**) which is resistant to further breakdown. Alternative minor pathways of formononetin degradation lead to 4′-*O*-methylequol (**395**) and 4′-*O*-demethylangolensin (**493**) (*47, 180*) both of which are oestrogenic. In sheep, more than 80% of ingested formononetin may ultimately be absorbed into the bloodstream as equol. Much of this is rapidly excreted in the urine as a water soluble metabolite produced in the liver, but some invariably persists in the blood plasma either "free" or as a sulphoconjugate thereby exerting a harmful influence on the female reproductive cycle for a prolonged period of time.

Coumestrol (**452**) appears to be much less amenable to degradation by rumen microflora and this difference in metabolism when compared with isoflavones may to some extent account for its much greater oestrogenic potency. Fortunately, coumestrol and other coumestans [e. g. 9-*O*-methylcoumestrol which can be converted to (**452**) by microbially-mediated demethylation] normally occur at extremely low concentrations in forage plants and thus, relatively speaking, are of less oestrogenic significance than simple isoflavones such as formononetin. Major problems may arise, however, when animals feed regularly on clover or lucerne foliage infected with pathogenic fungi (e. g. the leaf-spot causing organism *Pseudopeziza medicaginis* on lucerne) as these can cause a dramatic increase in the levels of coumestrol and related compounds (*73, 523, 524*). Aphid infestation and, significantly, "physiogenic" leaf/stem spotting (a condition which develops in the absence of disease-causing organisms as the plants age) similarly lead to a marked accumulation of coumestans in leaves of *Medicago* (*79, 262, 521, 522*) and *Trifolium* (*827*) species. The possibility of immunising farm animals against the harmful effects of coumestrol and other plant oestrogens is now under active investigation (*180, 748*). For a more detailed account of the chemical and biological properties of *Medicago* coumestans, the reader is referred to the comprehensive review published by BICKOFF et al. (*76*).

Some additional minor properties/uses of isoflavonoid compounds are summarised in Table 1.

Table 1. *Minor Properties and Uses of Some Isoflavonoid Compounds*

Property/Use	Compound(s) Involved	References
Anti-arthritic activity	(**11**)	(*533*)
Anti-haemolytic activity	(**3, 8, 9, 81, 87, 551**)	(*315, 349, 570*)
Anti-inflammatory activity	(**243**)	(*777*)
Anti-oxidative activity	(**9, 81**)	(*349, 570*)
Anti-pyretic activity	(**243**)	(*777*)
Anti-spasmodic activity	(**3**)	(*739*)
Anti-tubercular activity	(**276, 362**)	(*411, 487*)
Anti-tumour activity	([−]-**279**, [+]-**279, 625, 626**)	(*467*)
Anti-ulcer activity	Possibly one or more of the following: (**5, 16, 81, 279, 456**)	(*778*)
Diuretic activity	(**98, 590, 609**)	(*466*)
Enzyme inhibition		
Catechol-*o*-methyltransferase	(**50, 81, 87, 89, 99, 108, 116, 118**)	(*163, 792*)
DOPA decarboxylase	(**81, 87, 89, 99, 108, 116, 118**)	(*163, 782, 792*)
Dopamine β-hydroxylase	(**81, 99**)	(*792*)
β-Galactosidase	(**515, 517, 556, 558**)	(*19, 334*)
β-Glucosidase	Methyl ether of (**400**)	(*461*)
Histidine decarboxylase	(**81, 87, 89, 99, 108**)	(*792*)
Lipase	(**3, 81, 511, 551**)	(*604*)
Pectin hydrolase	(**12**), Methyl ethers of (**183**) and (**400**)	(*461, 688*)
Pectin lyase	(**12**), Methyl ether of (**183**)	(*688*)
Pectin transeliminase	Methyl ether of (**400**)	(*461*)
Hypocholesterolemic activity	(**5, 83**)	(*732*)
Hypolipidemic activity	(**5, 83, 94**)	(*733, 750*)
Hypotensive activity	(**87, 116, 118**)	(*163, 792*)
In bright nickel plating	(**11**)	(*52*)

IV. Introduction to Tables 2, 3 and 4

Table 2 lists the structures, molecular weights/molecular formulae, trivial names and sources of the more than 510 naturally occurring isoflavonoid aglycones and legume-derived 2-arylbenzofurans reported up to the end of December 1981. Glycosylated compounds are considered in Table 3, whilst Table 4 deals with partially identified isoflavonoids and some substances of leguminous origin for which very few structural details are available but which, nevertheless, may be of an isoflavonoid nature. Because of their length, Tables 2 and 3 are divided into sections each covering a particular class of isoflavonoid aglycone or glycoside. In addition, however, compounds within most sections of Table 2 are themselves subdivided in order to emphasise specific molecular substituents, the presence or absence of which can often be readily deduced by standard chemical and/or spectroscopic procedures. Thus, 5-deoxyisoflavones and isoflavanones are distinguished from their 5-oxy counterparts; similarly, 6a-hydroxypterocarpans and 12a-oxygenated (OH/OCH$_3$) rotenoids are, for the purposes of this review, considered separately from — but in the same section as — 6aH-pterocarpans and 12aH-rotenoids respectively. Care has also been taken to ensure that simple and complex isoflavonoids (see page 4) do not intermingle.

Compounds within the various sections and subsections of Table 2 have been arranged systematically on the initial basis of increasing molecular weight. Where this is the same for two or more compounds their order is determined respectively by a) the degree of substitution apparent in ring A (or the equivalent ring D for all rotenoid derivatives) with monosubstitution preceding disubstitution, and so on; b) the relative positions of these substituents as determined by the numbering system shown for the parent compound [e. g. the 6,7-substituted isoflavone (**52**) appears before its 7,8-substituted isomer (**53**)]; and c) the number and position of A-ring hydroxyl groups, these having priority over methoxy or other substituents [e. g. formononetin (**5**) is placed ahead of isoformononetin (**6**), and texasin (**17**) precedes kakkatin (**18**) which itself outranks glycitein (**19**)]. Isomeric isoflavonoids with identical A-rings [e. g. sophorapterocarpan A (**311**)/phaseollidin (**312**), and vestitol (**397**)/isovestitol (**398**)] can usually be assigned to their correct position by applying the above rules to ring B [or its A (rotenoid) or D (pterocarpan/coumestan) equivalent]. However, a number of complex compounds such as isoflavone (**137**)/wighteone (**138**), millettosin (**249**)/12a-hydroxyisomillettone (**250**), and dehydrotoxicarol (**270**)/villosol (**271**) can be allocated only by reference to the nature of their C-5 attachments, the relevant substituent priorities in decreasing order being α,α-dimethylallyl, γ,γ-dimethylallyl, *gem*-dimethylchromene and isopropenyldihydrofuran.

In Tables 2, 3 and 4, an attempt has been made to list as comprehensively as possible all the plant species and plant parts from which a given compound has been obtained[5]. However, when a publication reports the same substance(s) in many different species of the same genus [as for instance with the isoflavones of *Baptisia* (*540*), *Genista* (*318*) and *Thermopsis* (*204*)], unnecessary repetition has been avoided by use of the generic name (e. g. *Genista* spp.) only. Lastly, data in the "Plant Part(s) Examined" column are taken directly from the literature and although generally satisfactory, the reader should be aware that a few descriptions may not be botanically accurate. Others are annoyingly vague; for example, the terms "aerial parts" or "above ground parts" presumably mean stem plus leaf material, but whether flowers or pods and seeds (all known to be important sources of isoflavonoid compounds) are also included is a matter of opinion.

In the century and a quarter since HLASIWETZ (*344*) reported the isolation of ononin (**521**), a considerable number of review papers have dealt either wholly or in part with the rapidly bourgeoning subject of isoflavonoid compounds (*39, 76, 183, 193, 196, 402, 403, 608, 609, 610, 721, 805, 814, 824, 825*). Regrettably, many of these are rather limited in scope often covering, sometimes inadequately, only one particular type of isoflavonoid or else restricting the subject matter to information published during a specified period of time. From the viewpoint of the natural products chemist, both these methods of approach are unsatisfactory: first, because in order to obtain full information regarding the isoflavonoid constituents of a given species or genus it may be necessary for a researcher to consult several different publications, and secondly, constraints placed upon review authors with respect to the content and length of their contributions quite often mean that the coverage is far from comprehensive. In fact, remarkable as it may seem, no currently available reviews deal exhaustively with the entire range of naturally occurring isoflavonoid aglycones and glycosides now described in the literature.

The present contribution is an attempt to correct this situation, the term "naturally occurring" being used in a deliberately broad sense to encompass not only isoflavonoids of plant origin, but also novel substances produced for example by microbial modification of either natural or synthetic compounds. By judicious use of the various Tables, the Molecular Weight Index, the Trivial Name Index, and the Source Index, it is hoped that the review will prove valuable to chemists, botanists, plant pathologists and

---

[5] Apart from those species listed in Tables 2 – 4, a great many others are considered, on the basis of colour tests and fish-poisoning or insecticidal properties etc., to contain isoflavonoids (generally rotenone or similar compounds) but as yet the substances concerned have not been identified (*19a, 397, 410, 497, 592, 694, 695, 696, 821, 835a*).

taxonomists amongst others, by providing both an aid to isoflavonoid identification and an introduction to the distribution and chemical characteristics of this biologically fascinating group of natural products.

Note Added in Proof see p. 266.

Acknowledgements

I should like to thank P. M. Dewick for his comments on the manuscript and K. R. Markham for encouragement during its preparation. Thanks are also due to L. Crombie, F. M. Dean, D. M. X. Donnelly, L. Fellows, R. J. Grayer-Barkmeijer, J. B. Harborne, S. C. Jain, M. Jay, N. T. Keen, P. Khanna, H. Ohashi, P. M. Richardson, S. Tahara and A. Zaman, all of whom provided details of unpublished work, or helped in some other way. Financial assistance from the S.E.R.C. is gratefully acknowledged.

Table 2. *Naturally Occurring Isoflavonoid Aglycones*[a–d]

---

[a] Except where indicated all plant species belong to the subfamily Papilionoideae of the Leguminosae. Species from the legume subfamilies Caesalpinioideae and Mimosoideae are marked LC and LM respectively.

[b] Isoflavonoid producing species from non-leguminous plant families are marked with an asterisk (*), the relevant family being given in an abbreviated form as follows: (AMAR) = Amaranthaceae; (CARY) = Caryophyllaceae; (CHEN) = Chenopodiaceae; (COMP) = Compositae; (CRU)=Cruciferae; (CUPR)=Cupressaceae; (IRID)=Iridaceae; (MOR)= Moraceae; (MYRC) = Myricaceae; (MYR) = Myristicaceae; (POD) = Podocarpaceae; (ROS) = Rosaceae; (SCRO) = Scrophulariaceae; (STEM) = Stemonaceae.

[c] The symbol † indicates those plant parts in which a particular isoflavonoid may accumulate as a result of physiological stress. In most cases, the compounds concerned are thought to function as phytoalexins (see Section II and III of text).

[d] ? = Provisional source, or derivation; — = Data not available.

A: ISOFLAVONES

Plant and/or Other Sources	Plant Part(s) Examined	References

a) Simple 5-Deoxy Isoflavones

1   MW 238; $C_{15}H_{10}O_3$

Penicillium cyclopium
cultures [Metabolite]   (46)

Table 2 *(continued)*

Plant and/or Other Sources	Plant Part(s) Examined	References

2 MW 252; $C_{16}H_{12}O_3$

| *Glycyrrhiza glabra* | Root | *(61)* |

3 MW 254; $C_{15}H_{10}O_4$

DAIDZEIN

Plant Sources

Adenocarpus decorticans	Hydrolysed Leaf	*(302)*
A. foliolosus	Hydrolysed Leaf	*(318)*
LM *Albizia procera*	Heartwood/Stem Bark	*(210)*
Baphia spp.	Leaf	*(762a)*
Baptisia spp.	Leaf/Stem	*(538, 540)*
Cajanus cajan	Germinating Seed	*(733)*
Calicotome villosa	Hydrolysed Leaf	*(302)*
Chamaecytisus hirsutus	Hydrolysed Leaf	*(302)*
C. ratisbonensis	Hydrolysed Leaf	*(302)*
C. smyrnaeus	Hydrolysed Leaf	*(318)*
C. supinus	Hydrolysed Leaf	*(302)*
Chamaespartium sagittale	Hydrolysed Leaf	*(318)*
C. tridentatum	Hydrolysed Leaf	*(302)*
Cicer arietinum	Germinating Seed	*(828)*
Cytisus spp.	Hydrolysed Leaf	*(318)*
C. ardoini	Hydrolysed Leaf	*(302)*
C. ingramii	Hydrolysed Leaf	*(302)*
C. multiflorus	Hydrolysed Leaf	*(302)*
C. striatus	Hydrolysed Leaf	*(302)*
C. tribracteolatus	Hydrolysed Leaf	*(302)*
Dalbergia ecastophyllum	Trunk Wood	*(547)*
D. stevensonii	Heartwood	*(227)*
Echinospartum horridum	Hydrolysed Leaf/Stem	*(303)*
Erinacea anthyllis	Hydrolysed Leaf	*(318)*
Erythrina crista-galli	Trunk Bark/Trunk Wood	*(354)*
Genista spp.	Hydrolysed Leaf	*(318)*
G. clavata	Hydrolysed Leaf	*(302)*

Table 2 (continued)

Plant and/or Other Sources	Plant Part(s) Examined	References
Glycine max	Fermented Beans ("Tempeh")	(315)
	Hydrolysed Seed Meal	(571)
	†Hypocotyl	(445)
	†Leaf	(444)
	Seed	(602)
Laburnum anagyroides	Hydrolysed Leaf	(318)
Lespedeza cyrtobotrya	Heartwood	(562)
Lygos monosperma	Hydrolysed Leaf	(318)
L. raetam	Hydrolysed Leaf	(318)
Machaerium villosum	Heartwood	(95, 499)
Medicago sativa	†Leaf	(307, 607)
Ononis spp.	Leaf	(135a)
Phaseolus aureus	Callus Tissue/Cell Suspension Cultures	(53)
	Germinating Seed/ Seedlings	(214, 732, 733)
P. mungo	Germinating Seed	(732, 733)
P. vulgaris	†Pod	(831)
Piptanthus nepalensis	Twig	(625)
Pisum sativum	Silage	(76)
Psoralea corylifolia	Root	(311)
Pterodon apparicioi	Trunk Wood	(512)
Pueraria spp.	Root	(67, 569, 738)
P. thunbergiana (=lobata)	Flower	(491)
	Rhizome	(717)
Sophora subprostrata	Root	(502)
Stauracanthus boivinii	Hydrolysed Leaf	(302)
S. genistoides	Hydrolysed Leaf	(318)
Teline spp.	Hydrolysed Leaf	(318)
Thermopsis spp.	Leaf/Stem	(204)
T. fabacea	Root	(21)
Trifolium pratense	Leaf/Stem	(78, 307)
T. repens	Leaf	(307)
T. subterraneum	Burr Husk/Petiole/Seed/Stem	(260)
	Leaf	(260, 307)
Ulex spp.	Hydrolysed Leaf/Stem	(318)
U. micranthus	Hydrolysed Leaf/Stem	(302)
Vigna unguiculata	Germinating Seed	(828)
	Hypocotyl	(633, 634)

Other Sources

Micromonospora halophytica cultures [Metabolite]		(274)
Streptomyces xanthophaeus cultures [Metabolite]		(334)

Table 2 *(continued)*

Plant and/or Other Sources	Plant Part(s) Examined	References

4 MW 266; $C_{17}H_{14}O_3$

Glycyrrhiza glabra	Root	*(61)*

5 MW 268; $C_{16}H_{12}O_4$

FORMONONETIN (BIOCHANIN B; NEOCHANIN; PRATOL)

LM *Albizia procera*	Heartwood/Stem Bark	*(210)*
Amorpha fruticosa	Root	*(736)*
Baptisia spp.	Leaf/Stem	*(540)*
B. australis	Flower/Leaf/Root/Stem	*(510)*
Cajanus cajan	†Etiolated Stem	*(363)*
	Germinating Seed	*(733)*
Castanospermum australe	Wood	*(230)*
Centrosema spp.	†Leaf	*(537)*
Chamaecytisus ratisbonensis	Hydrolysed Leaf	*(302)*
Chamaespartium sagittale	Hydrolysed Leaf	*(302)*
C. tridentatum	Hydrolysed Leaf	*(302)*
Cicer spp.	Etiolated Stem	*(382)*
C. arietinum	Callus Tissue	*(716)*
	Etiolated Stem	*(361, 382)*
	Germinating Seed	*(85, 815, 828)*
	Leaf	*(356)*
	Root	*(359)*
Cladrastis lutea	Heartwood	*(726)*
C. platycarpa	Leaf	*(600)*
	Heartwood/Sapwood	*(601)*
	Trunk Bark	*(351, 352)*
C. shikokiana	Bark/Heartwood/Leaf/	
	Root Bark/Root Wood/	*(599)*
	Sapwood	
Cytisus proliferus	Hydrolysed Leaf	*(318)*
C. striatus	Hydrolysed Leaf	*(302)*
C. tribracteolatus	Hydrolysed Leaf	*(302)*
Dalbergia baroni	Heartwood	*(223)*
D. barretoana	Wood	*(90)*

Table 2 *(continued)*

Plant and/or Other Sources	Plant Part(s) Examined	References
D. cearensis	Softwood	*(308)*
D. ecastophyllum	Trunk Wood/Vine Wood	*(226, 547)*
D. inundata	Wood	*(511)*
D. nitidula	Heartwood	*(514)*
D. oliveri	Heartwood	*(225)*
D. paniculata	Bark	*(583, 631)*
D. spruceana	Sapwood	*(176)*
D. stevensonii	Heartwood	*(227)*
D. variabilis	Wood	*(497)*
D. volubilis	Leaf	*(146)*
Diplotropis purpurea	Trunk Wood	*(102)*
Echinospartum horridum	Hydrolysed Leaf/Stem	*(303)*
Ferreirea spectabilis	Wood	*(123)*
Genista spp.	Hydrolysed Leaf	*(318)*
G. patula	Aerial Parts	*(618)*
Glycyrrhiza sp.	Root Bark	*(709)*
G. echinata	Callus Tissue	*(26)*
G. glabra	Root	*(235, 458, 689)*
G. glabra var. *glandulifera*	Callus Tissue	*(270)*
G. glabra var. *typica*	Root	*(561)*
Hedysarum polybotrys	Root	*(485)*
Maackia amurensis		
var. *buergeri*	Heartwood	*(778)*
Machaerium kuhlmannii	Trunk Wood	*(611)*
M. nictitans	Trunk Wood	*(611)*
M. vestitum	Trunk Wood	*(496)*
M. villosum	Heartwood	*(91, 95, 499)*
Medicago sativa	†Leaf	*(307, 607)*
	Root	*(607)*
Myroxylon balsamum	Trunk Wood	*(96)*
Onobrychis spp.	Leaf	*(370)*
O. viciifolia	Leaf	*(370)*
	Seedlings	*(15)*
Ononis spp.	Leaf	*(135a)*
O. arvensis	Root	*(475)*
O. spinosa	Root	*(47a, 335)*
Pericopsis sp.	Heartwood	*(355)*
P. laxiflora	Heartwood	*(252)*
P. mooniana	Heartwood	*(252)*
Pickeringia montana	Hydrolysed Leaf	*(317)*
Piptanthus nepalensis	Hydrolysed Leaf	*(317)*
	Twig	*(625)*
Pisum sativum	†Seedlings	*(136)*
Pterocarpus indicus	Heartwood	*(66, 178)*
P. vidalianus	Heartwood	*(616)*
Pueraria thunbergiana	Flower	*(493)*
(=*lobata*)	Rhizome	*(717)*

References, pp. 229—265

Table 2 *(continued)*

Plant and/or Other Sources	Plant Part(s) Examined	References
Sophora flavescens	Root	*(501)*
S. tomentosa	Aerial Parts	*(470, 471)*
Spartium junceum	Aerial Parts	*(619)*
Teline canariensis	Hydrólysed Leaf	*(302)*
Thermopsis spp.	Leaf/Stem	*(204)*
T. fabacea	Root	*(21)*
Tipuana tipu	Root/Trunk Wood	*(94)*
Trifolium spp.	Hydrolysed Leaf/Leaf	*(27, 263, 438)*
T. alexandrinum	Leaf	*(734)*
T. hybridum	Root	*(259)*
T. incarnatum	Flower	*(704)*
T. pratense	Flower	*(649)*
	Leaf	*(45, 78, 307, 438)*
	Root	*(106, 258, 259, 648)*
	Stem	*(78)*
T. repens	Leaf	*(307, 832)*
T. subterraneum	Burr Husk/Petiole/Seed	*(260)*
	Leaf	*(48, 88, 260, 307)*
	Seedlings	*(661)*
	Stem	*(88, 260)*
Ulex micranthus	Hydrolysed Leaf/Stem	*(302)*
Vatairea heteroptera	Trunk Wood	*(253)*
* *Virola caducifolia* (MYR)	Trunk Wood	*(100)*
* *V. multinervia* (MYR)	Trunk Wood	*(100)*
Wisteria floribunda	Bark	*(596)*
	Wood	*(596, 781)*

6   MW 268; $C_{16}H_{12}O_4$

ISOFORMONONETIN

Glycine max	†Leaf	*(393)*
Machaerium villosum	Heartwood	*(95, 499)*

Table 2 *(continued)*

Plant and/or Other Sources	Plant Part(s) Examined	References

7   MW 270; $C_{15}H_{10}O_5$

2'-HYDROXYDAIDZEIN

Phaseolus vulgaris   †Pod   *(831)*

8   MW 270; $C_{15}H_{10}O_5$

3'-HYDROXYDAIDZEIN

Machaerium villosum   Heartwood   *(95, 499)*

9   MW 270; $C_{15}H_{10}O_5$

DEMETHYLTEXASIN (SOYBEAN FACTOR 2)

Centrosema haitiense   †Leaf   *(537)*
Glycine max   Fermented Beans ("Tempeh")   *(315)*

10   MW 282; $C_{16}H_{10}O_5$

CORYLINAL

Psoralea corylifolia   Seed   *(310)*

Table 2 *(continued)*

Plant and/or Other Sources	Plant Part(s) Examined	References

11   MW 282; $C_{16}H_{10}O_5$

PSEUDOBAPTIGENIN

Baptisia spp.	Leaf/Stem	*(538, 540)*
Cladrastis platycarpa	Bark/Leaf	*(599)*
	Wood	*(597)*
C. shikokiana	As given for **5**	*(599)*
Dalbergia assamica	Pod	*(155)*
	Seed	*(156)*
D. lanceolaria	Flower/Leaf	*(533)*
D. sericea	Bark	*(630)*
	Leaf	*(194)*
D. spruceana	Heartwood	*(176)*
D. stevensonii	Heartwood	*(227)*
Maackia amurensis		
var. *buergeri*	Heartwood	*(773)*
Pisum sativum	†Seedlings	*(136)*
Pterocarpus erinaceus	Heartwood	*(55)*
Trifolium hybridum	Root	*(259)*
T. pratense	Hydrolysed Leaf/Stem	*(78)*
	Root	*(258, 259, 648)*

12   MW 282; $C_{17}H_{14}O_4$

7,4'-DI-*O*-METHYLDAIDZEIN

Dalbergia violacea		
(= *miscolobium*)	Heartwood	*(305)*
Pterodon apparicioi	Trunk Wood	*(512)*

Table 2 *(continued)*

Plant and/or Other Sources	Plant Part(s) Examined	References

13   MW 284; $C_{16}H_{12}O_5$

2'-HYDROXYFORMONONETIN

Trifolium repens	†Leaf	*(832)*
* *Virola caducifolia* (MYR)	Trunk Wood	*(100)*
* *V. multinervia* (MYR)	Trunk Wood	*(100)*

14   MW 284; $C_{16}H_{12}O_5$

TERALIN (THERALIN)

Thermopsis alterniflora	Leaf	*(431)*

15   MW 284; $C_{16}H_{12}O_5$

CALYCOSIN

Baptisia spp.	Leaf/Stem	*(538, 539, 540)*
Bowdichia nitida	Heartwood	*(123)*
Cadia purpurea	Leaf	*(845)*
Cladrastis shikokiana	†Heartwood	*(595)*
Cyclolobium clausseni	Heartwood	*(295)*
Dalbergia baroni	Heartwood	*(223a)*
D. latifolia	Heartwood	*(223a)*
Machaerium mucronulatum	Heartwood	*(499)*
M. vestitum	Trunk Wood	*(295)*
M. villosum	Heartwood	*(499)*
Myroxylon balsamum	Trunk Wood	*(96)*
M. peruiferum	Trunk Wood	*(536)*

Table 2 *(continued)*

Plant and/or Other Sources	Plant Part(s) Examined	References
Pterocarpus dalbergioides	Heartwood	*(627)*
Sophora secundiflora	Aerial Parts (Mainly Stem)	*(558)*
Thermopsis ssp.	Leaf/Stem	*(204)*
T. fabacea	Root	*(21)*
Trifolium pratense	Hydrolysed Leaf/Stem	*(78)*
	Root	*(258, 648)*

16   MW 284; $C_{16}H_{12}O_5$

3'-METHOXYDAIDZEIN

Cyclolobium clausseni	Heartwood	*(93)*
Maackia amurensis var. *buergeri*	Heartwood	*(778)*
Machaerium villosum	Heartwood	*(95)*

17   MW 284; $C_{16}H_{12}O_5$

TEXASIN

Baptisia australis	Leaf/Stem	*(540)*
B. leucophaea	Leaf/Stem	*(540)*
Myroxylon peruiferum	Trunk Wood	*(536)*
Platymiscium praecox	Heartwood	*(92)*

18   MW 284; $C_{16}H_{12}O_5$

KAKKATIN

Pueraria sp.	Flower	*(488)*

Table 2 *(continued)*

Plant and/or Other Sources	Plant Part(s) Examined	References

19   MW 284; $C_{16}H_{12}O_5$

GLYCITEIN

Centrosema haitiense	†Leaf	*(537)*
C. pubescens	†Leaf	*(537)*
Glycine max	Hydrolysed Seed Meal	*(571)*
Mildbraedeodendron excelsa	Heartwood	*(550)*

20   MW 284; $C_{16}H_{12}O_5$

RETUSIN

Dalbergia retusa	Heartwood	*(414)*
Dipteryx odorata	Heartwood	*(332)*

21   MW 286; $C_{15}H_{10}O_6$

BAPTIGENIN

Baptisia tinctoria	Root	*(243, 251)*

22   MW 294; $C_{18}H_{14}O_4$

Glycyrrhiza glabra	Root	*(61)*

Table 2 *(continued)*

Plant and/or Other Sources	Plant Part(s) Examined	References

23   MW 294; $C_{18}H_{14}O_4$

GLYZARIN

Glycyrrhiza glabra   Root   *(62)*

24   MW 296; $C_{17}H_{12}O_5$

PSEUDOBAPTIGENIN METHYL ETHER

Calopogonium mucunoides   Seed   *(808)*

25   MW 298; $C_{17}H_{14}O_5$

CLADRIN

Cladrastis lutea   Heartwood   *(726)*
C. platycarpa   Bark/Leaf/Sapwood   *(599)*
                              Heartwood   *(601)*
C. shikokiana   As given for **5**   *(599)*

Table 2 *(continued)*

Plant and/or Other Sources	Plant Part(s) Examined	References

26 MW 298; $C_{16}H_{10}O_6$

GLYZAGLABRIN

Glycyrrhiza glabra   Root   *(63)*

27 MW 298; $C_{17}H_{14}O_5$

SAYANEDINE

Pisum sativum   Pod   *(398, 769)*

28 MW 298; $C_{17}H_{14}O_5$

AFRORMOSIN (AFROMOSIN; CASTANIN)

Amphimas pterocarpoides	Heartwood	*(55)*
Baptisia spp.	Leaf/Stem	*(540)*
B. australis	Flower/Leaf/Stem	*(510)*
Castanospermum australe	Wood	*(230)*
Centrosema spp.	†Leaf	*(537)*
Cladrastis lutea	Heartwood	*(726)*
C. platycarpa	Heartwood/Sapwood	*(601)*
	Leaf	*(600)*
	Trunk Bark	*(351, 352)*
C. shikokiana	As given for **5**	*(599)*
Dalbergia riparia	Wood	*(103)*
Myrocarpus fastigiatus	Wood	*(319)*

Table 2 *(continued)*

Plant and/or Other Sources	Plant Part(s) Examined	References
Myroxylon balsamum	Wood	*(319)*
M. peruiferum	Trunk Wood	*(536)*
Onobrychis spp.	Leaf	*(370)*
O. viciifolia	Leaf	*(370)*
	Seedlings	*(15, 213)*
Pericopsis sp.	Heartwood	*(355)*
P. elata (= *Afrormosia elata*)	Heartwood	*(252, 254, 549)*
P. mooniana	Heartwood	*(252)*
Pisum sativum	†Seedlings	*(136)*
Pterodon apparicioi	Trunk Wood	*(273)*
Wisteria floribunda	Bark	*(596, 740)*
	Wood	*(596, 740, 781)*

29 MW 298; $C_{17}H_{14}O_5$

8-*O*-METHYLRETUSIN

Cladrastis platycarpa	Leaf/Heartwood/Sapwood	*(599)*
	Trunk Bark	*(353)*
C. shikokiana	As given for **5**	*(599)*
Dalbergia retusa	Heartwood	*(304, 414)*
D. variabilis	Bark	*(497)*
Dipteryx odorata	Heartwood	*(332)*
Monopteryx uaucu	Trunk Wood	*(16)*
Pericopsis schliebenii	Heartwood	*(252)*
Xanthocercis zambesiaca	Heartwood	*(328)*

30 MW 300; $C_{16}H_{12}O_6$

KOPARIN

Castanospermum australe	Wood	*(54)*

Table 2 *(continued)*

Plant and/or Other Sources	Plant Part(s) Examined	References

31 MW 300; $C_{16}H_{12}O_6$

GLIRICIDIN

Gliricidia sepium Heartwood *(534)*

32 MW 310; $C_{17}H_{10}O_6$

MAXIMA-ISOFLAVONE A (MAXIMA SUBSTANCE A)

(Revised Structure)

Tephrosia maxima Root *(490, 672, 678)*

33 MW 312; $C_{17}H_{12}O_6$

CUNEATIN

Cicer cuneatum Etiolated Stem *(382)*

34 MW 312; $C_{18}H_{16}O_5$

CABREUVIN

Table 2 *(continued)*

Plant and/or Other Sources	Plant Part(s) Examined	References
Calopogonium mucunoides	Seed	*(808)*
Myrocarpus fastigiatus	Wood	*(297)*
Myroxylon balsamum	Wood	*(297)*
M. peruiferum	Trunk Wood	*(536)*

35   MW 312; $C_{17}H_{12}O_6$

FUJIKINETIN

Cladrastis platycarpa	Leaf	*(600)*
	Heartwood/Sapwood	*(601)*
	Trunk Bark	*(350, 351, 352)*
C. shikokiana	As given for **5**	*(599)*
Dalbergia riparia	Wood	*(103)*
D. sericea	Bark	*(630)*
Pterodon apparicioi	Trunk Wood	*(273)*

36   MW 314; $C_{17}H_{14}O_6$

ODORATIN

Dipteryx odorata	Heartwood	*(332)*
Pterodon apparicioi	Trunk Wood	*(273)*

37   MW 314; $C_{17}H_{14}O_6$

3'-HYDROXY-8-*O*-METHYLRETUSIN

Table 2 *(continued)*

Plant and/or Other Sources	Plant Part(s) Examined	References
Dipteryx odorata	Heartwood	*(332)*
Monopteryx uaucu	Trunk Wood	*(16)*
Myroxylon balsamum	Trunk Wood	*(96)*
Xanthocercis zambesiaca	Heartwood	*(328)*

38   MW 314; $C_{17}H_{14}O_6$

DIPTERYXIN

Dipteryx odorata	Heartwood	*(332)*

39   MW 326; $C_{18}H_{14}O_6$

CUNEATIN METHYL ETHER

Pterodon apparicioi	Trunk Wood	*(273)*

40   MW 326; $C_{18}H_{14}O_6$

FUJIKINETIN METHYL ETHER

Cordyla africana	Heartwood	*(134)*

41   MW 326; $C_{18}H_{14}O_6$

Xanthocercis zambesiaca	Heartwood	*(328)*

Table 2 *(continued)*

Plant and/or Other Sources	Plant Part(s) Examined	References

42 MW 328; $C_{18}H_{16}O_6$

Amorpha fruticosa	Seedlings	*(184)*
Dalbergia monetaria	Seed	*(223a)*

43 MW 328; $C_{18}H_{16}O_6$

Myroxylon peruiferum	Trunk Wood	*(536)*

44 MW 328; $C_{18}H_{16}O_6$

CLADRASTIN

Cladrastis lutea	Heartwood	*(726)*
C. platycarpa	Leaf	*(600)*
	Heartwood/Sapwood	*(601)*
	Trunk Bark	*(351, 352)*
C. shikokiana	As given for **5**	*(599)*

45 MW 328; $C_{18}H_{16}O_6$

Monopteryx uaucu	Trunk Wood	*(16)*
Xanthocercis zambesiaca	Heartwood	*(328)*

Table 2 *(continued)*

Plant and/or Other Sources	Plant Part(s) Examined	References

46 MW 342; $C_{19}H_{18}O_6$

Amorpha fruticosa	Fruit	*(594)*
Calopogonium mucunoides	Seed	*(808)*

47 MW 342; $C_{18}H_{14}O_7$

Cordyla africana	Heartwood	*(135)*
Dalbergia assamica	Pod	*(155)*
	Seed	*(156)*
Mildbraedeodendron excelsa	Heartwood	*(550)*

48 MW 342; $C_{18}H_{14}O_7$
DALPATEIN

Dalbergia paniculata	Seed	*(9, 662)*

49 MW 342; $C_{19}H_{18}O_6$

Cordyla africana	Heartwood	*(134)*
Pterodon apparicioi	Trunk Wood	*(512)*
P. pubescens	Trunk Wood	*(99)*

References, pp. 229—265

Table 2 *(continued)*

Plant and/or Other Sources	Plant Part(s) Examined	References

50   MW 344; $C_{18}H_{16}O_7$

Streptomyces sp.
   cultures [Metabolite] | | *(163)*

51   MW 356; $C_{19}H_{16}O_7$

MILLDURONE

Cordyla africana	Heartwood	*(134)*
Dalbergia paniculata	Seed	*(9)*
Mildbraedeodendron excelsa	Heartwood	*(550)*
Millettia dura	Seed	*(613)*
Pterodon apparicioi	Trunk Wood	*(273)*
P. pubescens	Trunk Wood	*(99)*

52   MW 356; $C_{19}H_{16}O_7$

Cordyla africana	Heartwood	*(134)*
Mildbraedeodendron excelsa	Heartwood	*(550)*

53   MW 356; $C_{19}H_{16}O_7$

Pterodon apparicioi	Trunk Wood	*(273)*

Table 2 *(continued)*

Plant and/or Other Sources	Plant Part(s) Examined	References

54 MW 356; $C_{19}H_{16}O_7$
PETALOSTETIN

| *Petalostemon candidum* | Whole Plant | *(783)* |

55 MW 356; $C_{19}H_{16}O_7$

| *Xanthocercis zambesiaca* | Heartwood | *(328)* |

56 MW 358; $C_{19}H_{18}O_7$

| *Pterodon apparicioi* | Trunk Wood | *(273)* |

57 MW 372; $C_{20}H_{20}O_7$

| *Pterodon apparicioi* | Trunk Wood | *(273)* |
| *P. pubescens* | Trunk Wood | *(99)* |

Table 2 *(continued)*

Plant and/or Other Sources	Plant Part(s) Examined	References

58   MW 372; $C_{20}H_{20}O_7$

Cordyla africana	Heartwood	(*134*)
Mildbraedeodendron excelsa	Heartwood	(*550*)
Pterodon apparicioi	Trunk Wood	(*273*)
P. pubescens	Trunk Wood	(*273*)

59   MW 372; $C_{20}H_{20}O_7$

Pterodon apparicioi	Trunk Wood	(*512*)

b) Complex 5-Deoxy Isoflavones

60   MW 320; $C_{20}H_{16}O_4$

CORYLIN

Psoralea corylifolia	Seed	(*400*)

61   MW 320; $C_{20}H_{16}O_4$

ERYTHRININ A

Erythrina variegata	Bark	(*209*)

Table 2 *(continued)*

Plant and/or Other Sources	Plant Part(s) Examined	References

62   MW 322; $C_{20}H_{18}O_4$

NEOBAVAISOFLAVONE

Psoralea corylifolia   Seed   *(34, 35, 310)*

63   MW 334; $C_{21}H_{18}O_4$

CALOPOGONIUM ISOFLAVONE A

Calopogonium mucunoides   Seed   *(807)*

64   MW 336; $C_{20}H_{16}O_5$

GLABRONE

Glycyrrhiza glabra   Root   *(458)*

65   MW 336; $C_{21}H_{20}O_4$

DURLETTONE

Millettia dura   Seed   *(613)*

Table 2 *(continued)*

Plant and/or Other Sources	Plant Part(s) Examined	References

66   MW 336; $C_{19}H_{12}O_6$

DEHYDRONEOTENONE (NORTON & HANSBERRY'S COMPOUND VI)

Neorautanenia amboensis	Root	*(119)*
N. edulis	Root	*(119)*
Pachyrrhizus erosus	Seed	*(185, 186, 416, 586)*

67   MW 338; $C_{20}H_{18}O_5$

PSORALENOL

Psoralea corylifolia	Seed	*(774)*

68   MW 348; $C_{21}H_{16}O_5$

CALOPOGONIUM ISOFLAVONE B

Calopogonium mucunoides	Seed	*(808)*

Table 2 *(continued)*

Plant and/or Other Sources	Plant Part(s) Examined	References

69   MW 350; $C_{21}H_{18}O_5$

MAXIMA-ISOFLAVONE B (MAXIMA SUBSTANCE B)

| *Tephrosia maxima* | Root | *(490, 672, 677, 678)* |

70   MW 378; $C_{22}H_{18}O_6$

JAMAICIN (PISCIDIA COMPOUND A.K. 6)

| *Piscidia erythrina* | Bark | *(125, 767)* |
| | Root Bark/Whole Root | *(239, 563)* |

71   MW 378; $C_{22}H_{18}O_6$

DURMILLONE

Millettia dura	Seed	*(613)*
M. ferruginea	Seed	*(342)*
M. rubiginosa	Root	*(208)*

Table 2 *(continued)*

Plant and/or Other Sources	Plant Part(s) Examined	References

72 MW 380; $C_{22}H_{20}O_6$

MAXIMA-ISOFLAVONE C (MAXIMIN; MAXIMA SUBSTANCE C)

Tephrosia maxima	Pod	*(680)*
(= *T. purpurea* var. *maxima*)	Root	*(672, 676, 678)*

73 MW 382; $C_{22}H_{22}O_6$

LICORICONE

Glycyrrhiza uralensis	Root	*(420)*

74 MW 394; $C_{23}H_{22}O_6$

BARBIGERONE

Tephrosia barbigera	Seed	*(806)*

Table 2 *(continued)*

Plant and/or Other Sources	Plant Part(s) Examined	References

75 MW 408; $C_{23}H_{20}O_7$
FERRUGONE

Millettia ferruginea   Seed   *(342)*

76 MW 408; $C_{23}H_{20}O_7$
ICHTHYNONE

Millettia rubiginosa	Root	*(208)*
Piscidia erythrina	Root Bark/Root Wood	*(706, 720)*
	Whole Root	*(239)*

77 MW 416; $C_{26}H_{24}O_5$
MUNETONE
(Revised Structure)

Mundulea sericea (= *suberosa*)   Root Bark   *(43, 228, 229, 282)*

Table 2 *(continued)*

Plant and/or Other Sources	Plant Part(s) Examined	References

78 MW 434; $C_{26}H_{26}O_6$

CAJAISOFLAVONE

Cajanus cajan Root Bark *(58)*

79 MW 434; $C_{26}H_{26}O_6$

MUNDULONE

Mundulea sericea (= *suberosa*) Bark *(130, 652)*
  Root Bark *(282)*

c) Simple 5-Oxy Isoflavones

80 MW 268; $C_{16}H_{12}O_4$

Derris robusta Seed *(159)*

81 MW 270; $C_{15}H_{10}O_5$

GENISTEIN (PRUNETOL; SOPHORICOL)

Table 2 *(continued)*

Plant and/or Other Sources	Plant Part(s) Examined	References
Plant Sources		
Adenocarpus decorticans	Hydrolysed Leaf	*(302)*
A. foliolosus	Hydrolysed Leaf	*(318)*
^{LM} *Albizia procera*	Heartwood/Stem Bark	*(210)*
Apios tuberosa	†Leaf	*(396)*
Argyrocytisus battandieri	†Leaf	*(356)*
Baptisia spp.	Leaf/Stem	*(538, 540)*
Bowdichia nitida	Heartwood	*(123)*
Cajanus cajan	†Etiolated Stem	*(363)*
	Root Bark	*(59)*
Calicotome spinosa	Hydrolysed Leaf	*(318)*
C. villosa	Hydrolysed Leaf	*(318)*
Canavalia ensiformis	†Etiolated Hypocotyl	*(356)*
Chamaecytisus albus	Hydrolysed Leaf	*(318)*
C. eriocarpus	Hydrolysed Leaf	*(302)*
C. hirsutus	Hydrolysed Leaf	*(302)*
C. ratisbonensis	Hydrolysed Leaf	*(302)*
C. smyrnaeus	Hydrolysed Leaf	*(318)*
C. supinus	Hydrolysed Leaf	*(302)*
Chamaespartium sagittale	Hydrolysed Leaf	*(318)*
C. tridentatum	Hydrolysed Leaf	*(302)*
Chronanthus biflorus	Hydrolysed Leaf	*(318)*
Crotalaria juncea	†Leaf	*(356)*
Cytisus spp.	Hydrolysed Leaf	*(318)*
C. ardoini	Hydrolysed Leaf	*(302)*
C. ingramii	Hydrolysed Leaf	*(302)*
C. multiflorus	Hydrolysed Flower/Leaf	*(302)*
C. scoparius	Flower	*(494)*
C. striatus	Hydrolysed Leaf	*(302)*
C. tribracteolatus	Hydrolysed Leaf	*(302)*
Dolichos biflorus	†Leaf/Stem	*(443)*
Echinospartum horridum	Hydrolysed Leaf/Stem	*(303)*
Erinacea anthyllis	Hydrolysed Leaf	*(318)*
Erythrina crista-galli	Trunk Bark/Trunk Wood	*(354)*
Genista spp.	Hydrolysed Leaf	*(318)*
G. clavata	Hydrolysed Leaf	*(302)*
G. germanica	Hydrolysed Leaf	*(302)*
G. lydia	Flower	*(791)*
G. ovata	Aerial Parts	*(575)*
G. patula	Aerial Parts	*(618)*
G. rumelica	Aerial Parts	*(576)*
Glycine max	Fermented Beans ("Tempeh")	*(315)*
	Hydrolysed Seed Meal	*(571)*
	Seed	*(602)*
Hardenbergia violacea	†Leaf	*(356)*
Lablab niger	†Etiolated Hypocotyl	*(367)*
Laburnum alpinum	Heartwood	*(237)*

Table 2 *(continued)*

Plant and/or Other Sources	Plant Part(s) Examined	References
L. anagyroides	Hydrolysed Leaf	*(318)*
	†Leaf	*(389)*
	Sapwood	*(165)*
Lembotropis emeriflora	Hydrolysed Leaf	*(302)*
Lespedeza cyrtobotrya	Heartwood	*(562)*
Lupinus albus	†Etiolated Hypocotyl	*(389)*
L. luteus	Hydrolysed Flower	*(505)*
	Senescent Leaf/Stem	*(507)*
	Root/Hydrolysed Root	*(505, 506)*
Lygos monosperma	Hydrolysed Leaf	*(318)*
L. raetam	Hydrolysed Leaf	*(318)*
Maackia amurensis var. *buergeri*	Heartwood	*(778)*
Medicago sativa	Leaf	*(307)*
Moghania macrophylla	Wood	*(482)*
Neonotonia wightii	†Stem	*(391)*
Pericopsis laxiflora	Heartwood	*(252)*
Phaseolus vulgaris	†Pod	*(77, 829)*
Piptanthus nepalensis	Twig	*(625)*
Pisum sativum	Silage	*(76)*
* *Podocarpus spicatus* (POD)	Heartwood	*(115)*
* *Prunus aequinoctialis* (ROS)	Wood	*(330)*
* *P. avium* (ROS)	Wood	*(330)*
* *P. mahaleb* (ROS)	Wood	*(620)*
* *P. maximowiczii* (ROS)	Wood	*(330)*
* *P. nipponica* (ROS)	Wood	*(330)*
* *P. verecunda* (ROS)	Wood	*(331)*
Pterocarpus angolensis	Heartwood	*(55a)*
Pueraria thunbergiana	Flower	*(491)*
Sophora japonica	Fruit	*(776)*
S. subprostrata	Root	*(468)*
Spartium junceum	Aerial Parts	*(619)*
Stauracanthus boivinii	Hydrolysed Leaf	*(302)*
S. genistoides	Hydrolysed Leaf	*(318)*
Stizolobium deeringianum	†Etiolated Stem	*(356)*
Teline spp.	Hydrolysed Leaf	*(318)*
Thermopsis spp.	Leaf/Stem	*(204)*
T. fabacea	Root	*(21)*
Trifolium spp.	Hydrolysed Leaf/Leaf	*(27, 192, 263)*
T. alexandrinum	Leaf	*(734)*
T. hybridum	Root	*(259)*
T. pratense	Leaf	*(192, 307)*
	Root	*(258, 259, 648)*
T. repens	Leaf	*(307)*
	Root?	*(259)*
T. subterraneum	Burr Husk/Petiole/Seed	*(260)*
	Leaf	*(48, 88, 192, 260, 307)*

Table 2 *(continued)*

Plant and/or Other Sources	Plant Part(s) Examined	References
	Seedlings	*(661)*
	Stem	*(88, 260)*
Ulex spp.	Hydrolysed Leaf/Stem	*(318)*
U. micranthus	Hydrolysed Leaf/Stem	*(302)*
Other Sources		
Aspergillus niger		
cultures [Metabolite]		*(792)*
Fusarium spp.		
cultures [Metabolite]		*(44a)*
Micromonospora halophytica		
cultures [Metabolite]		*(274)*
Mycobacterium phlei		
cultures [Metabolite]		*(348)*
Streptomyces griseus		
cultures [Metabolite]		*(474)*
S. xanthophaeus		
cultures [Metabolite]		*(334)*

82   MW 282; $C_{17}H_{14}O_4$

Arachis hypogaea	Seed	*(786)*

83   MW 284; $C_{16}H_{12}O_5$

BIOCHANIN A (PRATENSOL)

Plant Sources

[LM]*Albizia procera*	Heartwood/Stem Bark	*(210)*
Andira inermis	Heartwood	*(174)*
A. parviflora	Trunk Wood	*(102)*
Baptisia spp.	Leaf/Stem	*(540)*
Cicer spp.	Etiolated Stem	*(382)*
C. arietinum	Etiolated Stem	*(361, 382)*
	Germinating Seed	*(84, 815, 828)*
	Leaf	*(356)*
	Root	*(359)*

Table 2 *(continued)*

Plant and/or Other Sources	Plant Part(s) Examined	References
* *Cotoneaster pannosa* (ROS)	Fruit	*(177)*
* *C. serotina* (ROS)	Fruit	*(177)*
Dalbergia lanceolaria	Root	*(146a)*
D. nitidula	Heartwood	*(514)*
D. paniculata	Bark	*(583, 631)*
D. sissoo	Flower	*(42)*
	Leaf	*(41)*
	Mature Pod	*(727)*
D. spruceana	Sapwood	*(176)*
D. volubilis	Flower	*(147, 148, 149)*
	Leaf	*(146)*
Echinospartum horridum	Hydrolysed Leaf/Stem	*(303)*
Ferreirea spectabilis	Heartwood/Wood	*(123, 453)*
Lupinus termis	Seed	*(719)*
Medicago sativa	Leaf	*(307)*
Monopteryx inpae	Trunk Wood	*(16)*
M. uaucu	Trunk Wood	*(16)*
Myroxylon peruiferum	Trunk Wood	*(536)*
Pericopsis sp.	Heartwood	*(355)*
P. elata	Heartwood	*(252)*
P. laxiflora	Heartwood	*(252)*
P. mooniana	Heartwood	*(252)*
Pueraria thunbergiana	Flower	*(491, 493)*
Sophora japonica	Root	*(469)*
S. mollis	Heartwood	*(401)*
Thermopsis spp.	Leaf/Stem	*(204)*
Trifolium spp.	Leaf	*(27, 263, 434)*
T. alexandrinum	Leaf	*(734)*
T. pratense	Flower	*(83, 649)*
	Leaf	*(307)*
	Root	*(258, 648)*
T. subterraneum	Burr Husk/Petiole/ Seed/Stem	*(260)*
	Leaf	*(48, 260, 307)*
	Seedlings	*(661)*
* *Virola caducifolia* (MYR)	Trunk Wood	*(104)*
Other Source		
Fusarium spp. cultures [Metabolite]		*(44a)*

84   MW 284; $C_{16}H_{12}O_5$
PRUNETIN

Table 2 *(continued)*

Plant and/or Other Sources	Plant Part(s) Examined	References
Plant Sources		
Dalbergia violacea (= *miscolobium*)	Leaf	*(253)*
Genista carinalis	Aerial Parts	*(578)*
Glycyrrhiza glabra	Leaf	*(430)*
* *Prunus* sp. (*emarginata*?) (ROS)	Bark	*(250)*
* *P. aequinoctialis* (ROS)	Wood	*(330)*
* *P. avium* (ROS)	Wood	*(330)*
* *P. mahaleb* (ROS)	Bark	*(646)*
	Wood	*(620)*
* *P. maximowiczii* (ROS)	Wood	*(330)*
* *P. nipponica* (ROS)	Wood	*(330)*
* *P. puddum* (ROS)	Bark	*(579, 580, 671)*
	Heartwood	*(288)*
* *P. verecunda* (ROS)	Wood	*(331)*
Pterocarpus angolensis	Heartwood	*(454, 456)*
Other Source		
Mycobacterium phlei cultures [Metabolite]		*(348)*

85 MW 284; $C_{16}H_{12}O_5$

5-*O*-METHYLGENISTEIN (ISOPRUNETIN)

Adenocarpus decorticans	Hydrolysed Leaf	*(302)*
A. foliolosus	Hydrolysed Leaf	*(318)*
Calicotome spinosa	Hydrolysed Leaf	*(318)*
C. villosa	Hydrolysed Leaf	*(318)*
Chamaecytisus hirsutus	Hydrolysed Leaf	*(302)*
C. ratisbonensis	Hydrolysed Leaf	*(302)*
C. supinus	Hydrolysed Leaf	*(302)*
Chamaespartium sagittale	Hydrolysed Leaf	*(318)*
C. tridentatum	Hydrolysed Leaf	*(302)*
Chronanthus biflorus	Hydrolysed Leaf	*(318)*
Cytisus ardoini	Hydrolysed Leaf	*(302)*
C. ingramii	Hydrolysed Leaf	*(302)*
C. multiflorus	Hydrolysed Leaf	*(302)*
C. striatus	Hydrolysed Leaf	*(302)*
C. tribracteolatus	Hydrolysed Leaf	*(302)*

Table 2 *(continued)*

Plant and/or Other Sources	Plant Part(s) Examined	References
Echinospartum horridum	Hydrolysed Leaf/Stem	*(303)*
Erinacea anthyllis	Hydrolysed Leaf	*(318)*
Genista spp.	Hydrolysed Leaf	*(318)*
G. clavata	Hydrolysed Leaf	*(302)*
G. hispanica	Twig	*(624)*
Laburnum alpinum	Heartwood	*(237)*
L. anagyroides	Hydrolysed Leaf	*(318)*
	Sapwood	*(165)*
Lembotropis emeriflora	Hydrolysed Leaf	*(302)*
Lupinus luteus	Hydrolysed Root	*(505)*
Lygos monosperma	Hydrolysed Leaf	*(318)*
L. raetam	Hydrolysed Leaf	*(318)*
Ormosia excelsa	Wood	*(296)*
Spartium junceum	Aerial Parts	*(619)*
Stauracanthus boivinii	Hydrolysed Leaf	*(302)*
S. genistoides	Hydrolysed Leaf	*(318)*
Teline spp.	Hydrolysed Leaf	*(318)*
Thermopsis montana	Hydrolysed Leaf	*(302)*
Ulex spp.	Hydrolysed Leaf	*(318)*
U. micranthus	Hydrolysed Leaf	*(302)*

86   MW 286; $C_{15}H_{10}O_6$

2'-HYDROXYGENISTEIN

Apios tuberosa	† Leaf	*(396)*
Argyrocytisus battandieri	† Leaf	*(356)*
Cajanus cajan	† Etiolated Stem	*(363)*
Crotalaria juncea	† Leaf	*(356)*
Dolichos biflorus	† Leaf/Stem	*(443)*
Hardenbergia violacea	† Leaf	*(356)*
Lablab niger	† Etiolated Hypocotyl	*(367)*
Laburnum anagyroides	† Leaf	*(389)*
Lupinus albus	† Etiolated Hypocotyl	*(389)*
	Leaf	*(777a)*
Moghania macrophylla	Wood	*(482, 651)*
Neonotonia wightii	† Stem	*(391)*
Phaseolus vulgaris	† Pod	*(77, 829)*
Spartium junceum	† Leaf	*(389)*
Stizolobium deeringianum	† Etiolated Stem	*(356)*

Table 2 *(continued)*

Plant and/or Other Sources	Plant Part(s) Examined	References

87   MW 286; $C_{15}H_{10}O_6$

OROBOL (NORSANTAL; SANTOL)

Plant Sources

Baptisia spp.	Leaf/Stem	*(538, 540)*
Bolusanthus speciosus	Seed	*(244, 317)*
Cytisus scoparius	Flower	*(494)*
Lathyrus spp.	Hydrolysed Leaf	*(379)*
L. nissolia	Hydrolysed Phyllode	*(302)*
Thermopsis spp.	Leaf/Stem	*(204)*

Other Sources

Aspergillus niger		
cultures [Metabolite]		*(792)*
Fusarium spp.		
cultures [Metabolite]		*(44a)*
Stemphylium sp.		
cultures [Metabolite]		*(792)*
Streptomyces sp.		
cultures [Metabolite]		*(792)*

88   MW 286; $C_{15}H_{10}O_6$

6-HYDROXYGENISTEIN

Baptisia hirsuta	Leaf/Stem	*(539, 540)*
Centrosema plumieri	† Leaf	*(537)*

References, pp. 229—265

Table 2 *(continued)*

Plant and/or Other Sources	Plant Part(s) Examined	References

89 MW 286; $C_{15}H_{10}O_6$

8-HYDROXYGENISTEIN

Aspergillus niger
  cultures [Metabolite]   *(792)*

90 MW 298; $C_{16}H_{10}O_6$

5-HYDROXYPSEUDOBAPTIGENIN (3',4'-METHYLENEDIOXYOROBOL)

Lupinus luteus	Hydrolysed Flower	*(505)*
	Root/Hydrolysed Root	*(505, 506)*
	Senescent Leaf/Stem	*(507)*
Sophora japonica	Root	*(469)*

91 MW 298; $C_{17}H_{14}O_5$

5-*O*-METHYLBIOCHANIN A

Plant Source

  Echinospartum horridum   Hydrolysed Leaf/Stem   *(303)*

Other Source

  Fusarium spp.
    cultures [Metabolite]   *(44a)*

Table 2 *(continued)*

Plant and/or Other Sources	Plant Part(s) Examined	References

92   MW 298; $C_{16}H_{10}O_6$

IRILONE

* *Iris germanica* (IRID)	Rhizome	*(218, 621)*
Trifolium pratense	Root	*(256, 648)*

93   MW 300; $C_{16}H_{12}O_6$

2'-HYDROXYBIOCHANIN A

* *Virola caducifolia* (MYR)	Trunk Wood	*(104)*
* *V. multinervia* (MYR)	Trunk Wood	*(100)*

94   MW 300; $C_{16}H_{12}O_6$

PRATENSEIN

Cicer arietinum	Aerial Parts	*(347)*
	Leaf	*(437)*
	Germinating Seed	*(828)*
Cytisus scoparius	Flower?	*(494)*
Monopteryx inpae	Trunk Wood	*(16)*
Sophora japonica	Wood	*(779)*
Thermopsis spp.	Leaf/Stem	*(204)*
Trifolium pratense	Leaf/Stem	*(823)*
	Root	*(258)*
T. subterraneum	Leaf	*(826)*

References, pp. 229—265

Table 2 *(continued)*

Plant and/or Other Sources	Plant Part(s) Examined	References

95   MW 300; $C_{16}H_{12}O_6$

3'-O-METHYLOROBOL

Dalbergia inundata	Wood	*(511)*
Thermopsis spp.	Leaf/Stem	*(204)*
* *Wyethia helenioides* (COMP)	Aerial Parts	*(81a)*
* *W. mollis* (COMP)	Leaf/Stem	*(810)*

96   MW 300; $C_{16}H_{12}O_6$

CAJANIN

Cajanus cajan	† Etiolated Stem	*(363)*
Canavalia ensiformis	† Etiolated Hypocotyl	*(356)*
Centrosema pascuorum	† Leaf	*(537)*
C. pubescens	† Leaf	*(537)*

97   MW 300; $C_{16}H_{12}O_6$

SANTAL

Baphia nitida	Heartwood	*(615, 698)*
Pterocarpus osun	Heartwood	*(14)*
P. santalinus	Heartwood	*(687, 698)*
* *Wyethia helenioides* (COMP)	Aerial Parts	*(81a)*
* *W. mollis* (COMP)	Leaf/Stem	*(810)*

Table 2 *(continued)*

Plant and/or Other Sources	Plant Part(s) Examined	References

98   MW 300; $C_{16}H_{12}O_6$

TECTORIGENIN

Baptisia spp.	Leaf/Stem	*(540)*
Centrosema spp.	† Leaf	*(537)*
Dalbergia riparia	Wood	*(103)*
D. sissoo	Bark	*(220)*
	Flower	*(42)*
	Mature Pod	*(727)*
D. stevensonii	Heartwood	*(227)*
D. volubilis	Bark	*(450)*
	Leaf	*(146)*
* *Iris germanica* (IRID)	Rhizome	*(621)*
Ononis spinosa	Root	*(47a)*

99   MW 300; $C_{16}H_{12}O_6$

ISOTECTORIGENIN (PSEUDOTECTORIGENIN)

Plant Source

Dalbergia sissoo	Bark	*(220)*

Other Source

Aspergillus niger cultures [Metabolite]		*(792)*

100   MW 312; $C_{17}H_{12}O_6$

BETAVULGARIN

Table 2 *(continued)*

Plant and/or Other Sources	Plant Part(s) Examined	References
* *Beta corolliflora* (CHEN)	† Leaf	*(692)*
* *B. lomatogona* (CHEN)	† Leaf	*(692)*
* *B. procumbens* (CHEN)	† Leaf	*(692)*
* *B. trigyna* (CHEN)	† Leaf	*(692)*
* *B. vulgaris* (CHEN)	† Leaf	*(276)*
	† Stem	*(692)*
* *B. vulgaris* ssp. maritima (CHEN)	† Leaf	*(692)*
* *Dianthus* sp. (CARY)	† Leaf	*(691a)*

101 MW 312; $C_{17}H_{12}O_6$

IRISOLONE

* *Iris florentina* (IRID)	Rhizome	*(23, 565)*
* *I. germanica* (IRID)	Rhizome	*(217, 621)*
* *I. nepalensis* (IRID)	Rhizome	*(290, 291, 650)*

102 MW 314; $C_{17}H_{14}O_6$

2′-METHOXYBIOCHANIN A

* *Virola caducifolia* (MYR)	Trunk Wood	*(104)*
* *V. multinervia* (MYR)	Trunk Wood	*(100)*

103 MW 314; $C_{17}H_{14}O_6$

or

Cajanus cajan	† Etiolated Stem	*(363)*

Table 2 *(continued)*

Plant and/or Other Sources	Plant Part(s) Examined	References

104   MW 314; $C_{17}H_{14}O_6$

IRISOLIDONE

* *Iris germanica* (IRID)	Rhizome	*(621)*
* *I. kashmiriana* (IRID)	Rhizome	*(219)*
* *I. nepalensis* (IRID)	Rhizome	*(650)*
* *I. tingitana* (IRID)	Bulb	*(234)*
Pericopsis mooniana	Heartwood	*(252)*
Pueraria thunbergiana	Flower	*(491)*
Sophora japonica	Root	*(469)*

105   MW 314; $C_{17}H_{14}O_6$

7-*O*-METHYLTECTORIGENIN

Dalbergia sissoo	Flower	*(42)*
	Immature Pod	*(12)*
D. spruceana	—	*(825)*
D. volubilis	Bark	*(450)*
Pterocarpus angolensis	Heartwood	*(564)*

106   MW 314; $C_{17}H_{14}O_6$

MUNINGIN

Pterocarpus angolensis	Heartwood	*(455)*

References, pp. 229—265

Table 2 *(continued)*

Plant and/or Other Sources	Plant Part(s) Examined	References

107 MW 316; $C_{16}H_{12}O_7$

JUNIPEGENIN A

* *Juniperus macropoda* (CUPR)   Leaf   *(722)*

108 MW 316; $C_{16}H_{12}O_7$

8-METHOXYOROBOL

Aspergillus niger
cultures [Metabolite]   *(782, 792)*

109 MW 326; $C_{18}H_{14}O_6$

DERRUSTONE

Derris robusta   Root   *(231)*

110 MW 326; $C_{18}H_{14}O_6$

TLATLANCUAYIN

* *Iresine celosioides* (AMAR)   Whole Plant   *(181)*

Table 2 *(continued)*

Plant and/or Other Sources	Plant Part(s) Examined	References

111 MW 326; $C_{18}H_{14}O_6$

IRISOLONE METHYL ETHER

* *Iris tingitana* (IRID)   Bulb   *(234)*

112 MW 328; $C_{18}H_{16}O_6$

7,4'-DI-O-METHYLTECTORIGENIN

Dalbergia sissoo   Flower   *(40)*

113 MW 328; $C_{18}H_{16}O_6$

5-METHOXYAFRORMOSIN

Cladrastis platycarpa	Heartwood/Sapwood	*(601)*
	Leaf	*(600)*
	Trunk Bark	*(352)*
C. shikokiana	As given for **5**	*(599)*
Wisteria floribunda	Wood	*(596)*

114 MW 328; $C_{18}H_{16}O_6$

ISO-5-METHOXYAFRORMOSIN

Cladrastis platycarpa	Bark/Heartwood/Sapwood	*(599)*
C. shikokiana	Bark/Heartwood/Sapwood	*(599)*

Table 2 *(continued)*

Plant and/or Other Sources	Plant Part(s) Examined	References

115   MW 330; $C_{17}H_{14}O_7$

PODOSPICATIN

* *Podocarpus spicatus* (POD)   Heartwood   *(113, 115)*

116   MW 330; $C_{17}H_{14}O_7$

IRISTECTORIGENIN A

Plant Sources

* *Iris germanica* (IRID)   Rhizome   *(621)*
* *I. unguicularis* (IRID)   Rhizome   *(20, 22)*
  Monopteryx inpae   Trunk Wood   *(16)*

Other Source

  Streptomyces sp.
  cultures [Metabolite]   *(163)*

117   MW 330; $C_{17}H_{14}O_7$

IRISTECTORIGENIN B

* *Iris florentina* (IRID)   Rhizome   *(23, 565)*
* *I. germanica* (IRID)   Rhizome   *(621)*

Table 2 *(continued)*

Plant and/or Other Sources	Plant Part(s) Examined	References

118   MW 330; $C_{17}H_{14}O_7$

Plant Source

  Monopteryx inpae   Trunk Wood   *(16)*

Other Source

  Streptomyces sp.
  cultures [Metabolite]   *(163)*

119   MW 342; $C_{18}H_{14}O_7$

DIPTERYXINE (ISOPLATYCARPANETIN)

Cladrastis platycarpa	Bark/Heartwood/Sapwood	*(599)*
C. shikokiana	Bark/Heartwood/Sapwood	*(599)*
Dipteryx odorata	Bark	*(574)*

120   MW 342; $C_{18}H_{14}O_7$

IRISKUMAONIN

* *Iris kumaonensis* (IRID)	Whole Plant	*(415)*
* *I. tingitana* (IRID)	Bulb	*(234)*

References, pp. 229—265

Table 2 *(continued)*

Plant and/or Other Sources	Plant Part(s) Examined	References

121   MW 342; $C_{18}H_{14}O_7$
PLATYCARPANETIN

Cladrastis platycarpa	Heartwood/Sapwood	*(601)*
	Leaf	*(599)*
	Trunk Bark	*(352)*
C. shikokiana	As given for **5**	*(599)*

122   MW 344; $C_{18}H_{16}O_7$

DERRUGENIN

(Revised Structure)

| *Derris robusta* | Seed Shell | *(158, 784)* |

123   MW 344; $C_{18}H_{16}O_7$

JUNIPEGENIN B

| * *Juniperus macropoda* (CUPR) | Leaf | *(723)* |

124   MW 344; $C_{18}H_{16}O_7$

| *Monopteryx inpae* | Trunk Wood | *(16)* |

Table 2 *(continued)*

Plant and/or Other Sources	Plant Part(s) Examined	References

125   MW 356; $C_{19}H_{16}O_7$

ODORATINE

| *Cordyla africana* | Heartwood | *(135)* |
| *Dipteryx odorata* | Bark | *(574)* |

126   MW 356; $C_{19}H_{16}O_7$

IRISKUMAONIN METHYL ETHER

| * *Iris germanica* (IRID) | Rhizome | *(621)* |
| * *I. tingitana* (IRID) | Bulb | *(234)* |

127   MW 358; $C_{19}H_{18}O_7$

ROBUSTIGENIN

| *Derris robusta* | Seed Shell | *(157)* |

128   MW 360; $C_{18}H_{16}O_8$

IRIGENIN

Table 2 *(continued)*

Plant and/or Other Sources	Plant Part(s) Examined	References
* *Iris florentina* (IRID)	Rhizome	*(23, 565)*
* *I. germanica* (IRID)	Rhizome	*(236, 621)*
* *I. nepalensis* (IRID)	Rhizome	*(291, 650)*
* *I. tingitana* (IRID)	Bulb	*(234)*
* *I. unguicularis* (IRID)	Rhizome	*(20, 22)*
* *Juniperus macropoda* (CUPR)	Leaf	*(722)*

129   MW 372; $C_{20}H_{20}O_7$

ROBUSTIGENIN METHYL ETHER

Derris robusta	Seed Shell	*(161)*

130   MW 374; $C_{19}H_{18}O_8$

CAVIUNIN

Dalbergia barretoana	Wood	*(90)*
D. inundata	Wood	*(511)*
D. nigra	Heartwood/Sapwood	*(298)*
D. paniculata	Bark	*(632)*
	Flower	*(4, 7, 8)*
	Root	*(4, 663, 666)*
	Seed	*(4, 9, 662)*
	Wood	*(4)*
D. riparia	Wood	*(103)*
D. spruceana	Bark/Sapwood	*(176)*
D. villosa	Wood	*(90)*

131   MW 374; $C_{19}H_{18}O_8$

JUNIPEGENIN C

* *Juniperus macropoda* (CUPR)	Leaf	*(723)*

Table 2 (continued)

Plant and/or Other Sources	Plant Part(s) Examined	References

132   MW 374; $C_{19}H_{18}O_8$

ISOCAVIUNIN

Dalbergia sissoo	Mature Pod	(727)

133   MW 386; $C_{20}H_{18}O_8$

IRISFLORENTIN

* Iris florentina (IRID)	Rhizome	(23, 565)
* I. germanica (IRID)	Rhizome	(621)
* I. tingitana (IRID)	Bulb	(234)

134   MW 386; $C_{20}H_{18}O_8$

Cordyla africana	Heartwood	(135)

d) Complex 5-Oxy Isoflavones

135   MW 336; $C_{20}H_{16}O_5$

ALPINUMISOFLAVONE

Table 2 *(continued)*

Plant and/or Other Sources	Plant Part(s) Examined	References
Calopogonium mucunoides	Seed	*(808)*
Erythrina variegata	Bark	*(209)*
Laburnum alpinum	Twig	*(399)*
Millettia thonningii	Seed	*(607a)*

136   MW 336; $C_{20}H_{16}O_5$

DERRONE

Derris robusta	Seed	*(162)*

137   MW 338; $C_{20}H_{18}O_5$

Moghania macrophylla	Wood	*(482)*

138   MW 338; $C_{20}H_{18}O_5$

WIGHTEONE (ERYTHRININ B; LUPINUS COMPOUND LA-1)

Argyrocytisus battandieri	† Leaf	*(356)*
Erythrina variegata	Bark	*(209)*
Laburnum anagyroides	† Leaf	*(389)*
Lupinus albus	† Hypocotyl	*(391)*
	Leaf	*(321, 777a)*
L. angustifolius	Leaf	*(777a)*
L. polyphyllus	Leaf	*(321, 356)*
Neonotonia wightii	† Stem	*(391)*

Table 2 *(continued)*

Plant and/or Other Sources	Plant Part(s) Examined	References

139 MW 350; $C_{21}H_{18}O_5$

4'-O-METHYLALPINUMISOFLAVONE

Calopogonium mucunoides   Seed   *(807)*

140 MW 350; $C_{21}H_{18}O_5$

4'-O-METHYLDERRONE

Calopogonium mucunoides   Seed   *(807)*
Derris robusta   Seed   *(162a)*

141 MW 352; $C_{20}H_{16}O_6$

LICOISOFLAVONE B

Glycyrrhiza sp.   Root Bark   *(709)*

142 MW 352; $C_{20}H_{16}O_6$

PARVISOFLAVONE B

Poecilanthe parviflora   Trunk Wood   *(25)*

References, pp. 229—265

Table 2 *(continued)*

Plant and/or Other Sources	Plant Part(s) Examined	References

143 MW 352; $C_{20}H_{16}O_6$

PARVISOFLAVONE A

| *Poecilanthe parviflora* | Trunk Wood | *(25)* |

144 MW 354; $C_{20}H_{18}O_6$

LICOISOFLAVONE A (PHASEOLUTEONE)

Glycyrrhiza sp.	Root	*(460)*
Hardenbergia violacea	† Leaf	*(356)*
Phaseolus vulgaris	†Pod	*(829)*

145 MW 354; $C_{20}H_{18}O_6$

LUTEONE

Argyrocytisus battandieri	† Leaf	*(356)*
Hardenbergia violacea	† Leaf	*(356)*
Laburnum anagyroides	† Leaf	*(389)*
Lupinus spp.	Leaf	*(321)*
L. albus	† Etiolated Hypocotyl	*(389)*
L. angustifolius	Leaf	*(777a)*
L. luteus	Fruit	*(269)*

Table 2 *(continued)*

Plant and/or Other Sources	Plant Part(s) Examined	References

146   MW 354; $C_{20}H_{18}O_6$
ERYTHRININ C

Erythrina variegata   Bark   *(209)*

147   MW 354; $C_{20}H_{18}O_6$
2,3-DEHYDROKIEVITONE

Phaseolus vulgaris   † Pod   *(830)*

148   MW 364; $C_{21}H_{16}O_6$
ROBUSTONE

Derris robusta   Root   *(231)*
                             Seed   *(160)*

Table 2 *(continued)*

Plant and/or Other Sources	Plant Part(s) Examined	References

149   MW 364; $C_{22}H_{20}O_5$

ALPINUMISOFLAVONE DIMETHYL ETHER

Derris robusta   Seed   *(160)*

150   MW 366; $C_{21}H_{18}O_6$

DERRUBONE

Derris robusta   Root   *(231)*

150a MW 366; $C_{21}H_{18}O_6$

Millettia thonningii   Seed   *(607a)*

151   MW 368; $C_{21}H_{20}O_6$

AURMILLONE

Millettia auriculata   Seed   *(664, 665)*

Table 2 *(continued)*

Plant and/or Other Sources	Plant Part(s) Examined	References

152　MW 378; $C_{22}H_{18}O_6$

ROBUSTONE METHYL ETHER

Derris robusta　　　　　Root　　　　　　(230)
　　　　　　　　　　　　　Seed　　　　　　(160)

153　MW 378; $C_{22}H_{18}O_6$

GLABRESCIONE A

Derris glabrescens　　　Seed　　　　　　(202)

154　MW 384; $C_{21}H_{20}O_7$

PISCERYTHRONE

Piscidia erythrina　　　Whole Root　　　(239)

Table 2 *(continued)*

Plant and/or Other Sources	Plant Part(s) Examined	References

155 MW 384; $C_{21}H_{20}O_7$

PISCIDONE

Piscidia erythrina   Whole Root   *(239, 610)*

156 MW 396; $C_{22}H_{20}O_7$

ELONGATIN

Tephrosia elongata   Aerial Parts/Root   *(758)*

157 MW 404; $C_{25}H_{24}O_5$

CHANDALONE

Derris scandens   Root   *(238)*

158 MW 404; $C_{25}H_{24}O_5$

OSAJIN

Table 2 *(continued)*

Plant and/or Other Sources	Plant Part(s) Examined	References
Derris scandens	Root	*(636)*
Erythrina variegata	Bark	*(209)*
Maclura pomifera (MOR)	Fruit	*(822)*

159   MW 404; $C_{25}H_{24}O_5$

WARANGALONE (SCANDENONE; ISO-OSAJIN)

Derris scandens	Root	*(238, 636)*

160   MW 406; $C_{25}H_{26}O_5$

3'-γ,γ-DIMETHYLALLYL-WIGHTEONE

Millettia pachycarpa	Aerial Parts	*(755)*

161   MW 406; $C_{25}H_{26}O_5$

8-γ,γ-DIMETHYLALLYL-WIGHTEONE

Millettia pachycarpa	Aerial Parts	*(755)*

Table 2 *(continued)*

Plant and/or Other Sources	Plant Part(s) Examined	References

162 MW 410; $C_{23}H_{22}O_7$

TOXICAROL ISOFLAVONE

Derris malaccensis   Root   *(326, 329)*

163 MW 418; $C_{26}H_{26}O_5$

SCANDINONE (NALLANIN)

Derris scandens   Root   *(636, 682)*

164 MW 420; $C_{25}H_{24}O_6$

ISOAURICULATIN
(Revised Structure)

Millettia auriculata   Leaf   *(557)*
                                Root   *(724)*

Table 2 *(continued)*

Plant and/or Other Sources	Plant Part(s) Examined	References

165   MW 420; $C_{25}H_{24}O_6$
ISOAURICULASIN

Millettia auriculata   Leaf   *(557)*

166   MW 420; $C_{25}H_{24}O_6$
POMIFERIN

* *Maclura pomifera* (MOR)   Fruit   *(822)*

167   MW 420; $C_{25}H_{24}O_6$
AURICULATIN

Millettia auriculata   Root   *(725)*
                                    Seed   *(664, 665)*

Table 2 *(continued)*

Plant and/or Other Sources	Plant Part(s) Examined	References

168   MW 420; $C_{25}H_{24}O_6$

AURICULASIN

Millettia auriculata	Leaf	*(557)*
	Seed	*(664, 665)*

169   MW 422; $C_{25}H_{26}O_6$

3′-γ,γ-DIMETHYLALLYL-LUTEONE

Lupinus angustifolius	Root	*(707a)*

170   MW 422; $C_{25}H_{26}O_6$

6,8-DI-γ,γ-DIMETHYLALLYL-OROBOL

Millettia pachycarpa	Aerial Parts	*(755)*

Table 2 *(continued)*

Plant and/or Other Sources	Plant Part(s) Examined	References

171   MW 434; $C_{26}H_{26}O_6$
AURICULIN

Millettia auriculata   Root   *(724)*

172   MW 436; $C_{26}H_{28}O_6$

Millettia pachycarpa   Leaf   *(754)*

173   MW 436; $C_{26}H_{28}O_6$

or.

Millettia pachycarpa   Leaf   *(754)*

Table 2 *(continued)*

Plant and/or Other Sources	Plant Part(s) Examined	References

174 MW 450; $C_{27}H_{30}O_6$

GLABRESCIONE B

| *Derris glabrescens* | Seed | *(202)* |

175 MW 468; $C_{27}H_{32}O_7$

| *Millettia pachycarpa* | Leaf | *(754)* |

176 MW 468; $C_{27}H_{32}O_7$

| *Millettia pachycarpa* | Leaf | *(754)* |

Table 2 *(continued)*

Plant and/or Other Sources	Plant Part(s) Examined	References

e) Chloro Isoflavones

177   MW 304; $C_{15}H_9O_5Cl$

6-CHLOROGENISTEIN

Streptomyces griseus cultures [Metabolite]		*(474)*

178   MW 338; $C_{15}H_8O_5Cl_2$

6,3'-DICHLOROGENISTEIN

Streptomyces griseus cultures [Metabolite]		*(474)*

B: ISOFLAVONEQUINONE

Plant and/or Other Sources	Plant Part(s) Examined	References

179   MW 298; $C_{16}H_{10}O_6$

BOWDICHIONE

| *Bowdichia nitida* | Heartwood | *(123)* |

Table 2 *(continued)*

C: COUMARANOCHROMONES

Plant and/or Other Sources	Plant Part(s) Examined	References

180 MW 382; $C_{21}H_{18}O_7$

LISETIN

Piscidia communis	—	*(239)*
P. erythrina	Root Bark	*(563)*
	Whole Root	*(239)*

181 MW 418; $C_{25}H_{22}O_6$

MILLETTIN

Millettia auriculata	Seed	*(665)*

D: ISOFLAVANONES

Plant and/or Other Sources	Plant Part(s) Examined	References

a) Simple 5-Deoxy Isoflavanones

182 MW 256; $C_{15}H_{12}O_4$

DIHYDRODAIDZEIN

Pericopsis mooniana	Heartwood	*(252)*

Table 2 *(continued)*

Plant and/or Other Sources	Plant Part(s) Examined	References

183   MW 270; $C_{16}H_{14}O_4$
DIHYDROFORMONONETIN

Myroxylon balsamum   Trunk Wood   *(96)*

184   MW 272; $C_{15}H_{12}O_5$
2'-HYDROXYDIHYDRODAIDZEIN

Phaseolus vulgaris   † Pod   *(831)*

185   MW 286; $C_{16}H_{14}O_5$
VESTITONE

Plant Sources

Medicago rugosa	† Leaf	*(381)*
Onobrychis viciifolia	† Leaf	*(370)*
Tipuana tipu	† Leaf	*(356)*
Trifolium repens	† Leaf	*(832)*

Other Source

Nectria haematococca cultures [Metabolite]	*(205)*

Table 2 *(continued)*

Plant and/or Other Sources	Plant Part(s) Examined	References

186　MW 286; $C_{16}H_{14}O_5$

3'-HYDROXYDIHYDROFORMONONETIN

Myroxylon balsamum	Trunk Wood	*(96)*

187　MW 300; $C_{17}H_{16}O_5$

SATIVANONE

Dalbergia stevensonii	Heartwood	*(227)*
Medicago sativa	† Leaf	*(371)*

188　MW 300; $C_{17}H_{16}O_5$

3'-METHOXYDIHYDROFORMONONETIN

Cladrastis shikokiana	† Heartwood	*(595)*

189　MW 300; $C_{16}H_{12}O_6$

SOPHOROL

Table 2 *(continued)*

Plant and/or Other Sources	Plant Part(s) Examined	References
Plant Source		
Maackia amurensis var. *buergeri*	Heartwood	*(771)*
Other Source		
Nectria haematococca cultures [Metabolite]		*(205)*

190 MW 300; $C_{17}H_{16}O_5$

ISOSATIVANONE

Medicago rugosa	Leaf	*(381)*

191 MW 314; $C_{17}H_{14}O_6$

ONOGENIN

Dalbergia stevensonii	Heartwood	*(227)*
Ononis arvensis	Root	*(478)*

192 MW 316; $C_{17}H_{16}O_6$

VIOLANONE

Dalbergia oliveri	Heartwood	*(225)*
D. violacea	Heartwood	*(610)*

References, pp. 229—265

Table 2 *(continued)*

Plant and/or Other Sources	Plant Part(s) Examined	References

192a   MW 316; $C_{17}H_{16}O_6$

LESPEDEOL C

Lespedeza cyrtobotrya	Heartwood	*(562a)*

193   MW 328; $C_{18}H_{16}O_6$

Cordyla africana	Heartwood	*(134)*
Mildbraedeodendron excelsa	Heartwood	*(550)*

194   MW 330; $C_{18}H_{18}O_6$

3'-*O*-METHYLVIOLANONE

Dalbergia cearensis	Softwood	*(308)*

195   MW 330; $C_{18}H_{18}O_6$

3'-METHOXYISOSATIVANONE

Myroxylon peruiferum	Trunk Wood	*(536)*

Table 2 *(continued)*

Plant and/or Other Sources	Plant Part(s) Examined	References

b) Complex 5-Deoxy Isoflavanones

196   MW 324; $C_{19}H_{16}O_5$
       NEORAUNONE

| *Neorautanenia amboensis* | Root Bark | *(108)* |

197   MW 338; $C_{19}H_{14}O_6$
       NEOTENONE (NEORAUTENONE)

Neorautanenia amboensis	Root	*(119)*
N. edulis	Root/Tuber	*(119, 795)*
N. pseudopachyrrhiza	Root	*(185, 186)*
Pachyrrhizus erosus	Seed	*(185, 186, 416)*

198   MW 340; $C_{20}H_{20}O_5$
       5-DEOXYKIEVITONE

| *Phaseolus vulgaris* | † Pod | *(830)* |

References, pp. 229—265

Table 2 *(continued)*

Plant and/or Other Sources	Plant Part(s) Examined	References

199   MW 354; $C_{20}H_{18}O_6$

NEPSEUDIN

H_3CO OCH_3
OCH_3

Neorautanenia
  pseudopachyrrhiza   Root   *(186)*

200   MW 354; $C_{20}H_{18}O_6$

AMBONONE

OCH_3
H_3CO OCH_3

Neorautanenia amboensis   Root Bark   *(108)*

201   MW 354; $C_{19}H_{14}O_7$

EROSENONE

Pachyrrhizus erosus   Seed   *(416)*

202   MW 386; $C_{21}H_{22}O_7$

SECONDIFLORAN

HO OCH_3
OH

Sophora secundiflora   Aerial Parts
                (Mainly Stem)   *(559)*

Table 2 *(continued)*

Plant and/or Other Sources	Plant Part(s) Examined	References

c) Simple 5-Oxy Isoflavanones

203   MW 286; $C_{16}H_{14}O_5$

DIHYDROBIOCHANIN A

| *Andira parviflora* | Trunk Wood | *(102)* |

204   MW 288; $C_{15}H_{12}O_6$

DALBERGIOIDIN

Dolichos biflorus	† Leaf/Stem	*(443)*
Lablab niger	† Etiolated Hypocotyl	*(367)*
Lespedeza cyrtobotrya	Heartwood	*(562)*
Macrotyloma axillare	† Leaf	*(356)*
Ougeinia dalbergioides	Heartwood	*(38)*
	Leaf	*(13)*
Phaseolus vulgaris	† Pod	*(829)*
Stizolobium deeringianum	† Etiolated Stem	*(356)*

205   MW 302; $C_{16}H_{14}O_6$

FERREIRIN

| *Ferreirea spectabilis* | Heartwood | *(457)* |

Table 2 *(continued)*

Plant and/or Other Sources	Plant Part(s) Examined	References

206　MW 302; $C_{16}H_{14}O_6$

ISOFERREIRIN

Dolichos biflorus	† Leaf/Stem	*(443)*
Stizolobium deeringianum	† Etiolated Stem	*(356)*

207　MW 316; $C_{17}H_{16}O_6$

HOMOFERREIRIN

Argyrocytisus battandieri	† Leaf	*(356)*
Cicer arietinum	Root	*(44)*
Ferreirea spectabilis	Heartwood	*(457)*
Ougeinia dalbergioides	Heartwood	*(38)*
	Leaf	*(13)*

208　MW 316; $C_{17}H_{16}O_6$

CAJANOL

(Revised Structure)

Cajanus cajan	† Root	*(372)*
	† Etiolated Stem	*(363, 372)*
Stizolobium deeringianum	† Etiolated Stem	*(356)*

Table 2 *(continued)*

Plant and/or Other Sources	Plant Part(s) Examined	References

209   MW 332; $C_{17}H_{16}O_7$

PARVISOFLAVANONE

Poecilanthe parviflora   Trunk Wood   (25)

210   MW 346; $C_{18}H_{18}O_7$

OUGENIN

Ougeinia dalbergioides   Heartwood   (38)
                                     Leaf   (13)

d) Complex 5-Oxy Isoflavanones

211   MW 354; $C_{20}H_{18}O_6$

LICOISOFLAVANONE

Glycyrrhiza sp.   Root Bark   (709)

212   MW 354; $C_{20}H_{18}O_6$

1″,2″-DEHYDROCYCLOKIEVITONE

Phaseolus vulgaris   † Pod   (830)

References, pp. 229—265

Table 2 *(continued)*

Plant and/or Other Sources	Plant Part(s) Examined	References

213 MW 356; $C_{20}H_{20}O_6$

KIEVITONE (PHASEOLUS SUBSTANCE II; VIGNATIN)

Dolichos biflorus	† Leaf/Stem	*(443)*
Lablab niger	† Etiolated Hypocotyl	*(367)*
Macroptilium atropurpureum	† Leaf	*(356)*
Macrotyloma axillare	† Leaf	*(356)*
Mucuna utilis	† Germinating Seed	*(583a)*
Phaseolus aureus	† Etiolated Hypocotyl	*(356)*
P. calcaratus	† Germinating Seed/ Leaf/Root	*(356, 773a)*
P. lunatus	† Cotyledon	*(633)*
P. vulgaris	† Etiolated Hypocotyl/ Hypocotyl	*(127, 761)*
	† Germinating Seed	*(285)*
	† Leaf	*(286)*
	† Pod	*(829)*
	† Root	*(129)*
Stizolobium deeringianum	† Etiolated Stem	*(356)*
Vigna unguiculata	† Cotyledon	*(633)*
	† Germinating Seed	*(441, 545, 634)*
	† Etiolated Hypocotyl/ Hypocotyl	*(29, 633)*

214 MW 368; $C_{21}H_{20}O_6$

ISOSOPHORONOL

Sophora tomentosa	Root	*(203)*

Table 2 *(continued)*

Plant and/or Other Sources	Plant Part(s) Examined	References

215   MW 370; $C_{21}H_{22}O_6$
SOPHORAISOFLAVANONE A

Sophora tomentosa   Aerial Parts   *(471)*

216   MW 374; $C_{20}H_{22}O_7$
KIEVITONE HYDRATE

Fusarium solani f. sp.
phaseoli cultures
[Metabolite]   *(489)*

217   MW 422; $C_{25}H_{26}O_6$
CAJANONE

Cajanus cajan   Root   *(654)*

Table 2 *(continued)*

Plant and/or Other Sources	Plant Part(s) Examined	References

218   MW 422; $C_{25}H_{26}O_6$
LESPEDEOL B

Lespedeza homoloba   Bark   *(788)*

219   MW 424; $C_{25}H_{28}O_6$
LESPEDEOL A

Lespedeza homoloba   Bark   *(790)*

220   MW 436; $C_{26}H_{28}O_6$
2'-*O*-METHYLCAJANONE

Cajanus cajan   Root Bark   *(60)*

Table 2 *(continued)*

Plant and/or Other Sources	Plant Part(s) Examined	References

221   MW 438; $C_{26}H_{30}O_6$

ISOSOPHORANONE

Sophora tomentosa	Aerial Parts	*(471)*
	Root	*(203)*

222   MW 438; $C_{26}H_{30}O_6$

SOPHORAISOFLAVANONE B

| *Sophora franchetiana* | Root | *(472)* |

E: ROTENOIDS

Plant and/or Other Sources	Plant Part(s) Examined	References

a) Simple 12a-Deoxy Rotenoids

223   MW 342; $C_{19}H_{18}O_6$

MUNDUSERONE

Mundulea sericea		
(*=suberosa*)	Root Bark	*(249)*

Table 2 *(continued)*

Plant and/or Other Sources	Plant Part(s) Examined	References

224 MW 358; $C_{19}H_{18}O_7$

SERMUNDONE

Mundulea sericea
  (=suberosa)   Root Bark   *(229, 609)*

b) Complex 12a-Deoxy Rotenoids

225 MW 336; $C_{19}H_{12}O_6$

DOLINEONE (DOLICHONE)

Neorautanenia
  pseudopachyrrhiza   Root   *(185, 186)*
Pachyrrhizus erosus   Seed   *(185, 186, 416)*

226 MW 352; $C_{20}H_{16}O_6$

ELLIPTONE

Crotalaria burhia   Callus Tissue   *(787)*
                           Root/Seed   *(447)*
C. medicaginea   Whole Plant   *(404)*
Derris elliptica   Root   *(199, 324)*
D. malaccensis   Root   *(326)*

Table 2 *(continued)*

Plant and/or Other Sources	Plant Part(s) Examined	References
Tephrosia falciformis	Root	*(417)*
T. purpurea	Callus Tissue/	
	Leaf/Root/Stem	*(731)*
T. strigosa	Aerial Parts	*(418)*
T. vogelii	Callus Tissue	*(731)*

227   MW 352; $C_{20}H_{16}O_6$

EROSONE (ISO-ELLIPTONE; NORTON & HANSBERRY'S COMPOUND V)

Pachyrrhizus erosus   Seed   *(416, 586, 609)*

228   MW 366; $C_{20}H_{14}O_7$

PACHYRRHIZONE (NORTON & HANSBERRY'S COMPOUND II)

Pachyrrhizus erosus   Seed   *(68, 416, 552,*
                                                           586)

229   MW 368; $C_{20}H_{16}O_7$

MALACCOL

Derris malaccensis   Root   *(325, 326)*

Table 2 *(continued)*

Plant and/or Other Sources	Plant Part(s) Examined	References

230 MW 378; $C_{22}H_{18}O_6$

MILLETTONE

Lonchocarpus rugosus	—	*(613)*
Millettia dura	Seed	*(613)*
Piscidia erythrina	Whole Root	*(239)*

231 MW 378; $C_{22}H_{18}O_6$

ISOMILLETTONE

Piscidia erythrina	Whole Root	*(239)*

232 MW 394; $C_{23}H_{22}O_6$

DEGUELIN

Crotalaria burhia	Callus Tissue	*(787)*
	Root/Seed	*(447)*
Derris elliptica	Callus Tissue	*(465)*
	Root	*(131, 168, 199)*

Table 2 *(continued)*

Plant and/or Other Sources	Plant Part(s) Examined	References
D. malaccensis	Root	*(133, 326)*
Lonchocarpus chrysophyllus	Root	*(695)*
L. longifolius	Root	*(96a)*
L. nicou	Root	*(168, 199)*
L. spruceanus	Root	*(553a)*
L. unifoliatus	Seed	*(201)*
L. urucu	Root	*(749)*
Millettia ferruginea	Seed	*(280)*
M. pachycarpa	Seed	*(164)*
Mundulea sericea		
(=*suberosa*)	Bark	*(834)*
Piscidia mollis	Seed	*(553a)*
Tephrosia cinerea	Root	*(284)*
T. falciformis	Root	*(417)*
T. purpurea	Callus Tissue/Leaf/	
	Root/Stem	*(731)*
T. strigosa	Aerial Parts	*(418)*
T. toxicaria	Root	*(694)*
T. villosa	Root	*(191)*
T. virginiana	Root	*(289)*
T. vogelli	Callus Tissue	*(731)*
	Leaf	*(131, 168, 199, 200, 316)*
	Petiole/Stem	*(200, 316)*
	Root	*(191, 200, 316)*
	Seed	*(554, 675)*

233   MW 394; $C_{23}H_{22}O_6$

ROTENONE (TUBOTOXIN; NICOULINE)

Amorpha fruticosa	Seedlings	*(184a, 184b)*
Antheroporum pierrei	Seed	*(410)*
Crotalaria burhia	Callus Tissue	*(787)*
	Root/Seed	*(447)*
C. medicaginea	Whole Plant	*(404)*
Derris amazonica	Root	*(483)*
D. brevipes	Stem	*(208)*
D. chinensis	Root	*(410, 694)*
D. cuneifolia	Root	*(410)*

Table 2 *(continued)*

Plant and/or Other Sources	Plant Part(s) Examined	References
D. elliptica	Bark/Petiole/Stem	*(151)*
	Callus Tissue	*(465)*
	Root	*(151, 199, 275,*
		324, 423, 644)
	Seed	*(835)*
D. ferruginea	Root	*(681)*
D. fordii	Root ?	*(164)*
D. grandifolia	Aerial Parts	*(410)*
D. malaccensis	Root	*(133, 326)*
	Seed	*(835)*
D. negrensis	—	*(804)*
D. polyantha	Root	*(410, 694)*
D. uliginosa	Bark ?/Stem/Twig ?	*(408)*
	Root	*(410, 644)*
	Seed	*(835)*
Lonchocarpus chrysophyllus	Root	*(410, 483)*
	Stem	*(410)*
L. longifolius	Root	*(96a)*
L. martynii	Root	*(410, 483)*
	Stem	*(410)*
L. nicou	Branch	*(695)*
	Leaf	*(410)*
	Root/Root Bark	*(199, 408)*
	Stem	*(408)*
L. rariflorus	Root	*(483)*
L. sericeus	Leaf/Root	*(410)*
L. spruceanus	Root	*(553a)*
L. urucu (= Derris urucu)	Creeper	*(101)*
	Leaf/Stem	*(483)*
	Root	*(483, 749)*
L. utilis	Leaf	*(483, 546)*
	Root/Stem	*(483)*
L. velutinus	Root	*(408)*
Millettia dura	Seed	*(613)*
M. ferruginea	Leaf/Petiole/Stem	*(410)*
	Seed	*(170, 280)*
M. ichthyochtona	Seed	*(410)*
M. laurentii	Leaf/Root/Stem	*(410)*
M. mannii	Leaf/Root/Stem	*(410)*
M. pachycarpa	Root	*(281, 679)*
	Seed	*(164)*
M. rubiginosa	Root	*(208)*
M. taiwaniana	Root	*(423)*
M. versicolor	Leaf/Root/Stem	*(410)*
Mundulea pauciflora	Root/Stem Bark	*(410)*
M. sericea (=suberosa)	Pod/Seed/Stem	*(410, 694)*
	Stem Bark/Branch Bark	*(553, 834)*
Neorautanenia amboensis	Root	*(589)*

Table 2 *(continued)*

Plant and/or Other Sources	Plant Part(s) Examined	References
N. ficifolia	Tuber	*(713)*
Ormocarpum glabrum	—	*(410)*
Ostryoderris lucida	Bark	*(821)*
Pachyrrhizus erosus	Seed	*(151, 416, 585, 586)*
Piscidia erythrina	Bark	*(125, 422)*
	Root Bark	*(563, 706)*
	Root Wood	*(706, 720)*
	Whole Root	*(239)*
P. mollis	Seed	*(553a)*
Poiretia tetraphylla	Flower/Leaf	*(584)*
Spatholobus roxburghii	Root Bark	*(407)*
Tephrosia candida	Pod/Root/Root/Bark/ Seed/Stem Bark	*(410, 479)*
T. cinerea	Aerial Parts	*(410)*
	Root	*(284)*
T. ehrenbergiana	Seed	*(835)*
T. falciformis	Root	*(417)*
T. latidens	Root	*(410)*
T. lindheimeri	—	*(696)*
T. macropoda	Root	*(544)*
T. obovata	Whole Plant	*(150, 151)*
T. piscatoria	Root/Seed	*(644)*
T. purpurea	Callus Tissue/Leaf/Stem	*(731)*
	Root	*(191, 731)*
T. strigosa	Aerial Parts	*(418)*
T. toxicaria	Pod/Seed	*(410)*
	Root	*(326)*
T. villosa	Root	*(191)*
T. virginiana	Root	*(169, 289)*
T. vogelii	Callus Tissue	*(731)*
	Flower	*(835)*
	Leaf/Petiole/Root/Stem	*(200, 316)*
	Pod/Seed	*(410)*
* *Verbascum thapsus* (SCRO)	Leaf	*(587, 588)*

234 MW 394; $C_{23}H_{22}O_6$

MYRICONOL

* *Myrica nagi* (MYRC)   Stem Bark   *(480)*

Table 2 *(continued)*

Plant and/or Other Sources	Plant Part(s) Examined	References

234a   MW 396; $C_{23}H_{24}O_6$

Amorpha fruticosa   Seedlings   *(184a)*

235   MW 410; $C_{23}H_{22}O_7$
AMORPHIGENIN

Plant Sources

Amorpha spp.	Seed	*(426, 427)*
A. fruticosa	Root	*(736)*
	Seed	*(2, 3, 166, 425, 427)*
	Seedlings	*(184a)*
Dalbergia monetaria	Seed	*(223a)*

Other Source

Animal and Insect
Tissue Homogenates
[Metabolite]   *(268)*

Table 2 *(continued)*

Plant and/or Other Sources	Plant Part(s) Examined	References

236   MW 410; $C_{23}H_{22}O_7$
TOXICAROL

Crotalaria burhia	Callus Tissue	*(787)*
	Root/Seed	*(447)*
Derris elliptica	Root	*(131)*
D. malaccensis	Root	*(325, 326)*
Tephrosia obovata	Whole Plant	*(150)*
T. toxicaria	Root	*(167, 326)*

237   MW 410; $C_{23}H_{22}O_7$
SUMATROL

Crotalaria burhia	Callus Tissue	*(787)*
	Root/Seed	*(447)*
Derris malaccensis	Root	*(133, 326, 446)*
Millettia auriculata	Root	*(725)*
	Seed	*(664)*
Piscidia erythrina	Whole Root	*(239)*
Tephrosia toxicaria	Root	*(326)*

Table 2 *(continued)*

Plant and/or Other Sources	Plant Part(s) Examined	References

238   MW 412; $C_{23}H_{24}O_7$

DIHYDROAMORPHIGENIN

Amorpha fruticosa	Seed	*(425, 426)*

239   MW 412; $C_{23}H_{24}O_7$

DALPANOL

Amorpha fruticosa	Seedlings	*(184a)*
Dalbergia paniculata	Seed	*(5, 662)*

240   MW 426; $C_{23}H_{22}O_8$

VILLOSIN

Tephrosia villosa	Pod	*(484)*

Table 2 *(continued)*

Plant and/or Other Sources	Plant Part(s) Examined	References

241   MW 428; $C_{23}H_{24}O_8$
AMORPHIGENOL

Plant Sources

Amorpha spp.	Seed	(*426, 427*)
A. fruticosa	Seed	(*425, 427*)
	Seedlings	(*184b*)

Other Source

Animal and Insect Tissue Homogenates [Metabolite]		(*268*)

c) Simple 12a-Oxy Rotenoids

242   MW 358; $C_{19}H_{18}O_7$
12a-HYDROXYMUNDUSERONE

Pachyrrhizus erosus	Seed	(*416*)

References, pp. 229—265

Table 2 *(continued)*

Plant and/or Other Sources	Plant Part(s) Examined	References

243 MW 390; $C_{19}H_{18}O_9$
CLITORIACETAL

Clitoria macrophylla   Root   *(777)*

d) Complex 12a-Oxy Rotenoids

244 MW 352; $C_{19}H_{12}O_7$
12a-HYDROXYDOLINEONE

Neorautanenia amboensis   Root   *(589)*
Pachyrrhizus erosus   Seed   *(416, 481)*

245 MW 366; $C_{20}H_{14}O_7$
12a-METHOXYDOLINEONE

Neorautanenia amboensis   Root   *(589)*

Table 2 *(continued)*

Plant and/or Other Sources	Plant Part(s) Examined	References

246 MW 368; $C_{20}H_{16}O_7$

12a-HYDROXYEROSONE

Neorautanenia amboensis	Root Bark	*(108)*
Pachyrrhizus erosus	Seed	*(416)*

247 MW 382; $C_{21}H_{18}O_7$

NEOBANONE

Neorautanenia amboensis	Root	*(590)*

248 MW 382; $C_{20}H_{14}O_8$

12a-HYDROXYPACHYRRHIZONE

Pachyrrhizus erosus	Seed	*(416, 481)*

References, pp. 229—265

Table 2 *(continued)*

Plant and/or Other Sources	Plant Part(s) Examined	References

249　MW 394; $C_{22}H_{18}O_7$

MILLETTOSIN

Millettia dura	Seed	*(613)*

250　MW 394; $C_{22}H_{18}O_7$

12a-HYDROXYISOMILLETTONE

Neorautanenia amboensis	Root	*(590)*

251　MW 410; $C_{23}H_{22}O_7$

TEPHROSIN

Crotalaria burhia	Callus Tissue	*(787)*
	Root/Seed	*(447)*
C. medicaginea	Whole Plant	*(404)*
Derris elliptica	Root	*(168, 199)*

Table 2 *(continued)*

Plant and/or Other Sources	Plant Part(s) Examined	References
D. malaccensis	Root	*(133)*
Lonchocarpus chrysophyllus	Root	*(695)*
L. longifolius	Root	*(96a)*
L. nicou	Root	*(168, 199)*
L. spruceanus	Root	*(553a)*
L. unifoliatus	Seed	*(201)*
L. urucu (= *Derris urucu*)	Creeper	*(101)*
	Root	*(749)*
Millettia dura	Seed	*(613)*
M. ferruginea	Seed	*(170, 280)*
Mundulea sericea (= *suberosa*)	Branch Bark/Root Bark/ Stem Bark	*(553, 834)*
Piscidia mollis	Seed	*(553a)*
Tephrosia falciformis	Root	*(417)*
T. obovata	Whole Plant	*(150)*
T. praecana	Seed	*(133a)*
T. purpurea	Callus Tissue/Leaf/ Stem/Root	*(731)*
T. villosa	Root	*(191)*
T. virginiana	Root	*(169, 289)*
T. vogelii	Callus Tissue	*(731)*
	Leaf	*(131, 168, 199)*
	Pod/Stem	*(833)*
	Root	*(191, 833)*
	Seed	*(554, 675)*

252   MW 410; $C_{23}H_{22}O_7$

12a-HYDROXYROTENONE

Plant Sources

Derris elliptica	Root ?	*(199)*
Lonchocarpus longifolius	Root	*(96a)*
L. nicou	Root ?	*(199)*
L. spruceanus	Root	*(553a)*
L. urucu (= *Derris urucu*)	Creeper	*(101)*
	Root	*(749)*

Table 2 *(continued)*

Plant and/or Other Sources	Plant Part(s) Examined	References
Mundulea sericea		
(= *suberosa*)	Branch Bark/Root Bark/	
	Stem Bark	*(553)*
Neorautanenia amboensis	Root	*(589)*
Pachyrrhizus erosus	Seed	*(416)*
Piscidia mollis	Seed	*(553a)*
Tephrosia praecana	Seed	*(133a)*
T. purpurea	Root	*(191)*
T. villosa	Root	*(191)*
T. vogelii	Leaf ?	*(199)*

Other Source

Animal and Insect
Tissue Homogenates
[Metabolite]   *(268)*

253   MW 424; $C_{24}H_{24}O_7$

12a-METHOXYROTENONE

Neorautanenia amboensis   Root   *(589)*

254   MW 426; $C_{23}H_{22}O_8$

DALBINOL (12a-HYDROXYAMORPHIGENIN)

Table 2 *(continued)*

Plant and/or Other Sources	Plant Part(s) Examined	References
Plant Sources		
Amorpha fruticosa	Seedlings	*(184a)*
Dalbergia assamica	Seed	*(156)*
D. latifolia	Seed	*(153)*
D. monetaria	Seed	*(223a)*
Other Source		
Animal and Insect Tissue Homogenates [Metabolite]		*(268)*

255   MW 426; $C_{23}H_{22}O_8$
11-HYDROXYTEPHROSIN

Amorpha fruticosa   Fruit   *(560)*

256   MW 426; $C_{23}H_{22}O_8$
VILLOSINOL

Tephrosia villosa   Pod   *(712)*

Table 2 *(continued)*

Plant and/or Other Sources	Plant Part(s) Examined	References

256a MW 428; $C_{23}H_{24}O_8$

12-HYDROXYDALBINOL

Dalbergia monetaria   Seed   *(223a)*

257 MW 442; $C_{23}H_{22}O_9$

VILLOL

Tephrosia villosa   Pod   *(484)*

258 MW 444; $C_{23}H_{24}O_9$

12a-HYDROXYAMORPHIGENOL

Animal and Insect
Tissue Homogenates
[Metabolite]   *(268)*

Table 2 *(continued)*

F. DEHYDROROTENOIDS

Plant and/or Other Sources	Plant Part(s) Examined	References

a) Simple Dehydrorotenoids

259   MW 370; $C_{19}H_{14}O_8$
        STEMONONE

| *Stemona collinsae* (STEM) | Root | (743) |

260   MW 372; $C_{19}H_{16}O_8$
        STEMONAL

| *Stemona collinsae* (STEM) | Root | (743) |

261   MW 400; $C_{21}H_{20}O_8$
        STEMONACETAL

| *Clitoria macrophylla* | Root | (777) |
| *Stemona collinsae* (STEM) | Root | (743) |

References, pp. 229—265

Table 2 *(continued)*

Plant and/or Other Sources	Plant Part(s) Examined	References

b) Complex Dehydrorotenoids

262   MW 334; $C_{19}H_{10}O_6$
        DEHYDRODOLINEONE

Neorautanenia amboensis	Root	*(589)*

263   MW 364; $C_{20}H_{12}O_7$
        DEHYDROPACHYRRHIZONE

Pachyrrhizus erosus	Seed	*(416, 481)*

264   MW 376; $C_{22}H_{16}O_6$
        DEHYDROMILLETTONE

Piscidia erythrina	Whole Root	*(239)*

Table 2 *(continued)*

Plant and/or Other Sources	Plant Part(s) Examined	References

265   MW 392; $C_{23}H_{20}O_6$
DEHYDRODEGUELIN

Amorpha fruticosa	Fruit	*(594)*
Derris elliptica	Root	*(173, 694)*
Millettia dura	Seed	*(613)*
Tephrosia candida	Seed	*(152a)*
T. vogelii	Leaf	*(199)*
	Seed	*(554, 675)*

266   MW 392; $C_{23}H_{20}O_6$
DEHYDROROTENONE

Derris elliptica	Root ?	*(199)*
D. ferruginea	Root	*(681)*
D. negrensis	—	*(804)*
D. uliginosa	Root	*(86)*
Lonchocarpus longifolius	Root	*(96a)*
L. urucu (= *Derris urucu*)	Creeper	*(101)*
Millettia ferruginea	Seed	*(170)*
M. pachycarpa	Root	*(679)*
Mundulea sericea	Branch Bark/Root Bark/	
(= *suberosa*)	Stem Bark	*(553, 834)*
Neorautanenia amboensis	Root	*(589)*
Tephrosia falciformis	Root	*(278)*
T. virginiana	Root	*(169, 289)*

Table 2 *(continued)*

Plant and/or Other Sources	Plant Part(s) Examined	References

267 MW 406; $C_{23}H_{18}O_7$

ROTENONONE

Amorpha canescens	Root	*(645)*
Neorautanenia amboensis	Root	*(590)*

268 MW 408; $C_{23}H_{20}O_7$

AMORPHOLONE

Amorpha canescens	Root	*(645)*

269 MW 408; $C_{23}H_{20}O_7$

DEHYDROAMORPHIGENIN

Amorpha spp.	Seed	*(426, 427)*
A. fruticosa	Seed	*(425, 427)*

Table 2 *(continued)*

Plant and/or Other Sources	Plant Part(s) Examined	References

270   MW 408; $C_{23}H_{20}O_7$

DEHYDROTOXICAROL

Amorpha fruticosa	Fruit	*(594)*
	Seed	*(690)*
Derris elliptica	Root	*(173, 694)*

271   MW 408; $C_{23}H_{20}O_7$

VILLOSOL

| *Tephrosia villosa* | Pod | *(712)* |

272   MW 410; $C_{23}H_{22}O_7$

DEHYDRODALPANOL

| *Dalbergia paniculata* | Seed | *(9)* |

References, pp. 229—265

Table 2 *(continued)*

Plant and/or Other Sources	Plant Part(s) Examined	References

273 MW 422; $C_{23}H_{18}O_8$

VILLOSONE

Tephrosia villosa Pod (*484*)

274 MW 438; $C_{24}H_{22}O_8$

VILLINOL

Tephrosia villosa Pod (*484*)

G: PTEROCARPANS

Plant and/or Other Sources	Plant Part(s) Examined	References

a) Simple 6a-Deoxy Pterocarpans

275 MW 256; $C_{15}H_{12}O_4$

DEMETHYLMEDICARPIN

Table 2 *(continued)*

Plant and/or Other Sources	Plant Part(s) Examined	References
Plant Sources		
LM *Albizia procera*	Heartwood	*(210)*
Erythrina crista-galli	† Leaf	*(394)*
E. sandwicensis	† Leaf	*(378)*
Melilotus alba	Leaf (Fungal Metabolite)	*(360)*
Pachyrrhizus erosus	† Etiolated Stem	*(377)*
Phaseolus vulgaris	† Pod	*(831)*
Psophocarpus tetragonolobus	† Etiolated Stem	*(394)*
Trifolium repens	Leaf (Fungal Metabolite)	*(832)*
Other Sources		
Colletotrichum coccodes		
cultures [Metabolite]		*(339)*
Fusarium spp.		
cultures [Metabolite]		*(817, 817a)*
Gibberella saubinetti		
cultures [Metabolite]		*(817a)*

276   MW 270; $C_{16}H_{14}O_4$

MEDICARPIN (DEMETHYLHOMOPTEROCARPIN)

Plant Sources

Aldina heterophylla	Trunk Wood	*(102)*
Andira inermis	Heartwood	*(175)*
Canavalia ensiformis	Callus Tissue	*(314)*
	† Cotyledon	*(440)*
	† Etiolated Hypocotyl/	
	Hypocotyl	*(440, 508)*
	† Germinating Seed	*(441)*
Caragana spp.	† Leaf	*(374)*
Carmichaelia flagelliformis	† Leaf	*(356)*
Cicer spp.	† Etiolated Stem	*(382)*
C. arietinum	† Etiolated Stem	*(361, 382)*
	† Germinating Seed	*(441)*
Dalbergia cearensis	Softwood	*(308)*
D. decipularis	Heartwood/Wood	*(176, 195)*

Table 2 *(continued)*

Plant and/or Other Sources	Plant Part(s) Examined	References
D. ecastophyllum	Trunk Wood	(547)
D. nitidula	Heartwood	(514)
D. oliveri	Heartwood	(225)
D. riparia	Wood	(103, 176)
D. sericea	† Leaf	(376)
D. spruceana	Sapwood	(176)
D. stevensonii	Heartwood	(227)
D. variabilis	Bark/Wood	(497)
D. volubilis	Bark	(450)
Derris amazonica	Creeper	(101)
Factorovskya aschersoniana	† Leaf	(373)
Hedysarum polybotrys	Root	(487)
Lathyrus spp.	† Leaf	(701)
L. nissolia	† Phyllode	(702)
Maackia amurensis		
var. buergeri	Heartwood	(531)
Machaerium acutifolium	Softwood	(612)
M. kuhlmannii	Trunk Wood	(176, 611)
M. nictitans	Trunk Wood	(176, 611)
M. vestitum	Trunk Wood	(496)
Medicago spp.	† Leaf	(371, 381)
M. sativa	† Leaf	(762)
	† Seedlings	(216)
Melilotus spp.	† Leaf	(366)
	† Shoot	(343)
Myroxylon balsamum	Trunk Wood	(96)
Neorautanenia amboensis	Root Bark	(108)
Onobrychis spp.	† Leaf	(370)
* Osteophleum platyspermum		
(MYR)	Branch/Leaf	(100)
Parochetus communis	† Leaf	(373)
Pericopsis schliebenii	Heartwood	(252)
Platymiscium trinitatis	Wood	(182)
Sophora japonica	† Leaf	(356)
Stizolobium deeringianum	† Etiolated Stem	(356)
Swartzia madagascariensis	Heartwood	(327)
Tipuana tipu	† Leaf	(356)
Trifolium spp.	† Leaf	(369, 384)
T. hybridum	Root	(259)
T. pratense	† Leaf	(80, 197, 340)
	† Seedlings	(211, 212)
T. repens	† Callus Tissue	(313a)
	† Leaf	(832)
	Root	(259, 775)
Trigonella spp.	† Leaf	(383, 390)
Vicia faba	† Cotyledon/Leaf/Pod	(322, 323)
	† Root	(348a)
Vigna unguiculata	† Etiolated Hypocotyl	(508)

Table 2 *(continued)*

Plant and/or Other Sources	Plant Part(s) Examined	References

Other Source

Fusarium spp.
cultures [Metabolite]   *(817a)*

277   MW 270; $C_{16}H_{14}O_4$
ISOMEDICARPIN

Plant Source

Psophocarpus tetragonolobus   † Etiolated Stem   *(394)*
                                † Pod   *(655)*

Other Sources

Fusarium spp.
cultures [Metabolite]   *(817, 817a)*
Gibberella saubinetti
cultures [Metabolite]   *(817a)*

278   MW 272; $C_{15}H_{12}O_5$
4-HYDROXYDEMETHYLMEDICARPIN

Melilotus alba   Leaf (Fungal Metabolite)   *(360)*

279   MW 284; $C_{16}H_{12}O_5$
MAACKIAIN (INERMIN; DEMETHYLPTEROCARPIN)

Table 2 *(continued)*

Plant and/or Other Sources	Plant Part(s) Examined	References
Aldina heterophylla	Trunk Wood	*(102)*
Andira inermis	Heartwood	*(174)*
Canavalia ensiformis	† Etiolated Hypocotyl	*(356)*
Caragana spp.	† Leaf	*(374)*
Cicer spp.	† Etiolated Stem	*(382)*
C. arietinum	† Etiolated Stem	*(361, 382)*
	† Germinating Seed	*(441)*
Dalbergia oliveri	Heartwood	*(225)*
D. sericea	† Leaf	*(376)*
D. spruceana	Heartwood/Sapwood	*(176)*
D. stevensonii	Heartwood	*(227)*
Derris elliptica	Root	*(586a)*
Diplotropis purpurea	Trunk Wood	*(102)*
Euchresta japonica	Root	*(745a)*
Lathyrus ssp.	† Leaf	*(701)*
Maackia amurensis		
var. *buergeri*	Heartwood	*(531, 772, 778)*
Millettia pendula	Heartwood	*(333)*
* *Osteophleum platyspermum* (MYR)	Branch/Leaf	*(100)*
Pericopsis schliebenii	Heartwood	*(252)*
Pisum fulvum	† Leaf	*(701)*
P. sativum	† Cotyledon	*(699)*
	† Leaf	*(701)*
	† Petal/Stipule	*(375)*
	† Pod	*(769)*
Pterocarpus dalbergioides	Heartwood	*(627)*
Sophora angustifolia	Callus Tissue	*(271)*
S. flavescens	Root	*(467)*
S. franchetiana	Root	*(472)*
S. japonica	† Leaf	*(356)*
	Root	*(469, 741)*
S. microphylla	Bark/Heartwood/Sapwood	*(114)*
S. subprostrata	Root	*(468, 473)*
S. tetraptera	Bark/Heartwood/Sapwood	*(114)*
S. tomentosa	Aerial Parts	*(470, 471)*
	Root	*(737)*
Stizolobium deeringianum	† Etiolated Stem	*(356)*
Swartzia madagascariensis	Heartwood	*(327)*
Tephrosia bidwilli	† Leaf	*(356)*
Thermopsis fabacea	Root	*(21)*
Tipuana tipu	† Leaf	*(356)*
Trifolium spp.	† Leaf	*(369, 384)*
T. hybridum	Root	*(259)*
T. pratense	† Leaf	*(80, 197, 340)*
	Root	*(259, 648)*
	† Seedlings	*(211, 212)*
Trigonella spp.	† Leaf	*(383, 390)*

Table 2 *(continued)*

Plant and/or Other Sources	Plant Part(s) Examined	References

280 MW 284; $C_{17}H_{16}O_4$

HOMOPTEROCARPIN (BAPHINITONE)

Baphia nitida	Heartwood	*(18, 548)*
Bowdichia nitida	Heartwood	*(123)*
Maackia amurensis		
var. *buergeri*	Heartwood	*(531)*
Machaerium villosum	Heartwood	*(91, 499)*
Pericopsis angolensis	Heartwood	*(221, 252, 327)*
P. schliebenii	Heartwood	*(252)*
Pterocarpus dalbergioides	Bark	*(715)*
	Sapwood	*(714)*
P. indicus	Heartwood	*(66, 122)*
P. macrocarpus	Bark	*(715)*
	Heartwood/Sapwood	*(714)*
P. osun	Heartwood	*(14)*
P. santalinus	Heartwood	*(138, 687, 763)*
P. soyauxii	Heartwood	*(452)*
P. vidalianus	Heartwood	*(616)*
Swartzia madagascariensis	Heartwood	*(327)*
Trifolium hybridum	Root	*(259)*

281 MW 286; $C_{16}H_{14}O_5$

NISSOLIN

Lathyrus nissolia	† Phyllode	*(702)*

Table 2 *(continued)*

Plant and/or Other Sources	Plant Part(s) Examined	References

282 MW 286; $C_{16}H_{14}O_5$

VESTICARPAN

Machaerium vestitum	Trunk Wood	*(496)*
Platymiscium trinitatis	Wood	*(182)*

283 MW 286; $C_{16}H_{14}O_5$

4-HYDROXYMEDICARPIN

Melilotus alba	Leaf (Fungal Metabolite)	*(360)*

284 MW 298; $C_{17}H_{14}O_5$

PTEROCARPIN

Baphia nitida	Heartwood	*(548, 615)*
Pterocarpus dalbergioides	Heartwood	*(452, 627, 714)*
P. indicus	Heartwood	*(66, 122, 178)*
P. macrocarpus	Heartwood	*(452, 714)*
P. osun	Heartwood	*(14)*
P. santalinus	Heartwood	*(138, 687)*
Sophora angustifolia	Callus Tissue	*(271)*
S. subprostrata	Root	*(473, 741)*
Swartzia madagascariensis	Heartwood	*(327)*

Table 2 *(continued)*

Plant and/or Other Sources	Plant Part(s) Examined	References

285   MW 300; $C_{17}H_{16}O_5$

METHYLNISSOLIN

Lathyrus nissolia   † Phyllode   *(702)*

286   MW 300; $C_{17}H_{16}O_5$

2-METHOXYMEDICARPIN

Pisum sativum   † Epicotyl/Root Crown   *(657, 658)*

287   MW 300; $C_{17}H_{16}O_5$

SPARTICARPIN

Spartium junceum   † Leaf   *(389)*

288   MW 300; $C_{16}H_{12}O_6$

4-HYDROXYMAACKIAIN

Dalbergia spruceana   Heartwood/Sapwood   *(176)*

References, pp. 229—265

Table 2 *(continued)*

Plant and/or Other Sources	Plant Part(s) Examined	References

289 MW 300; $C_{17}H_{16}O_5$

4-METHOXYMEDICARPIN

Swartzia madagascariensis	Heartwood	*(327)*
Trifolium cherleri	† Leaf	*(386)*
T. pallescens	† Leaf	*(356)*
T. repens	Root	*(259)*

290 MW 300; $C_{17}H_{16}O_5$

4-HYDROXYHOMOPTEROCARPIN

Melilotus alba	Leaf (Fungal Metabolite)	*(360)*
Trifolium hybridum	† Leaf	*(359)*
T. pallescens	† Leaf	*(356)*

291 MW 314; $C_{17}H_{14}O_6$

2-HYDROXYPTEROCARPIN

Neorautanenia edulis	Root Bark	*(669)*
Swartzia leiocalycina	Heartwood	*(224)*

292 MW 314; $C_{18}H_{18}O_5$

2-METHOXYHOMOPTEROCARPIN

Pisum sativum	† Epicotyl/Root Crown	*(657, 658)*

Table 2 *(continued)*

Plant and/or Other Sources	Plant Part(s) Examined	References

293 MW 314; $C_{17}H_{14}O_6$

4-METHOXYMAACKIAIN

Amorpha californica	† Leaf	*(356)*
Baptisia australis	† Leaf	*(356)*
Dalbergia spruceana	Sapwood	*(176)*
Sophora franchetiana	Root	*(467a)*
Swartzia madagascariensis	Heartwood	*(327)*
Tephrosia bidwilli	† Leaf	*(356)*
Trifolium hybridum	† Leaf	*(362, 369)*

294 MW 314; $C_{17}H_{14}O_6$

4-HYDROXYPTEROCARPIN

Dalbergia spruceana	Heartwood/Sapwood	*(176)*

295 MW 314; $C_{18}H_{18}O_5$

4-METHOXYHOMOPTEROCARPIN

Myroxylon peruiferum	Trunk Wood	*(536)*
Swartzia madagascariensis	Heartwood	*(327)*

Table 2 *(continued)*

Plant and/or Other Sources	Plant Part(s) Examined	References

296   MW 316; $C_{17}H_{16}O_6$
PHILENOPTERAN

Lonchocarpus laxiflorus   Debarked Root   *(635)*

297   MW 316; $C_{17}H_{16}O_6$

Swartzia laevicarpa   Trunk Wood   *(97)*

298   MW 316; $C_{17}H_{16}O_6$
MUCRONUCARPAN

Machaerium mucronulatum   Heartwood   *(499)*

299   MW 328; $C_{18}H_{16}O_6$
2-METHOXYPTEROCARPIN

Neorautanenia edulis   Root Bark   *(667)*

Table 2 *(continued)*

Plant and/or Other Sources	Plant Part(s) Examined	References

300   MW 328; $C_{18}H_{16}O_6$
4-METHOXYPTEROCARPIN

Neorautanenia ficifolia	Tuber	*(87)*
Swartzia madagascariensis	Heartwood	*(327)*

301   MW 330; $C_{18}H_{18}O_6$
9-*O*-METHYLPHILENOPTERAN

Lonchocarpus laxiflorus	Debarked Root	*(635)*

302   MW 330; $C_{17}H_{14}O_7$
TRIFOLIAN

Trifolium pratense	Root	*(257, 259)*

303   MW 330; $C_{18}H_{18}O_6$

Pisum sativum	† Epicotyl/Root Crown	*(657, 658)*

Table 2 *(continued)*

Plant and/or Other Sources	Plant Part(s) Examined	References

304 MW 346; $C_{18}H_{18}O_7$

Swartzia laevicarpa   Trunk Wood   (*97*)

305 MW 360; $C_{19}H_{20}O_7$

Swartzia laevicarpa   Trunk Wood   (*97*)

306 MW 376; $C_{19}H_{20}O_8$

Swartzia laevicarpa   Trunk Wood   (*97*)

b) Complex 6a-Deoxy Pterocarpans

307 MW 280; $C_{17}H_{12}O_4$

NEODUNOL

Neorautanenia amboensis	Root Bark	(*108*)
N. edulis	Root Bark	(*116, 118*)
Pachyrrhizus erosus	† Etiolated Stem	(*377*)

Table 2 *(continued)*

Plant and/or Other Sources	Plant Part(s) Examined	References

308   MW 308; $C_{18}H_{12}O_5$
NEODULIN (EDULIN)

Neorautanenia edulis	Tuber	*(795)*

309   MW 322; $C_{20}H_{18}O_4$
PHASEOLLIN (PHASEOLIN)

Erythrina abyssinica	Root	*(419)*
Phaseolus spp.	† Pod	*(190)*
P. vulgaris	Cell Suspension Cultures	*(222)*
	† Etiolated Hypocotyl/Hypocotyl	*(31, 685)*
	† Germinating Seed	*(285, 441)*
	† Leaf	*(31, 190, 286)*
	† Pod	*(190, 639)*
	† Root	*(129, 775)*
	† Stem	*(190)*
Vigna unguiculata	† Etiolated Hypocotyl	*(29)*

310   MW 322; $C_{20}H_{18}O_4$
NEORAUTENOL

Neorautanenia edulis	Root Bark	*(116, 118)*

References, pp. 229—265

Table 2 *(continued)*

Plant and/or Other Sources	Plant Part(s) Examined	References

311   MW 324; $C_{20}H_{20}O_4$
SOPHORAPTEROCARPAN A

Sophora franchetiana	Root	*(472)*

312   MW 324; $C_{20}H_{20}O_4$
PHASEOLLIDIN

Dolichos biflorus	† Leaf/Stem	*(443)*
Erythrina abyssinica	Root	*(419)*
E. crista-galli	† Leaf	*(394)*
E. sandwicensis	† Leaf	*(378)*
Lablab niger	† Etiolated Hypocotyl	*(367)*
Macroptilium atropurpureum	† Leaf	*(356)*
Phaseolus aureus	† Etiolated Hypocotyl	*(356)*
P. calcaratus	† Germinating Seed/Leaf/ Root	*(356, 773a)*
P. vulgaris	† Etiolated Hypocotyl	*(127)*
	† Germinating Seed	*(285)*
	† Leaf	*(286)*
	† Pod	*(640, 643)*
	† Root	*(129)*
Psophocarpus tetragonolobus	† Leaf/Pod	*(655)*
	† Etiolated Stem	*(394)*
Vigna unguiculata	† Etiolated Hypocotyl	*(29)*
	† Stem	*(653)*

Table 2 *(continued)*

Plant and/or Other Sources	Plant Part(s) Examined	References

313   MW 324; $C_{19}H_{16}O_5$
AMBONANE

Neorautanenia amboensis	Root Bark	*(108)*

314   MW 324; $C_{20}H_{20}O_4$
HOMOEDUDIOL

Neorautanenia edulis	Root Bark	*(116, 118)*
† *Pachyrrhizus erosus*	Etiolated Stem?	*(377)*

315   MW 336; $C_{21}H_{20}O_4$
HEMILEIOCARPIN

Dalbergia nitidula	Bark	*(801)*

References, pp. 229—265

Table 2 *(continued)*

Plant and/or Other Sources	Plant Part(s) Examined	References

316   MW 338; $C_{21}H_{22}O_4$
SANDWICENSIN

Erythrina sandwicensis   † Leaf   *(378)*

317   MW 338; $C_{20}H_{18}O_5$
APIOCARPIN

Apios tuberosa   † Leaf   *(396)*

318   MW 338; $C_{19}H_{14}O_6$
FICININ

Neorautanenia ficifolia   Root   *(121)*

Table 2 *(continued)*

Plant and/or Other Sources	Plant Part(s) Examined	References

319   MW 340; $C_{20}H_{20}O_5$

DOLICHIN A

Dolichos biflorus   † Leaf   (*392*)

320   MW 340; $C_{20}H_{20}O_5$

DOLICHIN B

Dolichos biflorus   † Leaf   (*392*)

321   MW 342; $C_{20}H_{22}O_5$

PHASEOLLIDIN HYDRATE

Fusarium solani f. sp.
phaseoli cultures
[Metabolite]   (*760*)

Table 2 *(continued)*

Plant and/or Other Sources	Plant Part(s) Examined	References

322 MW 350; $C_{21}H_{18}O_5$

NEORAUTENANE

Neorautanenia amboensis   Root Bark   *(108)*

323 MW 350; $C_{21}H_{18}O_5$

LEIOCARPIN

LC *Apuleia leiocarpa*   Bark   *(98)*
Dalbergia nitidula   Bark   *(801)*

324 MW 352; $C_{21}H_{20}O_5$

EDUNOL

Neorautanenia edulis   Root Bark   *(670)*

325 MW 352; $C_{21}H_{20}O_5$

NEORAUTANE

Neorautanenia edulis   Root Bark   *(667)*

Table 2 *(continued)*

Plant and/or Other Sources	Plant Part(s) Examined	References

326   MW 352; $C_{21}H_{20}O_5$

EDULENANOL

Neorautanenia amboensis   Root Bark   (*108*)

327   MW 354; $C_{21}H_{22}O_5$

1-METHOXYPHASEOLLIDIN

Psophocarpus tetragonolobus   † Leaf/Pod   (*655*)

328   MW 354; $C_{21}H_{22}O_5$

EDUDIOL

Neorautanenia edulis   Root Bark   (*116*)

329   MW 356; $C_{20}H_{20}O_6$

or

Septoria nodorum
cultures [Metabolite 1]   (*30*)

Table 2 *(continued)*

Plant and/or Other Sources	Plant Part(s) Examined	References

330   MW 356; $C_{20}H_{20}O_6$

Septoria nodorum
cultures [Metabolite 2]   *(30)*

331   MW 366; $C_{21}H_{18}O_6$
NEORAUTENANOL

Neorautanenia amboensis   Root Bark   *(108)*

332   MW 366; $C_{22}H_{22}O_5$
EDULENANE (EDULAAN)

Neorautanenia amboensis   Root Bark   *(108, 117)*

333   MW 368; $C_{21}H_{20}O_6$
NEORAUTANOL

Neorautanenia amboensis   Root Bark   *(108)*

Table 2 *(continued)*

Plant and/or Other Sources	Plant Part(s) Examined	References

334   MW 368; $C_{22}H_{24}O_5$
EDULENOL

| *Neorautanenia amboensis* | Root Bark | *(108, 116)* |
| *N. edulis* | Root Bark | *(668)* |

335   MW 368; $C_{22}H_{24}O_5$
EDULANE

| *Neorautanenia edulis* | Root Bark | *(116)* |

336   MW 382; $C_{22}H_{22}O_6$
DESMODIN

| *Desmodium gangeticum* | Root | *(659)* |

337   MW 382; $C_{22}H_{22}O_6$
NEORAUTANIN

| *Neorautanenia edulis* | Root Bark | *(116)* |

Table 2 *(continued)*

Plant and/or Other Sources	Plant Part(s) Examined	References

338 MW 382; $C_{22}H_{22}O_6$
NEORAUCARPANOL

Neorautanenia amboensis   Root Bark   *(108, 116)*

339 MW 390; $C_{25}H_{26}O_4$
FOLITENOL

Neorautanenia ficifolia   Root Bark   *(120)*

340 MW 390; $C_{25}H_{26}O_4$
FOLININ

Neorautanenia ficifolia   Root Bark   *(120)*

Table 2 *(continued)*

Plant and/or Other Sources	Plant Part(s) Examined	References

341 MW 392; $C_{25}H_{28}O_4$
LESPEIN

Lespedeza homoloba   Bark   (*789*)

342 MW 392; $C_{25}H_{28}O_4$
LESPEDEZIN

Lespedeza homoloba   Bark   (*789*)

343 MW 392; $C_{25}H_{28}O_4$
FICIFOLINOL

Neorautanenia ficifolia   Root Bark   (*120*)

Table 2 *(continued)*

Plant and/or Other Sources	Plant Part(s) Examined	References

344 MW 392; $C_{25}H_{28}O_4$

ERYTHRABYSSIN II

Erythrina abyssinica   Root   *(419)*

345 MW 396; $C_{23}H_{24}O_6$

NEORAUCARPAN

Neorautanenia edulis   Root Bark   *(116)*

346 MW 418; $C_{26}H_{26}O_5$

NITIDUCARPIN

Dalbergia nitidula   Bark   *(801)*

Table 2 *(continued)*

Plant and/or Other Sources	Plant Part(s) Examined	References

347 MW 418; $C_{26}H_{26}O_5$

GANGETININ

Desmodium gangeticum   Root   *(659)*

348 MW 420; $C_{26}H_{28}O_5$

NITIDUCOL

Dalbergia nitidula   Bark   *(801)*

349 MW 420; $C_{26}H_{28}O_5$

GANGETIN

Desmodium gangeticum   Root   *(660)*

Table 2 *(continued)*

Plant and/or Other Sources	Plant Part(s) Examined	References

c) Simple 6a-Oxy (OH) Pterocarpans

350   MW 272; $C_{15}H_{12}O_5$

GLYCINOL

Erythrina sandwicensis	† Leaf	*(378)*
Glycine max	† Cotyledon	*(393, 527, 816)*
Pueraria thunbergiana		
(=lobata)	† Leaf	*(356)*

351   MW 286; $C_{16}H_{14}O_5$

6a-HYDROXYMEDICARPIN

Plant Sources

Melilotus alba	Leaf (Fungal Metabolite)	*(360)*
Trifolium pratense	Leaf (Fungal Metabolite)	*(197, 360)*

Other Sources

Botrytis cinerea		
cultures [Metabolite]		*(197, 360)*
Nectria haematococca		
cultures [Metabolite]		*(205)*
Sclerotinia trifoliorum		
cultures [Metabolite]		*(80, 197)*

352   MW 286; $C_{16}H_{14}O_5$

6a-HYDROXYISOMEDICARPIN

Melilotus alba	Leaf (Fungal Metabolite)	*(360)*

Table 2 *(continued)*

Plant and/or Other Sources	Plant Part(s) Examined	References

353   MW 300; $C_{16}H_{12}O_6$

6a-HYDROXYMAACKIAIN (6a-HYDROXYINERMIN)

Plant Source

Trifolium pratense	Leaf (Fungal Metabolite)	*(197, 360)*

Other Sources

Ascochyta pisi	
cultures [Metabolite]	*(803)*
Botrytis cinerea	
cultures [Metabolite]	*(197)*
Fusarium anguioides	
cultures [Metabolite]	*(509)*
F. avenaceum	
cultures [Metabolite]	*(509)*
F. oxysporum f. sp. *pisi*	
cultures [Metabolite]	*(266)*
F. solani f. sp. *pisi*	
cultures [Metabolite]	*(800)*
Nectria haematococca	
cultures [Metabolite]	*(205)*
Sclerotinia trifoliorum	
cultures [Metabolite]	*(80, 197)*
Stemphylium botryosum	
cultures [Metabolite]	*(338a)*

354   MW 300; $C_{17}H_{16}O_5$

VARIABILIN (HOMOPISATIN)

Plant Sources

Caragana spp.	† Leaf	*(374)*
Dalbergia variabilis	Wood	*(497)*
Lathyrus spp.	† Leaf	*(701)*

Table 2 *(continued)*

Plant and/or Other Sources	Plant Part(s) Examined	References
Lens culinaris	† Cotyledon/Epicotyl/Leaf	*(699, 700, 701)*
L. nigricans	† Cotyledon	*(700)*
Parochetus communis	† Leaf ?	*(379)*
Trifolium pratense	Leaf (Fungal Metabolite ?)	*(80, 197)*

Other Source

| *Botrytis cinerea* cultures [Metabolite] | | *(360)* |

355   MW 302; $C_{16}H_{14}O_6$

6a,7-DIHYDROXYMEDICARPIN

| *Melilotus alba* | Leaf (Fungal Metabolite) | *(360)* |
| *Trifolium pratense* | Leaf (Fungal Metabolite) | *(360)* |

356   MW 314; $C_{17}H_{14}O_6$

PISATIN

Caragana spp.	† Leaf	*(374)*
Lathyrus spp.	† Cotyledon/Leaf	*(699, 701)*
Pisum spp.	† Pod	*(189)*
P. fulvum	† Leaf	*(701)*
	† Pod	*(189)*
P. sativum	† Cotyledon	*(699)*
	Callus Tissue	*(28)*
	† Epicotyl/Root Crown	*(657, 658)*
	† Germinating Seed/ Seedlings	*(136, 441)*
	† Leaf	*(29, 701, 745)*
	† Pod	*(189, 641, 769)*
	† Petal/Sepal/Stipule	*(375)*
	† Root	*(129)*
	† Stem	*(745)*
Tephrosia bidwilli	† Leaf	*(356)*
Trifolium pratense	† Leaf (Fungal Metabolite ?)	*(80, 197)*

Table 2 *(continued)*

Plant and/or Other Sources	Plant Part(s) Examined	References

357   MW 316; $C_{16}H_{12}O_7$

6a,7-DIHYDROXYMAACKIAIN

Trifolium pratense   Leaf (Fungal Metabolite)   *(360)*

358   MW 328; $C_{17}H_{12}O_7$

ACANTHOCARPAN

(Revised Structure)

| *Caragana acanthophylla* | † Leaf | *(374)* |
| *Tephrosia bidwilli* | † Leaf | *(356)* |

359   MW 330; $C_{17}H_{14}O_7$

TEPHROCARPIN

Tephrosia bidwilli   † Leaf   *(356)*

359a   MW 344; $C_{18}H_{16}O_7$

LATHYCARPIN

Lathyrus sativus   † Leaf   *(356)*

Table 2 *(continued)*

Plant and/or Other Sources	Plant Part(s) Examined	References

d) Complex 6a-Oxy (OH) Pterocarpans

360 MW 324; $C_{18}H_{12}O_6$
NEOBANOL

| *Neorautanenia amboensis* | Root | *(590)* |

361 MW 336; $C_{20}H_{16}O_5$
CLANDESTACARPIN

Glycine clandestina	† Leaf	*(528)*
G. tabacina	† Leaf	*(439)*
G. tomentella	† Leaf	*(439)*

362 MW 338; $C_{20}H_{18}O_5$
TUBEROSIN

Pueraria thunbergiana		
(= *lobata*)	† Leaf	*(356)*
P. tuberosa	Tuber	*(411)*

Table 2 *(continued)*

Plant and/or Other Sources	Plant Part(s) Examined	References

363   MW 338; $C_{20}H_{18}O_5$

6a-HYDROXYPHASEOLLIN

Plant Source

  Phaseolus vulgaris   Pod (Fungal Metabolite ?)   *(831)*

Other Sources

  Botrytis cinerea
    cultures [Metabolite]   *(793a)*
  Colletotrichum lindemuthianum
    cultures [Metabolite]   *(128)*

364   MW 338; $C_{20}H_{18}O_5$

GLYCEOLLIN II

Glycine canescens	† Leaf	*(439)*
G. falcata	† Leaf	*(439)*
G. gracilis	† Leaf	*(439)*
G. latrobeana	† Leaf	*(439)*
G. max	† Cotyledon	*(393, 529)*
	† Leaf	*(393)*
G. soja	† Leaf	*(439)*

365   MW 338; $C_{20}H_{18}O_5$

GLYCEOLLIN III

Table 2 *(continued)*

Plant and/or Other Sources	Plant Part(s) Examined	References
Glycine gracilis	† Leaf	*(439)*
G. latrobeana	† Leaf	*(439)*
G. max	† Cotyledon	*(393, 529)*
	† Leaf	*(393)*
G. soja	† Leaf	*(439)*

366   MW 338; $C_{20}H_{18}O_5$

CANESCACARPIN

Glycine canescens	† Leaf	*(528)*

367   MW 338; $C_{20}H_{18}O_5$

GLYCEOLLIN I (Formerly 6a-HYDROXYPHASEOLLIN)
(Revised Structure)

Glycine canescens	† Leaf	*(439)*
G. falcata	† Leaf	*(439)*
G. latrobeana	† Leaf	*(439)*
G. max	† Callus Tissue/Pod/Root	*(442)*
	† Cell Suspension Cultures	*(232)*
	† Cotyledon	*(126, 393, 442)*
	† Germinating Seed	*(441)*
	† Hypocotyl	*(442, 445, 753)*
	† Leaf	*(393, 444)*
G. soja	† Leaf	*(439)*
Psoralea acaulis	† Leaf	*(356)*
P. bituminosa	† Leaf	*(356)*
P. onobrychis	† Leaf	*(356)*

Table 2 *(continued)*

Plant and/or Other Sources	Plant Part(s) Examined	References

368   MW 340; $C_{20}H_{20}O_5$

SANDWICARPIN

Erythrina sandwicensis	† Leaf	*(378)*

369   MW 340; $C_{20}H_{20}O_5$

GLYCEOCARPIN (GLYCEOLLIDIN II)

Glycine max	† Cell Suspension Cultures	*(837)*
	† Cotyledon	*(393, 837)*
	† Leaf	*(393)*

370   MW 340; $C_{20}H_{20}O_5$

GLYCEOLLIDIN I

Glycine max	† Cell Suspension Cultures/ Cotyledon	*(837)*

Table 2 *(continued)*

Plant and/or Other Sources	Plant Part(s) Examined	References

371 MW 354; $C_{21}H_{22}O_5$

CRISTACARPIN (ERYTHRABYSSIN I)

Erythrina abyssinica	Root	*(419)*
E. crista-galli	† Leaf	*(394)*
E. sandwicensis	† Leaf	*(378)*
Psophocarpus tetragonolobus	† Etiolated Stem	*(394)*

372 MW 354; $C_{20}H_{18}O_6$

6a,7-DIHYDROXYPHASEOLLIN

Colletotrichum lindemuthianum cultures [Metabolite]		*(128)*

373 MW 354; $C_{21}H_{22}O_5$

GLYCEOLLIN IV

Glycine max	† Cotyledon	*(527)*
G. soja	† Leaf	*(439)*

Table 2 *(continued)*

Plant and/or Other Sources	Plant Part(s) Examined	References

374   MW 354; $C_{20}H_{18}O_6$
GLYCEOFURAN

| *Glycine max* | † Cotyledon/Leaf | *(393)* |

375   MW 368; $C_{21}H_{20}O_6$
9-*O*-METHYLGLYCEOFURAN

| *Glycine max* | † Leaf | *(393)* |

H: PTEROCARPANONES

Plant and/or Other Sources	Plant Part(s) Examined	References

a) Simple Pterocarpanones

376   MW 286; $C_{16}H_{14}O_5$

| *Nectria haematococca* cultures [Metabolite] | | *(205)* |

377   MW 300; $C_{16}H_{12}O_6$

| *Nectria haematococca* cultures [Metabolite] | | *(205)* |

Table 2 *(continued)*

Plant and/or Other Sources	Plant Part(s) Examined	References

b) Complex Pterocarpanones

378   MW 338; $C_{20}H_{18}O_5$

1a-HYDROXYPHASEOLLONE

Cladosporium herbarum cultures [Metabolite]		*(793a)*
Fusarium solani f. sp. *phaseoli* cultures [Metabolite]		*(794)*

379   MW 338; $C_{20}H_{18}O_5$

Glycine max	Cotyledons	*(837)*

I: PTEROCARPENES

Plant and/or Other Sources	Plant Part(s) Examined	References

a) Simple Pterocarpenes

380   MW 254; $C_{15}H_{10}O_4$

ANHYDROGLYCINOL

Lespedeza cyrtobotrya	Heartwood	*(562)*
Tetragonolobus maritimus	† Leaf	*(385)*

Table 2 *(continued)*

Plant and/or Other Sources	Plant Part(s) Examined	References

381   MW 282; $C_{17}H_{14}O_4$

ANHYDROVARIABILIN

Dalbergia decipularis	Heartwood	*(195)*
Swartzia madagascariensis	Heartwood	*(327)*

382   MW 296; $C_{17}H_{12}O_5$

ANHYDROPISATIN (FLEMICHAPPARIN B)

Flemingia chappar	Root	*(10)*
Lonchocarpus urucu		
(= *Derris urucu*)	Creeper	*(101)*
Pisum sativum	† Seedlings	*(136)*
Sophora japonica	Root	*(469)*

383   MW 312; $C_{18}H_{16}O_5$

BRYACARPENE-5

Brya ebenus	Heartwood	*(247)*

384   MW 312; $C_{17}H_{12}O_6$

Swartzia ulei	Trunk Wood	*(253)*

Table 2 *(continued)*

Plant and/or Other Sources	Plant Part(s) Examined	References

385　MW 328; $C_{18}H_{16}O_6$

BRYACARPENE-2

Brya ebenus　　　　Heartwood　　　　*(247)*

386　MW 328; $C_{18}H_{16}O_6$

BRYACARPENE-4

Brya ebenus　　　　Heartwood　　　　*(247)*

387　MW 342; $C_{19}H_{18}O_6$

BRYACARPENE-3

Brya ebenus　　　　Heartwood　　　　*(247)*

388　MW 342; $C_{18}H_{14}O_7$

LEIOCALYCIN

Swartzia leiocalycina　　　　Heartwood　　　　*(224)*

Table 2 *(continued)*

Plant and/or Other Sources	Plant Part(s) Examined	References

389   MW 344; $C_{18}H_{16}O_7$
BRYACARPENE-1

Brya ebenus   Heartwood   *(247)*

b) Complex Pterocarpenes

390   MW 278; $C_{17}H_{10}O_4$
NEORAUTEEN

Neorautanenia edulis   Root Bark   *(118)*

391   MW 306; $C_{18}H_{10}O_5$
NEODULEEN

Neorautanenia edulis   Root Bark   *(118)*

Table 2 *(continued)*

J: PTEROCARPENEQUINONES

Plant and/or Other Sources	Plant Part(s) Examined	References

392 MW 312; $C_{17}H_{12}O_6$

4-DEOXYBRYAQUINONE

Brya ebenus	Heartwood	*(248)*

393 MW 328; $C_{17}H_{12}O_7$

BRYAQUINONE

Brya ebenus	Heartwood	*(248)*

K: ISOFLAVANS

Plant and/or Other Sources	Plant Part(s) Examined	References

a) Simple Isoflavans

394 MW 242; $C_{15}H_{14}O_3$

EQUOL

Plant Source

Millettia pendula	Heartwood	*(333)*

Table 2 *(continued)*

Plant and/or Other Sources	Plant Part(s) Examined	References

Other Sources

    Cow urine [Metabolite]   *(463)*
    Goat urine [Metabolite]   *(462)*
    Hen urine [Metabolite]   *(336)*
    Horse urine [Metabolite]   *(17, 542, 543)*
    Sheep urine [Metabolite]   *(47, 89)*

395   MW 256; $C_{16}H_{16}O_3$
      4'-*O*-METHYLEQUOL

Sheep urine [Metabolite]   *(180)*

396   MW 258; $C_{15}H_{14}O_4$
      DEMETHYLVESTITOL

Plant Sources

Anthyllis vulneraria	† Leaf	*(364)*
Erythrina sandwicensis	† Leaf	*(378)*
Hosackia americana	† Leaf	*(356)*
Lablab niger	† Etiolated Hypocotyl	*(367)*
Lotus spp.	† Leaf	*(364, 386, 388)*
Phaseolus vulgaris	† Pod	*(831)*
Tetragonolobus spp.	† Leaf	*(364)*

Other Source

    Fusarium proliferatum
      cultures [Metabolite]   *(817)*

Table 2 *(continued)*

Plant and/or Other Sources	Plant Part(s) Examined	References

397 MW 272; $C_{16}H_{16}O_4$

VESTITOL

Plant Sources

Canavalia ensiformis	† Etiolated Hypocotyl	*(356)*
Carmichaelia flagelliformis	† Leaf	*(356)*
Cyclolobium clausseni	Heartwood	*(93, 295)*
Dalbergia ecastophyllum	Trunk Wood	*(547)*
D. sericea	† Leaf	*(376)*
D. variabilis	Wood	*(497)*
Factorovskya aschersoniana	† Leaf	*(373)*
Hosackia americana	† Leaf	*(356)*
Lotus spp.	† Leaf	*(364, 386, 388)*
L. corniculatus	† Leaf	*(82, 364)*
L. pedunculatus	Leaf/Root	*(707)*
Machaerium vestitum	Trunk Wood	*(496)*
Medicago spp.	† Leaf	*(371, 381)*
M. sativa	† Seedlings	*(216)*
Melilotus alba	Leaf (Fungal Metabolite)	*(360)*
Onobrychis spp.	† Leaf	*(370)*
Tetragonolobus requienii	† Leaf	*(364)*
Tipuana tipu	† Leaf	*(356)*
Trifolium spp.	† Leaf	*(369, 384)*
T. hybridum	Root	*(255, 259)*
Trigonella spp.	† Leaf	*(383, 390)*

Other Sources

Fusarium avenaceum		
cultures [Metabolite]		*(817a)*
Stemphylium botryosum		
cultures [Metabolite]		*(768)*

398 MW 272; $C_{16}H_{16}O_4$

ISOVESTITOL

Table 2 *(continued)*

Plant and/or Other Sources	Plant Part(s) Examined	References
Anthyllis vulneraria	† Leaf	*(364)*
Erythrina sandwicensis	† Leaf	*(378)*
Hosackia americana	† Leaf	*(356)*
Lablab niger	† Etiolated Hypocotyl	*(367)*
Tetragonolobus spp.	† Leaf	*(364)*
Trifolium spp.	† Leaf	*(369, 384)*

399 MW 272; $C_{16}H_{16}O_4$

NEOVESTITOL

Dalbergia sericea	† Leaf	*(376)*

400 MW 286; $C_{17}H_{18}O_4$

SATIVAN (SATIVIN)

Derris amazonica	Creeper	*(101)*
Lotus corniculatus	† Leaf	*(82, 364)*
L. hispidus	† Leaf	*(386)*
L. pedunculatus	Root	*(707)*
Medicago spp.	† Leaf	*(371)*
M. sativa	† Leaf	*(395)*
	† Seedlings	*(216)*
Trifolium spp.	† Leaf	*(369, 384)*
Trigonella spp.	† Leaf	*(383, 390)*

401 MW 286; $C_{17}H_{18}O_4$

ISOSATIVAN

Dalbergia ecastophyllum	Trunk Wood	*(547)*
Medicago rugosa	† Leaf	*(381)*
M. scutellata	† Leaf	*(371)*
Trifolium spp.	† Leaf	*(362, 369)*

Table 2 *(continued)*

Plant and/or Other Sources	Plant Part(s) Examined	References

402 MW 286; $C_{17}H_{18}O_4$

ARVENSAN

Trifolium arvense	† Leaf	*(384)*
T. stellatum	† Leaf	*(369)*

403 MW 286; $C_{16}H_{14}O_5$

MAACKIAINISOFLAVAN (DIHYDROMAACKIAIN)

Stemphylium botryosum cultures [Metabolite]		*(337)*

404 MW 300; $C_{17}H_{16}O_5$

ASTRACICERAN

Astragalus cicer	† Leaf	*(387)*
A. pyrenaicus	† Leaf	*(356)*

405 MW 302; $C_{17}H_{18}O_5$

ISOMUCRONULATOL

Astragalus glycyphyllos	† Leaf	*(356)*
A. penduliflorus	† Leaf	*(356)*
Carmichaelia flagelliformis	† Leaf	*(356)*

Table 2 *(continued)*

Plant and/or Other Sources	Plant Part(s) Examined	References
Colutea arborescens	† Leaf	*(356)*
	Seedlings	*(210a)*
Gliricidia sepium	Heartwood	*(413, 534)*
Glycyrrhiza glabra	† Leaf	*(365)*

406 MW 302; $C_{17}H_{18}O_5$

MUCRONULATOL

Astragalus cicer	† Leaf	*(387)*
A. gummifer	† Leaf	*(356)*
A. lusitanicus	Root	*(206)*
A. pyrenaicus	† Leaf	*(356)*
Dalbergia cearensis	Softwood	*(308)*
D. ecastophyllum	Vine Wood	*(226)*
D. oliveri	Heartwood	*(225)*
D. variabilis	Wood	*(497)*
Machaerium acutifolium	Heartwood	*(495)*
M. mucronulatum	Heartwood	*(499)*
	Root	*(498)*
M. opacum	Heartwood	*(614)*
M. vestitum	Trunk Wood	*(496)*
M. villosum	Heartwood	*(91, 499)*

407 MW 302; $C_{17}H_{18}O_5$

LAXIFLORAN (SPHAEROSIN)

Lablab niger	† Etiolated Hypocotyl	*(367)*
Lonchocarpus laxiflorus	Debarked Root	*(635)*
Sphaerophysa salsula	Root	*(432)*

Table 2 *(continued)*

Plant and/or Other Sources	Plant Part(s) Examined	References

408 MW 302; $C_{16}H_{14}O_6$

3-HYDROXYMAACKIAINISOFLAVAN
(3-HYDROXYINERMINISOFLAVAN)

Fusarium oxysporum f. sp. *pisi*
cultures [Metabolite]   (*265, 266*)

409 MW 302; $C_{17}H_{18}O_5$

5-METHOXYVESTITOL

Lotus edulis	† Leaf	(*388*)
L. hispidus	† Leaf	(*386*)

410 MW 302; $C_{17}H_{18}O_5$

LOTISOFLAVAN

Lotus angustissimus	† Leaf	(*388*)
L. edulis	† Leaf	(*388*)

411 MW 302; $C_{17}H_{18}O_5$

6-METHOXYVESTITOL

Fusarium solani f. sp. *pisi*
cultures [Metabolite]   (*658*)

Table 2 *(continued)*

Plant and/or Other Sources	Plant Part(s) Examined	References

412 MW 318; $C_{17}H_{18}O_6$

BRYAFLAVAN

Brya ebenus   Heartwood   *(247)*

413 MW 318; $C_{17}H_{18}O_6$

8-DEMETHYLDUARTIN

Dalbergia ecastophyllum   Trunk Wood   *(547)*
     Vine Wood   *(226)*

414 MW 332; $C_{18}H_{20}O_6$

LONCHOCARPAN

Lonchocarpus laxiflorus   Debarked Root   *(635)*

415 MW 332; $C_{18}H_{20}O_6$

DUARTIN

Table 2 *(continued)*

Plant and/or Other Sources	Plant Part(s) Examined	References
Machaerium acutifolium	Heartwood	*(495)*
M. mucronulatum	Heartwood	*(499)*
M. opacum	Heartwood	*(614)*
M. villosum	Heartwood	*(91, 499)*

416 MW 348; $C_{18}H_{20}O_7$

($=$ MACHAEROL C ?)

Machaerium sp.	Heartwood	*(595)*
M. pedicellatum	Heartwood	*(593)*

417 MW 362; $C_{19}H_{22}O_7$

($=$ MACHAEROL B ?)

Machaerium sp.	Heartwood	*(595)*
M. pedicellatum	Heartwood	*(593)*

b) Complex Isoflavans

418 MW 324; $C_{20}H_{20}O_4$

PHASEOLLINISOFLAVAN

Table 2 *(continued)*

Plant and/or Other Sources	Plant Part(s) Examined	References
Plant Sources		
Glycyrrhiza glabra		
var. *typica*	Root	*(561)*
Phaseolus vulgaris	† Cell Suspension Cultures	*(222)*
	† Etiolated Hypocotyl	*(127)*
	† Germinating Seed	*(285)*
	† Leaf	*(286)*
	† Root	*(129, 775)*
Other Source		
Stemphylium botryosum		
cultures [Metabolite]		*(341)*

419   MW 324; $C_{20}H_{20}O_4$

GLABRIDIN

Glycyrrhiza glabra	Root	*(708)*
G. glabra var. *typica*	Root	*(561)*

420   MW 338; $C_{21}H_{22}O_4$
2′-*O*-METHYLPHASEOLLINISOFLAVAN
(2′-METHOXYPHASEOLLINISOFLAVAN)

Phaseolus vulgaris	† Hypocotyl	*(797)*
	Root	*(775)*

References, pp. 229—265

Table 2 *(continued)*

Plant and/or Other Sources	Plant Part(s) Examined	References

421 MW 338; $C_{21}H_{22}O_4$

4'-*O*-METHYLGLABRIDIN

Glycyrrhiza glabra var. *typica*   Root   *(561)*

422 MW 340; $C_{21}H_{24}O_4$

2'-*O*-METHYLPHASEOLLIDINISOFLAVAN

Vigna unguiculata   † Stem   *(653)*

423 MW 352; $C_{21}H_{20}O_5$

LEIOCIN

Dalbergia nitidula   Bark   *(801)*

Table 2 *(continued)*

Plant and/or Other Sources	Plant Part(s) Examined	References

424   MW 352; $C_{21}H_{20}O_5$
LEIOCINOL

Dalbergia nitidula   Bark   *(801)*

425   MW 354; $C_{21}H_{22}O_5$
3′-METHOXYGLABRIDIN

Glycyrrhiza glabra var.
    typica   Root   *(561)*

426   MW 354; $C_{21}H_{22}O_5$
NEORAUFLAVANE

Neorautanenia edulis   Bark   *(117)*

Table 2 *(continued)*

Plant and/or Other Sources	Plant Part(s) Examined	References

427   MW 356; $C_{21}H_{24}O_5$
α,α-DIMETHYLALLYLCYCLOLOBIN

Cyclolobium clausseni   Heartwood   *(295)*

428   MW 368; $C_{22}H_{24}O_5$
SPHAEROSININ

Sphaerophysa salsula   Root   *(432)*

429   MW 370; $C_{22}H_{26}O_5$
UNANISOFLAVAN

Sophora secundiflora   Aerial Parts
                                (Mainly Stem)   *(558)*

Table 2 *(continued)*

Plant and/or Other Sources	Plant Part(s) Examined	References

430 MW 390; $C_{25}H_{26}O_4$

HISPAGLABRIDIN B

Glycyrrhiza glabra var.
  typica   Root   *(561)*

431 MW 392; $C_{25}H_{28}O_4$

HISPAGLABRIDIN A

Glycyrrhiza glabra var.
  typica   Root   *(561)*

432 MW 406; $C_{26}H_{30}O_4$

HEMINITIDULAN

Dalbergia nitidula   Bark   *(801)*

Table 2 *(continued)*

Plant and/or Other Sources	Plant Part(s) Examined	References

433　MW 420; $C_{26}H_{28}O_5$
NITIDULAN

Dalbergia nitidula　　　　　Bark　　　　　　　*(801)*

434　MW 422; $C_{26}H_{30}O_5$
NITIDULIN

Dalbergia nitidula　　　　　Bark　　　　　　　*(801)*

435　MW 424; $C_{26}H_{32}O_5$
LICORICIDIN

Glycyrrhiza glabra　　　　　Root　　　　　　　*(742)*

Table 2 *(continued)*

Plant and/or Other Sources	Plant Part(s) Examined	References

c) Isoflavan Dimer

436   MW 558; $C_{32}H_{30}O_9$
BISCYCLOLOBIN

Cyclolobium clausseni	Heartwood	*(295)*

L: ISOFLAVANOL

Plant and/or Other Sources	Plant Part(s) Examined	References

437   MW 340; $C_{19}H_{16}O_6$
AMBANOL

Neorautanenia amboensis	Root	*(591)*

M: ISOFLAVANQUINONES

Plant and/or Other Sources	Plant Part(s) Examined	References

438   MW 286; $C_{16}H_{14}O_5$
CLAUSSEQUINONE

Table 2 *(continued)*

Plant and/or Other Sources	Plant Part(s) Examined	References
Cyclolobium clausseni	Heartwood	*(93, 295)*
C. vecchii	Heartwood	*(93, 295)*
Millettia pendula	Heartwood	*(333)*

439   MW 316; $C_{17}H_{16}O_6$

PENDULONE

Millettia pendula	Heartwood	*(333)*

440   MW 316; $C_{17}H_{16}O_6$

MUCROQUINONE

Cyclolobium clausseni	Heartwood	*(93, 295)*
Machaerium mucronulatum	Heartwood	*(499)*

441   MW 346; $C_{18}H_{18}O_7$

AMORPHAQUINONE

Amorpha fruticosa	Root	*(736)*

Table 2 *(continued)*

Plant and/or Other Sources	Plant Part(s) Examined	References

442   MW 360; $C_{19}H_{20}O_7$
ABRUQUINONE A

Abrus precatorius   Root   (525)

443   MW 376; $C_{19}H_{20}O_8$
ABRUQUINONE C

Abrus precatorius   Root   (525)

444   MW 390; $C_{20}H_{22}O_8$
ABRUQUINONE B

Abrus precatorius   Root   (525)

Table 2 *(continued)*

N: ISOFLAVENES

Plant and/or Other Sources	Plant Part(s) Examined	References

a) Simple Isoflavenes

444a   MW 256; $C_{15}H_{12}O_4$

HAGININ D

| *Lespedeza cyrtobotrya* | Heartwood | *(562a)* |

445   MW 270; $C_{16}H_{14}O_4$

HAGININ B

| *Lespedeza cyrtobotrya* | Heartwood | *(562)* |

446   MW 286; $C_{16}H_{14}O_5$

SEPIOL

| *Gliricidia sepium* | Heartwood | *(412, 413)* |

446a   MW 286; $C_{16}H_{14}O_5$

HAGININ C

| *Lespedeza cyrtobotrya* | Heartwood | *(562a)* |

12*

Table 2 *(continued)*

Plant and/or Other Sources	Plant Part(s) Examined	References

447   MW 300; $C_{17}H_{16}O_5$

2'-*O*-METHYLSEPIOL

| *Gliricidia sepium* | Heartwood | *(413)* |

448   MW 300; $C_{17}H_{16}O_5$

HAGININ A

| *Lespedeza cyrtobotrya* | Heartwood | *(562)* |

449   MW 316; $C_{17}H_{16}O_6$

or

| *Baphia nitida* | Heartwood | *(24)* |

b) Complex Isoflavenes

450   MW 322; $C_{20}H_{18}O_4$

GLABRENE

| *Glycyrrhiza glabra* | Root | *(458)* |
| *G. glabra* var. *typica* | Root | *(561)* |

Table 2 *(continued)*

Plant and/or Other Sources	Plant Part(s) Examined	References

451   MW 352; $C_{21}H_{20}O_5$

NEORAUFLAVENE

Neorautanenia edulis	Bark	*(117)*

O: COUMESTANS

Plant and/or Other Sources	Plant Part(s) Examined	References

a) Simple Coumestans

452   MW 268; $C_{15}H_8O_5$

COUMESTROL

Astragalus sinicus	Leaf	*(76, 809)*
* *Brassica oleracea* (CRU)	Axillary Bud	*(464)*
Centrosema pubescens	† Germinating Seed/Leaf	*(773a)*
Dolichos biflorus	Leaf/Stem	*(443)*
Glycine max	Cotyledon/Seed Hull	*(520a)*
	† Hypocotyl	*(445)*
	† Leaf	*(444)*
	Seed/Seed Meal/Sprout	*(464)*
Medicago spp.	Burr/Stem	*(261, 262)*
	Flower/Petiole	*(261)*
	Leaf	*(261, 262, 526, 751)*

Table 2 *(continued)*

Plant and/or Other Sources	Plant Part(s) Examined	References
M. sativa	† Leaf/Leaf Meal	*(69, 71, 74, 307, 523, 526, 607)*
	Sprout	*(464)*
	Stem/Stem Tip	*(523)*
	Root	*(607)*
Melilotus alba	Leaf	*(751)*
Phaseolus aureus	Callus Tissue/Cell Suspension Cultures	*(53)*
	Root	*(843)*
	Seedlings	*(214)*
P. calcaratus	† Germinating Seed/Leaf/ Root	*(356, 773a)*
P. lunatus	† Root	*(691)*
P. vulgaris	† Cell Suspension Cultures	*(222)*
	† Germinating Seed	*(285)*
	† Hypocotyl	*(685)*
	† Leaf	*(286, 530)*
	† Pod	*(831)*
Pisum sativum	Commercial Frozen Peas	*(76, 464)*
	Silage	*(76)*
Psoralea corylifolia	Root	*(311)*
* *Spinacia oleracea* (CHEN)	Leaf	*(464)*
Trifolium alexandrinum	Leaf	*(734)*
T. fragiferum	Leaf	*(69, 263, 526)*
T. incarnatum	Leaf	*(51)*
T. nigrescens	Leaf	*(263)*
T. pratense	Leaf	*(307, 526)*
T. repens	Leaf	*(69, 70, 263, 307, 526, 827)*
T. subterraneum	Burr Husk/Petiole/Seed/ Stem	*(260)*
	Leaf	*(260, 307, 526)*
Trigonella corniculata	Leaf	*(751)*
Vigna unguiculata	Hypocotyl	*(633, 634)*

453 MW 282; $C_{16}H_{10}O_5$
9-*O*-METHYLCOUMESTROL (4'-*O*-METHYLCOUMESTROL)

| *Cicer arietinum* | Root | *(844)* |
| *Dalbergia oliveri* | Heartwood | *(225)* |

Table 2 *(continued)*

Plant and/or Other Sources	Plant Part(s) Examined	References
D. stevensonii	Heartwood	*(227)*
Medicago spp.	Burr/Leaf/Stem	*(262)*
M. sativa	† Leaf/Leaf Meal	*(72, 607)*
Myroxylon balsamum	Trunk Wood	*(96)*
Pisum sativum	Silage	*(76)*
Trifolium pratense	† Seedlings	*(212)*
T. repens	† Leaf	*(827)*

454   MW 284; $C_{15}H_8O_6$

REPENSOL

Trifolium repens	† Leaf	*(827)*

455   MW 284; $C_{15}H_8O_6$

LUCERNOL

Medicago sativa	Leaf Meal	*(765)*

456   MW 296; $C_{16}H_8O_6$

MEDICAGOL

Cicer arietinum	Root	*(844)*
Dalbergia oliveri	Heartwood	*(225)*
D. stevensonii	Heartwood	*(227)*
Euchresta japonica	Root	*(745a)*
Maackia amurensis var. *buergeri*	Heartwood	*(778)*
Medicago sativa	† Leaf/Leaf Meal	*(519, 607)*
Sophora tomentosa	Aerial Parts	*(470, 471)*
Trifolium pratense	Hydrolysed Leaf/Stem	*(78)*
	† Seedlings	*(212)*

Table 2 *(continued)*

Plant and/or Other Sources	Plant Part(s) Examined	References

457 MW 296; $C_{17}H_{12}O_5$
COUMESTROL DIMETHYL ETHER

Dalbergia decipularis	Heartwood	*(195)*
Swartzia madagascariensis	Heartwood	*(327)*

458 MW 298; $C_{16}H_{10}O_6$

TRIFOLIOL

Medicago sativa	Leaf Meal	*(71)*
Trifolium repens	Leaf/Leaf Meal	*(518, 827)*

459 MW 298; $C_{16}H_{10}O_6$

8-METHOXYCOUMESTROL (3′-METHOXYCOUMESTROL)

Medicago spp.	Burr/Leaf/Stem	*(262)*
M. sativa	Leaf Meal	*(75)*

460 MW 298; $C_{16}H_{10}O_6$

SATIVOL

Medicago sativa	† Leaf/Leaf Meal	*(607, 765)*

References, pp. 229—265

Table 2 *(continued)*

Plant and/or Other Sources	Plant Part(s) Examined	References

461 MW 300; $C_{15}H_8O_7$

DEMETHYLWEDELOLACTONE (NORWEDELOLACTONE)

* *Eclipta alba* (COMP)	Leaf	*(64, 65)*
* *Wedelia calendulacea* (COMP)	Leaf	*(721)*

462 MW 310; $C_{17}H_{10}O_6$

FLEMICHAPPARIN C

Flemingia chappar	Root	*(10)*

463 MW 312; $C_{17}H_{12}O_6$

WAIROL

Medicago sativa	† Leaf	*(79)*

464 MW 312; $C_{17}H_{12}O_6$

Medicago sativa	† Leaf/Leaf + Stem/Root	*(79, 607, 766)*
Myroxylon balsamum	Trunk Wood	*(96)*

Table 2 *(continued)*

Plant and/or Other Sources	Plant Part(s) Examined	References

465   MW 314; $C_{16}H_{10}O_7$

WEDELOLACTONE

* *Eclipta alba* (COMP)	Leaf	*(64, 65, 299)*
* *Wedelia calendulacea* (COMP)	Leaf	*(300, 301)*

466   MW 326; $C_{17}H_{10}O_7$

2-HYDROXYFLEMICHAPPARIN C

Swartzia leiocalycina	Heartwood	*(224)*

467   MW 326; $C_{17}H_{10}O_7$

TEPHROSOL

Tephrosia villosa	Root	*(684)*

467a   MW 326; $C_{17}H_{10}O_7$

SOPHORACOUMESTAN B

Sophora franchetiana	Root	*(467a)*

Table 2 *(continued)*

Plant and/or Other Sources	Plant Part(s) Examined	References

468   MW 356; $C_{18}H_{12}O_8$

Swartzia leiocalycina   Heartwood   (*224*)

b) Complex Coumestans

469   MW 320; $C_{18}H_8O_6$
EROSNIN (NORTON & HANSBERRY'S COMPOUND I)

Pachyrrhizus erosus   Seed   (*233, 416, 586*)

470   MW 334; $C_{20}H_{14}O_5$
SOPHORACOUMESTAN A

Sophora franchetiana   Root   (*472*)

471   MW 336; $C_{20}H_{16}O_5$
SOJAGOL

Glycine max   † Hypocotyl   (*445*)
    † Leaf   (*444*)

Table 2 *(continued)*

Plant and/or Other Sources	Plant Part(s) Examined	References
Phaseolus aureus	Callus Tissue/Cell Suspension Cultures	*(53)*
	Root	*(843)*

472   MW 336; $C_{20}H_{16}O_5$
PSORALIDIN

Dolichos biflorus	Leaf/Stem	*(443)*
Phaseolus lunatus	† Root	*(691)*
Psoralea corylifolia	Pericarp	*(34, 35, 140)*
	Kernel	*(448)*

473   MW 336; $C_{20}H_{16}O_5$
ISOPSORALIDIN

Psoralea corylifolia	Seed	*(400)*

474   MW 352; $C_{20}H_{16}O_6$
PSORALIDIN OXIDE

Psoralea corylifolia	Seed	*(312)*

References, pp. 229—265

Table 2 *(continued)*

Plant and/or Other Sources	Plant Part(s) Examined	References

475 MW 366; $C_{21}H_{18}O_6$

GLYCYROL

Glycyrrhiza sp.	Root	*(710)*

476 MW 366; $C_{21}H_{18}O_6$

ISOGLYCYROL

Glycyrrhiza sp.	Root	*(710)*

477 MW 368; $C_{20}H_{16}O_7$

CORYLIDIN

Psoralea corylifolia	Seed	*(309)*

478 MW 380; $C_{22}H_{20}O_6$

1-*O*-METHYLGLYCYROL (5-*O*-METHYLGLYCYROL)

Glycyrrhiza sp.	Root	*(710)*

Table 2 *(continued)*

P: 3-ARYLCOUMARINS

Plant and/or Other Sources	Plant Part(s) Examined	References

a) Simple 4-Deoxy 3-Arylcoumarins

479   MW 284; $C_{16}H_{12}O_5$

Dalbergia oliveri	Heartwood	*(225)*

480   MW 298; $C_{16}H_{10}O_6$

Dalbergia oliveri	Heartwood	*(225)*

b) Complex 4-Deoxy 3-Arylcoumarins

481   MW 336; $C_{19}H_{12}O_6$

PACHYRRHIZIN
(NEORAUTONE; NORTON & HANSBERRY'S COMPOUND III)

Neorautanenia amboensis	Root	*(119)*
N. edulis	Root/Tuber	*(1, 119, 795)*
N. pseudopachyrrhiza	Root	*(185, 186)*
Pachyrrhizus erosus	Seed	*(416, 586, 752)*

Table 2 *(continued)*

Plant and/or Other Sources	Plant Part(s) Examined	References

482   MW 366; $C_{20}H_{14}O_7$
NEOFOLIN

Neorautanenia ficifolia	Root	*(121)*

483   MW 382; $C_{22}H_{22}O_6$
GLYCYRIN

Glycyrrhiza sp.	Root	*(459)*

c) Simple 4-Oxy 3-Arylcoumarin

484   MW 356; $C_{19}H_{16}O_7$
DERRUSNIN

Derris glabrescens	Seed	*(202)*
D. robusta	Root	*(231)*
	Seed	*(160)*

Table 2 *(continued)*

Plant and/or Other Sources	Plant Part(s) Examined	References

d) Complex 4-Oxy 3-Arylcoumarins

485   MW 380; $C_{22}H_{20}O_6$

ROBUSTIC ACID

Derris robusta	Root	*(231, 405, 683)*
D. scandens	Root	*(171)*

486   MW 394; $C_{23}H_{22}O_6$

ROBUSTIC ACID METHYL ETHER (METHYL ROBUSTATE)

Derris robusta	Root	*(231, 405)*

487   MW 394; $C_{22}H_{18}O_7$

ROBUSTIN

Derris robusta	Root	*(231)*

488   MW 408; $C_{23}H_{20}O_7$

ROBUSTIN METHYL ETHER

Derris robusta	Root	*(231)*
	Seed	*(160)*

Table 2 *(continued)*

Plant and/or Other Sources	Plant Part(s) Examined	References

489 MW 408; $C_{23}H_{20}O_7$

GLABRESCIN

Derris glabrescens   Seed   *(202)*

490 MW 434; $C_{26}H_{26}O_6$
SCANDENIN

Derris scandens   Root   *(171, 238, 406, 682)*

491 MW 434; $C_{26}H_{26}O_6$
LONCHOCARPIC ACID (CHANDANIN)

Derris scandens   Root   *(171, 238, 406, 682)*

Lonchocarpus sp.   Root   *(409)*

Table 2 *(continued)*

Plant and/or Other Sources	Plant Part(s) Examined	References

492   MW 448; $C_{27}H_{28}O_6$
LONCHOCARPENIN

Derris scandens	Root	*(238)*

Q: α-METHYLDEOXYBENZOINS

Plant and/or Other Sources	Plant Part(s) Examined	References

a) Simple α-Methyldeoxybenzoins

493   MW 258; $C_{15}H_{14}O_4$
DEMETHYLANGOLENSIN

Sheep urine [Metabolite]		*(47)*

Table 2 *(continued)*

Plant and/or Other Sources	Plant Part(s) Examined	References

494 MW 272; $C_{16}H_{16}O_4$

ANGOLENSIN

Pericopsis sp.	Heartwood	*(355)*
P. elata (= Afrormosia elata)	Heartwood	*(252, 254)*
P. mooniana	Heartwood	*(252)*
Pterocarpus angolensis	Heartwood	*(56, 456)*
P. erinaceus	Heartwood	*(14)*
P. indicus	Bark/Sapwood	*(313)*
	Heartwood	*(178, 313)*

495 MW 286; $C_{17}H_{18}O_4$

2-*O*-METHYLANGOLENSIN

Pericopsis elata	Heartwood	*(252)*

496 MW 286; $C_{17}H_{18}O_4$

4-*O*-METHYLANGOLENSIN

Pterocarpus angolensis	Heartwood	*(56)*

13*

Table 2 *(continued)*

Plant and/or Other Sources	Plant Part(s) Examined	References

b) Complex α-Methyldeoxybenzoins

497   MW 476; $C_{31}H_{40}O_4$

4-*O*-α-CADINYLANGOLENSIN

Pterocarpus angolensis	Heartwood	*(56)*

498   MW 476; $C_{31}H_{40}O_4$

4-*O*-T-CADINYLANGOLENSIN

Pterocarpus angolensis	Heartwood	*(56)*

Table 2 *(continued)*

R: 2-ARYLBENZOFURANS OF LEGUMINOUS ORIGIN

Plant and/or Other Sources	Plant Part(s) Examined	References

a) Simple 2-Arylbenzofurans

498a   MW 242; $C_{14}H_{10}O_4$

| *Lespedeza cyrtobotrya* | Heartwood | *(562a)* |

499   MW 254; $C_{16}H_{14}O_3$
PARVIFURAN

| *Dalbergia parviflora* | Heartwood | *(568)* |

500   MW 256; $C_{15}H_{12}O_4$
6-DEMETHYLVIGNAFURAN

Anthyllis vulneraria	† Leaf	*(385)*
Coronilla emerus	† Leaf	*(215)*
Tetragonolobus maritimus	† Leaf	*(385)*

Table 2 *(continued)*

Plant and/or Other Sources	Plant Part(s) Examined	References

501   MW 270; $C_{16}H_{14}O_4$

VIGNAFURAN

Lablab niger	† Etiolated Hypocotyl	*(367)*
Vigna unguiculata	† Leaf	*(656)*
	† Seedlings	*(545)*

502   MW 270; $C_{15}H_{10}O_5$

Sophora tomentosa	Aerial Parts	*(470)*

503   MW 284; $C_{16}H_{12}O_5$

Sophora tomentosa	Aerial Parts	*(470)*

504   MW 286; $C_{16}H_{14}O_5$

PTEROFURAN

Pterocarpus indicus	Heartwood	*(178)*

References, pp. 229—265

Table 2 *(continued)*

Plant and/or Other Sources	Plant Part(s) Examined	References

505   MW 286; $C_{16}H_{14}O_5$

ISOPTEROFURAN

Coronilla emerus   † Leaf   (215)

506   MW 286; $C_{16}H_{14}O_5$

Myroxylon balsamum   Trunk Wood   (96)

506a   MW 300; $C_{16}H_{12}O_6$

SOPHORAFURAN A

Sophora franchetiana   Root   (467a)

507   MW 374; $C_{19}H_{18}O_8$

BRYEBINAL

or

Brya ebenus   Heartwood   (248)

Table 2 *(continued)*

Plant and/or Other Sources	Plant Part(s) Examined	References

b) Complex 2-Arylbenzofurans

508   MW 338; $C_{20}H_{18}O_5$
NEORAUFURANE

Neorautanenia edulis	Bark	*(117)*

509   MW 354; $C_{21}H_{22}O_5$
AMBOFURANOL

Neorautanenia amboensis	Bulb Bark	*(109)*

S: 2-ARYLBENZOFURANQUINONE

Plant and/or Other Sources	Plant Part(s) Examined	References

510   MW 358; $C_{18}H_{14}O_8$
BRYEBINALQUINONE

or

Brya ebenus	Heartwood	*(248)*

Table 3. *Naturally Occurring Isoflavonoid Glycosides*[a, b]

---

[a] Glycosides are arranged in groups according to the molecular weight of the aglycone, and its sectional/sub-sectional position as given in Table 2. The symbol $\neq$ indicates those glycosides based on aglycones not yet reported as natural products.

[b] For other abbreviations and symbols see Table 2.

A: ISOFLAVONE GLYCOSIDES

Plant and/or Other Sources	Plant Part(s) Examined	References
511 DAIDZEIN [3]-7-*O*-GLUCOSIDE (DAIDZIN)		
Baptisia spp.	Leaf/Stem	(*538, 540*)
Erythrina crista-galli	Trunk Bark/Trunk Wood	(*354*)
Glycine max	Seed/Seed Cake/Seed Meal	(*436, 572, 602, 812*)
Medicago sativa	Leaf	(*607*)
Piptanthus nepalensis	Twig	(*625*)
Psoralea acaulis	Root	(*839*)
Pueraria spp.	Root	(*67, 569, 738*)
Thermopsis spp.	Leaf/Stem	(*204*)
T. fabacea	Root	(*21*)
Trifolium pratense	Leaf/Stem	(*780*)
512 6″-*O*-ACETYLDAIDZIN		
Glycine max	Seed	(*602*)
513 DAIDZEIN-4′-*O*-GLUCOSIDE		
Piptanthus nepalensis	Twig	(*625*)
514 DAIDZEIN-7,4′-DI-*O*-GLUCOSIDE		
Piptanthus nepalensis	Twig	(*625*)
515 DAIDZEIN-7-*O*-RHAMNOSIDE		
Streptomyces xanthophaeus cultures [Metabolite]		(*19, 334*)
516 DAIDZEIN-7-*O*-RHAMNOSYLGLUCOSIDE		
Baptisia spp.	Leaf/Stem	(*538, 540*)
517 DAIDZEIN-7,4′-DI-*O*-RHAMNOSIDE		
Streptomyces xanthophaeus cultures [Metabolite]		(*334*)
518 DAIDZEIN-8-*C*-GLUCOSIDE (PUERARIN)		
Pueraria spp.	Root	(*67, 569*)

Table 3 *(continued)*

Plant and/or Other Sources	Plant Part(s) Examined	References
519 4′,6″-DI-O-ACETYLPUERARIN		
Pueraria tuberosa	Root	*(67)*
520 PUERARIN XYLOSIDE		
Pueraria spp.	Root	*(569)*
521 FORMONONETIN [5]-7-O-GLUCOSIDE (ONONIN)		
Amorpha fruticosa	Root	*(736)*
Baptisia spp.	Leaf/Stem	*(540)*
B. australis	Leaf/Root	*(510, 617)*
Cicer arietinum	Leaf/Stem	*(437)*
Cladrastis platycarpa	Bark	*(352, 353)*
	Heartwood/Sapwood	*(601)*
	Leaf	*(599)*
C. shikokiana	Bark/Heartwood/Leaf/ Root Bark/Root Wood/ Sapwood	*(599)*
Dalbergia paniculata	Bark	*(628, 631)*
Genista patula	Aerial Parts	*(618)*
Medicago sativa	Leaf/Root	*(607)*
Ononis antiquorum	Aerial Parts	*(793)*
O. arvensis	Root	*(245, 246, 476)*
O. spinosa	Root	*(245, 246, 335, 344)*
Piptanthus nanus	Aerial Parts	*(793)*
P. nepalensis	Twig	*(625)*
Pueraria thunbergiana	Flower	*(493)*
Spartium junceum	Aerial Parts	*(619)*
Thermopsis spp.	Leaf/Stem	*(204)*
T. fabacea	Root	*(21)*
T. rhombifolia	Aerial Parts	*(50, 204)*
Trifolium spp.	Leaf	*(263, 438, 718)*
T. pratense	Leaf	*(49)*
	Root	*(258)*
T. subterraneum	Leaf	*(49)*
Wisteria floribunda	Bark	*(596)*
	Wood	*(596, 781)*
522 ACYLATED ONONIN DERIVATIVE (=**527**?)		
Trifolium spp.	Leaf	*(263)*
523 FORMONONETIN-7-O-RHAMNOSYLGLUCOSIDE		
Baptisia spp.	Leaf/Stem	*(540)*
524 FORMONONETIN-7-O-RUTINOSIDE		
Dalbergia paniculata	Bark	*(629, 631)*

Table 3 *(continued)*

Plant and/or Other Sources	Plant Part(s) Examined	References
525 FORMONONETIN-7-*O*-LAMINARABIOSIDE		
Cladrastis platycarpa	As given for **521**	*(352, 353, 598, 599)*
C. shikokiana	As given for **521**	*(599)*
526 FORMONONETIN DIGLYCOSIDE		
Trifolium spp.	Leaf	*(263)*
527 FORMONONETIN-7-*O*-GLUCOSIDE-6″-MALONATE		
Trifolium pratense	Leaf	*(49)*
T. subterraneum	Leaf	*(49)*
528 FORMONONETIN-7-*O*-GLUCOSIDE-6″-MALONATE, METHYL ESTER		
Trifolium pratense	Leaf	*(49)*
T. subterraneum	Leaf	*(49)*
529 DEMETHYLTEXASIN [9]-4′-*O*-GLUCOSIDE		
Glycine max	Seed Flakes	*(264)*
530 PSEUDOBAPTIGENIN [11]-7-*O*-GLUCOSIDE (ROTHINDIN)		
Cladrastis platycarpa	As given for **521**	*(599)*
C. shikokiana	As given for **521**	*(599)*
Ononis spinosa	Root	*(335)*
Rothia indica	Aerial Parts	*(573)*
Trifolium pratense	Root	*(258)*
531 PSEUDOBAPTIGENIN-7-*O*-RHAMNOSYLGLUCOSIDE (PSEUDOBAPTISIN)		
Baptisia spp.	Leaf/Stem	*(538, 540)*
B. tinctoria	Root	*(294, 764)*
532 PSEUDOBAPTIGENIN-7-*O*-LAMINARABIOSIDE ?		
Cladrastis platycarpa	As given for **521**	*(599)*
C. shikokiana	As given for **521**	*(599)*
533 CALYCOSIN [15]-7-*O*-GLUCOSIDE		
Baptisia spp.	Leaf/Stem	*(538, 539, 540)*
Thermopsis spp.	Leaf/Stem	*(204)*
534 CALYCOSIN-7-*O*-RHAMNOSYLGLUCOSIDE		
Baptisia spp.	Leaf/Stem	*(538, 539, 540)*
535 TEXASIN [17]-7-*O*-GLUCOSIDE		
Baptisia australis	Leaf/Stem	*(541)*

Table 3 *(continued)*

Plant and/or Other Sources	Plant Part(s) Examined	References
536 GLYCITEIN [**19**]-7-*O*-GLUCOSIDE		
Glycine max	Seed Meal	*(572)*
537 RETUSIN [**20**]-7-*O*-NEOHESPERIDOSIDE		
[LM] *Prosopsis juliflora*	Stem Bark	*(747)*
538 BAPTIGENIN [**21**]-*O*-RHAMNOSIDE (BAPTISIN)		
Baptisia tinctoria	Root	*(243, 293)*
539 CLADRIN [**25**]-7-*O*-GLUCOSIDE		
Cladrastis platycarpa	As given for **521**	*(599)*
C. shikokiana	As given for **521**	*(599)*
540 CLADRIN-7-*O*-LAMINARABIOSIDE?		
Cladrastis platycarpa	As given for **521**	*(599)*
C. shikokiana	As given for **521**	*(599)*
541 AFRORMOSIN [**28**]-7-*O*-GLUCOSIDE (WISTIN)		
Amorpha fruticosa	Root	*(736)*
Baptisia spp.	Leaf/Stem	*(510, 540)*
Cladrastis platycarpa	As given for **521**	*(351, 352, 600, 601)*
C. shikokiana	As given for **521**	*(599)*
Wisteria brachybotrys	Bark	*(740)*
W. floribunda	Bark/Heartwood	*(596, 740, 781)*
	Root	*(740)*
542 AFRORMOSIN-7-*O*-RHAMNOSYLGLUCOSIDE		
Baptisia spp.	Leaf/Stem	*(540)*
543 AFRORMOSIN-7-*O*-LAMINARABIOSIDE?		
Cladrastis platycarpa	As given for **521**	*(599)*
C. shikokiana	As given for **521**	*(599)*
544 8-*O*-METHYLRETUSIN [**29**]-7-*O*-GLUCOSIDE		
Cladrastis platycarpa	As given for **521**	*(599)*
C. shikokiana	As given for **521**	*(599)*
Wisteria floribunda	Bark	*(596)*
545 8-*O*-METHYLRETUSIN-7-*O*-LAMINARABIOSIDE		
Cladrastis platycarpa	As given for **521**	*(352, 353, 599)*
C. shikokiana	As given for **521**	*(599)*

Table 3 *(continued)*

Plant and/or Other Sources	Plant Part(s) Examined	References

546 FUJIKINETIN [**35**]-7-*O*-GLUCOSIDE (FUJIKININ)

Cladrastis platycarpa	Bark	*(350, 351, 352, 598)*
	Heartwood/Sapwood	*(597, 599)*
	Leaf	*(600)*
C. shikokiana	As given for **521**	*(599)*

547 FUJIKINETIN-7-*O*-LAMINARABIOSIDE ?

Cladrastis platycarpa	As given for **521**	*(599)*
C. shikokiana	As given for **521**	*(599)*

547a 7-HYDROXY-2′,4′,5′-TRIMETHOXYISOFLAVONE [**42**]-7-*O*-GLUCOSIDE

Dalbergia monetaria	Seed	*(223a)*

548 CLADRASTIN [**44**]-7-*O*-GLUCOSIDE

Cladrastis platycarpa	As given for **521**	*(597, 598, 599, 600)*
C. shikokiana	As given for **521**	*(599)*

549 CLADRASTIN-7-*O*-LAMINARABIOSIDE ?

Cladrastis platycarpa	As given for **521**	*(599)*
C. shikokiana	As given for **521**	*(599)*

550 DALPATEIN [**48**]-7-*O*-GLUCOSIDE (DALPATIN)

Dalbergia paniculata	Seed	*(6)*

551 GENISTEIN [**81**]-7-*O*-GLUCOSIDE (GENISTIN; GENISTOSIDE)

Adenocarpus complicatus	Twig	*(624)*
Baptisia spp.	Leaf/Stem	*(540)*
Cytisus scoparius	Flower	*(345)*
Erythrina crista-galli	Trunk Bark/Trunk Wood	*(354)*
Genista lydia	Flower	*(791)*
G. ovata	Aerial Parts	*(575)*
G. rumelica	Aerial Parts	*(576, 577)*
G. tinctoria	Flower	*(145)*
Glycine max	Seed/Seed Cake/Seed Flake/Seed Meal	*(436, 572, 602,* 811, 812)
Lupinus angustifolius	—	*(513)*
L. luteus	Receptacle/Root	*(506, 507)*
L. polyphyllus	Leaf/Stem	*(345, 346, 513)*
Lygos raetam	Flower	*(623)*
Piptanthus nepalensis	Twig	*(625)*
* *Prunus aequinoctialis* (ROS)	Wood	*(330)*

Table 3 *(continued)*

Plant and/or Other Sources	Plant Part(s) Examined	References
* *P. avium* (ROS)	Wood	*(330)*
* *P. nipponica* (ROS)	Wood	*(330)*
Pueraria thunbergiana	Flower	*(493)*
Spartium junceum	Aerial Parts	*(619)*
Thermopsis spp.	Leaf/Stem	*(204)*
T. alterniflora	—	*(429)*
T. fabacea	Root	*(21)*
T. lanceolata	—	*(429)*
Trifolium spp.	Leaf/Stem	*(49, 263, 718)*
T. pratense	Root	*(258)*
Ulex nanus	Twig	*(624)*

552   6″-*O*-ACETYLGENISTIN

Glycine max	Seed	*(603)*

553   GENISTEIN-4′-*O*-GLUCOSIDE (SOPHORICOSIDE)

Piptanthus nepalensis	Twig	*(625)*
Sophora japonica	Fruit	*(142, 143, 841)*

554   GENISTEIN-7-*O*-GLUCOSYLGLUCOSIDE

Lupinus angustifolius	—	*(513)*
L. luteus	Bud/Cotyledon/Leaf/ Petal/Pod/	
	Receptacle	*(507)*
	Root	*(506, 507)*
L. polyphyllus	—	*(513)*

555   GENISTEIN-7,4′-DI-*O*-GLUCOSIDE

Piptanthus nepalensis	Twig	*(625)*
Thermopsis spp.	Leaf/Stem	*(204)*

556   GENISTEIN-7-*O*-RHAMNOSIDE

Streptomyces xanthophaeus cultures [Metabolite]		*(334)*

557   GENISTEIN-7-*O*-RHAMNOSYLGLUCOSIDE (SPHAEROBIOSIDE)

Baptisia spp.	Leaf/Stem	*(538, 540)*
B. sphaerocarpa	Flower	*(705)*

558   GENISTEIN-7,4′-DI-*O*-RHAMNOSIDE

Streptomyces xanthophaeus cultures [Metabolite]		*(334)*

559   GENISTEIN-4′-*O*-NEOHESPERIDOSIDE (SOPHORABIOSIDE)

Sophora japonica	Fruit	*(240, 242, 840)*

Table 3 *(continued)*

Plant and/or Other Sources	Plant Part(s) Examined	References
560 GENISTEIN-7,4'-DI-*O*-APIOSYLGLUCOSIDE (SAROTHAMNOSIDE)		
Cytisus (Sarothamnus) patens	Seed	*(124)*
C. scoparius	Seed	*(124)*
561 GENISTEIN-7-*O*-GLUCOSIDE-6''-MALONATE		
Trifolium subterraneum	Leaf	*(49)*
562 GENISTEIN-7-*O*-GLUCOSIDE-6''-MALONATE, METHYL ESTER ?		
Trifolium subterraneum	Leaf	*(49)*
563 GENISTEIN-8-*C*-GLUCOSIDE		
Dalbergia nitidula	Bark	*(802)*
Lupinus luteus	Bud/Cotyledon/Leaf/ Petal/Pod/Receptacle/ Root/Stem	*(507)*
	Flower	*(838)*
564 GENISTEIN-6,8-DI-*C*-GLUCOSIDE (PANICULATIN)		
Dalbergia nitidula	Bark	*(802)*
D. paniculata	Bark	*(583, 631)*
565 *ISOGENISTEIN-7-*O*-GLUCOSIDE		
Cajanus cajan	Root Bark	*(59)*
566 BIOCHANIN A [83]-7-*O*-GLUCOSIDE (ASTROSIDE; SISSOTRIN)		
Astragalus austriacus	Leaf/Stem	*(207)*
Baptisia leucantha	Leaf/Stem	*(540)*
Cicer arietinum	Leaf/Stem	*(437)*
	Germinating Seed	*(828)*
* *Cotoneaster pannosa* (ROS)	Flower	*(177)*
* *C. serotina* (ROS)	Flower/Fruit	*(177)*
Dalbergía paniculata	Bark	*(628, 631)*
D. sissoo	Flower	*(146a)*
	Leaf	*(12, 41)*
Pueraria thunbergiana	Flower	*(493)*
Sophora japonica	Wood	*(779)*
Thermopsis spp.	Leaf/Stem	*(204)*
Trifolium spp.	Leaf/Stem	*(49, 263, 434, 718)*
567 BIOCHANIN A-7-*O*-RHAMNOSYLGLUCOSIDE		
Baptisia spp.	Leaf/Stem	*(540)*

Table 3 *(continued)*

Plant and/or Other Sources	Plant Part(s) Examined	References

568 BIOCHANIN A-7-*O*-RUTINOSIDE

Dalbergia paniculata   Bark   *(629, 631)*

569 BIOCHANIN A-7-*O*-GENTIOBIOSIDE

Sophora japonica   Wood   *(779)*

570 BIOCHANIN A-7-*O*-XYLOSYLGLUCOSIDE

Sophora japonica   Wood   *(779)*

571 BIOCHANIN A-7-*O*-GLUCOSYLAPIOSIDE (LANCEOLARIN)

Dalbergia lanceolaria   Root Bark   *(532)*

572 BIOCHANIN A-7-*O*-GLUCOSIDE-6″-MALONATE
(SISSOTRIN-5-MALONATE)
(Revised Structure)

Trifolium pratense   Leaf   *(49, 780)*
T. subterraneum   Leaf   *(49)*

573 BIOCHANIN A-7-*O*-GLUCOSIDE-6″-MALONATE, METHYL ESTER

T. pratense   Leaf   *(49)*
T. subterraneum   Leaf   *(49)*

574 PRUNETIN [84]-4′-*O*-GLUCOSIDE
(PRUNITRIN; PRUNITROSIDE; TRIFOSIDE)

Genista carinalis   Aerial Parts   *(578)*
* *Prunus* sp.
   (*emarginata* ?) (ROS)   Bark   *(250)*
* *P. mahaleb* (ROS)   Bark   *(646)*
* *P. puddum* (ROS)   Bark   *(241, 580)*
Trifolium medium   Leaf/Stem   *(435)*
T. pratense   Leaf/Stem   *(428, 435)*

575 PRUNETIN-8-*C*-GLUCOSIDE

Dalbergia paniculata   Bark   *(628, 631)*

576 OROBOL [87]-7-*O*-GLUCOSIDE (= OROBOSIDE ?)

Baptisia spp.   Leaf/Stem   *(540)*
Thermopsis spp.   Leaf/Stem   *(204)*

577 OROBOL-*O*-GLUCOSIDE (OROBOSIDE)

Lathyrus macrorrhizus   Root   *(556)*
L. montanus (= Orobus
   tuberosus)   Leaf/Aerial Parts   *(110, 111, 112,*
                                                            144)

Table 3 *(continued)*

Plant and/or Other Sources	Plant Part(s) Examined	References
578 OROBOL-7-*O*-RHAMNOSYLGLUCOSIDE		
Baptisia spp.	Leaf/Stem	*(538, 540)*
579 OROBOL-8-*C*-GLUCOSIDE		
Dalbergia nitidula	Bark	*(802)*
Lupinus luteus	Flower	*(838)*
580 OROBOL-6,8-DI-*C*-GLUCOSIDE		
Dalbergia nitidula	Bark	*(802)*
581 6-HYDROXYGENISTEIN [**88**]-7-*O*-RHAMNOSYLGLUCOSIDE		
Baptisia hirsuta	Leaf/Stem	*(539, 540)*
582 5-HYDROXYPSEUDOBAPTIGENIN [**90**]-7-*O*-GLUCOSIDE		
Lupinus angustifolius	—	*(513)*
L. luteus	Root	*(506, 507)*
L. polyphyllus	—	*(513)*
583 5-HYDROXYPSEUDOBAPTIGENIN-7-*O*-GLUCOSYLGLUCOSIDE		
Lupinus luteus	Root	*(506, 507)*
584 ⁺7-*O*-METHYLBIOCHANIN A-6-*C*-RHAMNOSIDE (ISOVOLUBILIN)		
Dalbergia volubilis	Flower	*(148, 149)*
585 ⁺7-*O*-METHYLBIOCHANIN A-8-*C*-RHAMNOSIDE (VOLUBILIN)		
Dalbergia volubilis	Flower	*(147, 148, 149)*
586 IRILONE [**92**]-4′-*O*-GLUCOSIDE		
* *Iris florentina* (IRID)	Rhizome	*(785)*
Trifolium pratense	Root	*(256)*
587 PRATENSEIN [**94**]-7-*O*-GLUCOSIDE		
Thermopsis spp.	Leaf/Stem	*(204)*
Trifolium spp.	Leaf	*(718)*
588 3′-*O*-METHYLOROBOL [**95**]-7-*O*-GLUCOSIDE		
Thermopsis spp.	Leaf/Stem	*(204)*
589 3′-*O*-METHYLOROBOL-8-*C*-GLUCOSIDE (DALPANITIN)		
Dalbergia paniculata	Seed	*(6)*

Table 3 *(continued)*

Plant and/or Other Sources	Plant Part(s) Examined	References

590 TECTORIGENIN [**98**]-7-*O*-GLUCOSIDE
(TECTORIDIN; SHEKANIN)

Baptisia spp.	Leaf/Stem	*(540)*
* *Belamcanda chinensis* (IRID)	Rhizome	*(535)*
Dalbergia riparia	Wood	*(103)*
D. sissoo	Flower/Root/Root Bark	*(146a)*
	Pod	*(729)*
D. volubilis	Bark	*(450, 451)*
* *Iris germanica* (IRID)	Rhizome	*(433)*
* *I. tectorum* (IRID)	Rhizome	*(566, 735)*
Pueraria montana	Flower	*(486)*

591 TECTORIGENIN-7-*O*-GENTIOBIOSIDE

Dalbergia volubilis	Bark	*(449)*

592 IRISOLONE [**101**]-4′-*O*-BIOSIDE

* *Iris florentina* (IRID)	Rhizome	*(785)*

593 IRISOLIDONE [**104**]-7-*O*-GLUCOSIDE (KAKKALIDONE)

Pueraria thunbergiana		
(=lobata)	Flower	*(486, 492)*
Sophora japonica	Wood	*(779)*

594 IRISOLIDONE-7-*O*-XYLOSYLGLUCOSIDE (KAKKALIDE)

Pueraria thunbergiana	Flower	*(492)*

595 IRISOLIDONE-8-*C*-GLUCOSIDE (VOLUBILININ)

Dalbergia volubilis	Flower	*(149)*

596 7-*O*-METHYLTECTORIGENIN [**105**]-4′-*O*-GLUCOSIDE

Dalbergia volubilis	Bark	*(451)*

597 7-*O*-METHYLTECTORIGENIN-4′-*O*-RHAMNOSYLGLUCOSIDE

Dalbergia sissoo	Immature Pod	*(12)*

598 7-*O*-METHYLTECTORIGENIN-4′-*O*-GENTIOBIOSIDE

Dalbergia volubilis	Bark	*(451)*

599 ⁺IRIFLOGENIN-4′-*O*-GLUCOSIDE (IRIFLOSIDE)

* *Iris florentina* (IRID)	Rhizome	*(23)*

Table 3 *(continued)*

Plant and/or Other Sources	Plant Part(s) Examined	References

600   5-METHOXYAFRORMOSIN [113]-7-*O*-GLUCOSIDE

Cladrastis platycarpa	As given for **521**	*(598, 599, 600)*
C. shikokiana	As given for **521**	*(599)*

601   5-METHOXYAFRORMOSIN-7-*O*-LAMINARABIOSIDE ?

Cladrastis platycarpa	As given for **521**	*(599)*
C. shikokiana	As given for **521**	*(599)*

602   ISO-5-METHOXYAFRORMOSIN [114]-7-*O*-GLUCOSIDE

Cladrastis platycarpa	Bark/Heartwood/Sapwood	*(599)*
C. shikokiana	Bark/Heartwood/Sapwood	*(599)*

603   IRISTECTORIGENIN A [116]-7-*O*-GLUCOSIDE
(IRISTECTORIN A)

* *Iris tectorum* (IRID)	Rhizome	*(566)*

604   IRISTECTORIGENIN B [117]-7 (or 4′)-*O*-GLUCOSIDE
(IRISTECTORIN B)

* *Iris tectorum* (IRID)	Rhizome	*(567)*

605   ⁺HOMOTECTORIGENIN-7-*O*-GLUCOSIDE (HOMOTECTORIDIN)

* *Iris germanica* (IRID)	Rhizome	*(433)*

606   DIPTERYXINE [119]-7-*O*-GLUCOSIDE (ISOPLATYCARPANETIN-7-*O*-
GLUCOSIDE)

Cladrastis platycarpa	Bark/Heartwood/Sapwood	*(599)*
C. shikokiana	Bark/Heartwood/Sapwood	*(599)*

607   PLATYCARPANETIN [121]-7-*O*-GLUCOSIDE

Cladrastis platycarpa	Bark	*(352, 353, 598)*
	Heartwood/Sapwood	*(597, 599)*
	Leaf	*(599)*
C. shikokiana	As given for **521**	*(599)*

608   PLATYCARPANETIN-7-*O*-LAMINARABIOSIDE ?

Cladrastis platycarpa	As given for **521**	*(599)*
C. shikokiana	As given for **521**	*(599)*

609   IRIGENIN [128]-7-*O*-GLUCOSIDE (IRIDIN)

* *Belamcanda chinensis* (IRID)	Rhizome	*(500)*
* *Iris florentina* (IRID)	Rhizome	*(23, 36, 198)*
* *I. germanica* (IRID)	Rhizome	*(36, 236)*

Table 3 *(continued)*

Plant and/or Other Sources	Plant Part(s) Examined	References
* *I. kumaonensis* (IRID)	Rhizome	(279)
	Whole Plant	(217, 415)
* *I. pallida* (IRID)	Rhizome	(36)
* *I. unguicularis* (IRID)	Rhizome	(20)
* *Juniperus macropoda* (CUPR)	Leaf	(722)
610 CAVIUNIN [**130**]-7-*O*-GLUCOSIDE		
Dalbergia paniculata	Root	(663)
D. sissoo	Pod	(729)
611 CAVIUNIN-7-*O*-RHAMNOSYLGLUCOSIDE		
Dalbergia paniculata	Root	(666)
612 CAVIUNIN-7-*O*-GENTIOBIOSIDE		
Dalbergia sissoo	Pod	(728)
613 CAVIUNIN-7-*O*-GLYCOSIDE		
Dalbergia riparia	Wood	(103)
614 ISOCAVIUNIN [**132**]-7-*O*-GLUCOSIDE (ISOCAVIUDIN)		
Dalbergia paniculata	Bark	(632)
D. sissoo	Pod	(729)
615 ISOCAVIUNIN-7-O-GENTIOBIOSIDE		
Dalbergia sissoo	Pod	(730)

B: ISOFLAVANONE GLYCOSIDES

Plant and/or Other Sources	Plant Part(s) Examined	References
616 DIHYDROFORMONONETIN [**183**]-7-*O*-GLUCOSIDE (2,3-DIHYDRO-ONONIN)		
Ononis spinosa	Root	(335)
617 ‡DALPANIN		
Dalbergia paniculata	Flower	(6, 7, 8)
	Seed	(6)

Table 3 *(continued)*

C: ROTENOID GLYCOSIDES

Plant and/or Other Sources	Plant Part(s) Examined	References
618 AMORPHIGENIN [235]-*O*-GLUCOSIDE		
Amorpha spp.	Seed	*(424, 426, 427)*
A. fruticosa	Seed	*(424, 427)*
Dalbergia monetaria	Seed	*(223a)*
619 AMORPHIGENIN-*O*-VICIANOSIDE (AMORPHIN)		
Amorpha spp.	Seed	*(426, 427)*
A. fruticosa	Seed	*(2, 3, 166, 427)*
620 DALPANOL [239]-*O*-GLUCOSIDE		
Dalbergia paniculata	Seed	*(662)*
621 AMORPHIGENOL [241]-*O*-GLUCOSIDE		
Amorpha spp.	Seed	*(426, 427)*
A. fruticosa	Seed	*(427)*
	Seedlings	*(184b)*
622 AMORPHIGENOL-*O*-VICIANOSIDE (AMORPHOL)		
Amorpha spp.	Seed	*(426)*
A. fruticosa	Seed	*(426)*
623 DALBINOL [254]-*O*-GLUCOSIDE (DALBIN)		
Dalbergia assamica	Seed	*(156)*
D. latifolia	Seed	*(154)*
D. monetaria	Seed	*(223a)*
D. nitidula	Bark	*(802)*
623a 12-HYDROXYDALBIN		
D. monetaria	Seed	*(223a)*

D: PTEROCARPAN GLYCOSIDES

Plant and/or Other Sources	Plant Part(s) Examined	References
624 MEDICARPIN [276]-3-*O*-GLUCOSIDE		
Medicago sativa	Root	*(711)*
Ononis spinosa	Root	*(335)*
Trifolium hybridum	Root	*(259)*
T. repens	Root	*(259)*

Table 3 *(continued)*

Plant and/or Other Sources	Plant Part(s) Examined	References
625 (+)-MAACKIAIN [**279**]-3-*O*-GLUCOSIDE (SOPHOJAPONICIN)		
Sophora japonica	Root	*(467, 741)*
626 (−)-MAACKIAIN-3-*O*-GLUCOSIDE (TRIFOLIRHIZIN)		
Baptisia australis	Root	*(510, 617)*
Euchresta japonica	Root	*(745a)*
Ononis arvensis	Root	*(477)*
O. spinosa	Root	*(267, 335)*
Sophora flavescens	Root	*(467, 741)*
S. subprostrata	Root	*(473, 741)*
Thermopsis fabacea	Root	*(21)*
Trifolium hybridum	Root	*(259)*
T. pratense	Leaf/Stem	*(780)*
	Root	*(105, 106, 259)*
627 6'-*O*-ACETYLTRIFOLIRHIZIN (TRIFOLIRHIZIN-6'-MONOACETATE)		
Sophora subprostrata	Root	*(473)*

E. COUMESTAN GLYCOSIDES

Plant and/or Other Sources	Plant Part(s) Examined	References
628 COUMESTROL [**452**]-*O*-GLUCOSIDE		
Medicago sativa	† Leaf/Root	*(607)*
629 DEMETHYLWEDELOLACTONE [**461**]-3-*O*-GLUCOSIDE		
* *Eclipta alba* (COMP)	Leaf	*(64, 65)*

F: DEOXYBENZOIN GLYCOSIDE

Plant and/or Other Sources	Plant Part(s) Examined	References
630 ⁺ONONETIN-4-*O*-GLUCOSIDE (ONOSPIN)		
Ononis spinosa	Root	*(344, 818)*

Table 4. *Partially Identified Isoflavonoids, and Some Substances Possibly of an Isoflavonoid Nature Found in the Family Leguminosae*

Plant Sources	Plant Part(s) Examined	References
631 BELAMCANGENIN-*O*-GLUCOSIDE (BELAMCANDIN)		
* *Belamcanda chinensis* (IRID)	Root	*(813)*
632 CYTIFOLIOSIDE		
Cytisophyllum sessilifolium (= *Cytisus sessilifolius*)	Fruit/Leaf	*(626)*
633 DISAIN (ISOFLAVONE DIMER ?)		
* *Maclura aurantiaca* (MOR)	—	*(39)*
634 HIRTA SUBSTANCE A		
Tephrosia villosa (= *hirta*)	Root Bark	*(673)*
635 HIRTA SUBSTANCE C		
Tephrosia candida	Seed	*(674)*
T. villosa (= *hirta*)	Root Bark	*(673)*
636 ISOSHEHKANGENIN-*O*-GLUCOSIDE (ISOSHEHKANIN)		
* *Iris wattii* (IRID)	Rhizome ?	*(152, 283, 515)*
637 ISOTEPHROSIN (12a-HYDROXY ROTENOID)		
Lonchocarpus nicou	Root	*(172)*
Millettia ferruginea	Seed	*(170)*
638 MACHAEROL A (ISOFLAVAN)		
Machaerium pedicellatum	Heartwood	*(593)*
639 MACHAEROL B (ISOFLAVAN; =**417** ?)		
Machaerium pedicellatum	Heartwood	*(593)*
640 MACHAEROL C (ISOFLAVAN; =**416** ?)		
Machaerium pedicellatum	Heartwood	*(593)*
641 MUNDULEA SUBSTANCE A (=**79** ?)		
Mundulea sericea (= *suberosa*)	Seed	*(582)*
642 PURPURANIN A		
Tephrosia purpurea var. *maxima*	Pod	*(680)*

Table 4 *(continued)*

Plant and/or Other Sources	Plant Part(s) Examined	References
643 PURPURANIN B		
Tephrosia purpurea var. *maxima*	Pod	*(680)*
644 ROBUSTENIN		
Derris robusta	Root	*(683)*

Molecular Weight Index

Molecular Weight[1]	Compound Number
238	**(1)**
242	**(394)**, **(498a)**
252	**(2)**
254	**(3)**, **(380)**, **(499)**
256	**(182)**, **(275)**, **(395)**, **(444a)**, **(500)**
258	**(396)**, **(493)**
266	**(4)**
268	**(5)**, **(6)**, **(80)**, **(452)**
270	**(7)**, **(8)**, **(9)**, **(81)**, **(183)**, **(276)**, **(277)**, **(445)**, **(501)**, **(502)**
272	**(184)**, **(278)**, **(350)**, **(397)**, **(398)**, **(399)**, **(494)**
278	**(390)**
280	**(307)**
282	**(10)**, **(11)**, **(12)**, **(82)**, **(381)**, **(453)**
284	**(13)**, **(14)**, **(15)**, **(16)**, **(17)**, **(18)**, **(19)**, **(20)**, **(83)**, **(84)**, **(85)**, **(279)**, **(280)**, **(454)**, **(455)**, **(479)**, **(503)**
286	**(21)**, **(86)**, **(87)**, **(88)**, **(89)**, **(185)**, **(186)**, **(203)**, **(281)**, **(282)**, **(283)**, **(351)**, **(352)**, **(376)**, **(400)**, **(401)**, **(402)**, **(403)**, **(438)**, **(446)**, **(446a)**, **(495)**, **(496)**, **(504)**, **(505)**, **(506)**
288	**(204)**
294	**(22)**, **(23)**
296	**(24)**, **(382)**, **(456)**, **(457)**
298	**(25)**, **(26)**, **(27)**, **(28)**, **(29)**, **(90)**, **(91)**, **(92)**, **(179)**, **(284)**, **(458)**, **(459)**, **(460)**, **(480)**
300	**(30)**, **(31)**, **(93)**, **(94)**, **(95)**, **(96)**, **(97)**, **(98)**, **(99)**, **(187)**, **(188)**, **(189)**, **(190)**, **(285)**, **(286)**, **(287)**, **(288)**, **(289)**, **(290)**, **(353)**, **(354)**, **(377)**, **(404)**, **(447)**, **(448)**, **(461)**, **(506a)**
302	**(205)**, **(206)**, **(355)**, **(405)**, **(406)**, **(407)**, **(408)**, **(409)**, **(410)**, **(411)**
304	**(177)**
306	**(391)**
308	**(308)**
310	**(32)**, **(462)**
312	**(33)**, **(34)**, **(35)**, **(100)**, **(101)**, **(383)**, **(384)**, **(392)**, **(463)**, **(464)**

[1] Data applies to aglycones only.

Molecular Weight Index *(continued)*

Molecular Weight[1]	Compound Number
314	(36), (37), (38), (102), (103), (104), (105), (106), (191), (291), (292), (293), (294), (295), (356), (465)
316	(107), (108), (192), (192a), (207), (208), (296), (297), (298), (357), (439), (440), (449)
318	(412), (413)
320	(60), (61), (469)
322	(62), (309), (310), (450)
324	(196), (311), (312), (313), (314), (360), (418), (419)
326	(39), (40), (41), (109), (110), (111), (466), (467), (467a)
328	(42), (43), (44), (45), (112), (113), (114), (193), (299), (300), (358), (385), (386), (393)
330	(115), (116), (117), (118), (194), (195), (301), (302), (303), (359)
332	(209), (414), (415)
334	(63), (262), (470)
336	(64), (65), (66), (135), (136), (225), (315), (361), (471), (472), (473), (481)
338	(67), (137), (138), (178), (197), (316), (317), (318), (362), (363), (364), (365), (366), (367), (378), (379), (420), (421), (508)
340	(198), (319), (320), (368), (369), (370), (422), (437)
342	(46), (47), (48), (49), (119), (120), (121), (223), (321), (387), (388)
344	(50), (122), (123), (124), (359a), (389)
346	(210), (304), (441)
348	(68), (416)
350	(69), (139), (140), (322), (323)
352	(141), (142), (143), (226), (227), (244), (324), (325), (326), (423), (424), (451), (474)
354	(144), (145), (146), (147), (199), (200), (201), (211), (212), (327), (328), (371), (372), (373), (374), (425), (426), (509)
356	(51), (52), (53), (54), (55), (125), (126), (213), (329), (330), (427), (468), (484)
358	(56), (127), (224), (242), (510)
360	(128), (305), (442)
362	(417)
364	(148), (149), (263)
366	(150), (150a), (228), (245), (331), (332), (475), (476), (482)
368	(151), (214), (229), (246), (333), (334), (335), (375), (428), (477)
370	(215), (259), (429)
372	(57), (58), (59), (129), (260)
374	(130), (131), (132), (216), (507)
376	(264), (306), (443)
378	(70), (71), (152), (153), (230), (231)
380	(72), (478), (485)
382	(73), (180), (247), (248), (336), (337), (338), (483)
384	(154), (155)
386	(133), (134), (202)
390	(243), (339), (340), (430), (444)
392	(265), (266), (341), (342), (343), (344), (431)
394	(74), (232), (233), (234), (249), (250), (486), (487)
396	(156), (234a), (345)
400	(261)
404	(157), (158), (159)
406	(160), (161), (267), (432)

Molecular Weight Index *(continued)*

Trivial Name Index

[1] Derived trivial names such as 2'-hydroxydaidzein (7) and 8-*O*-methylretusin (29) are not included.

Trivial Name Index *(continued)*

Trivial Name Index *(continued)*

Name[1]	Compound Number	Name[1]	Compound Number
Disain	(633)	Glabrescin	(489)
Dolichin A	(319)	Glabrescione A	(153)
Dolichin B	(320)	Glabrescione B	(174)
Dolichone	(225)	Glabridin	(419)
Dolineone	(225)	Glabrone	(64)
Duartin	(415)	Gliricidin	(31)
Durlettone	(65)	Glyceocarpin	(369)
Durmillone	(71)	Glyceofuran	(374)
		Glyceollidin I	(370)
Edudiol	(328)	Glyceollidin II	(369)
Edulaan	(332)	Glyceollin I	(367)
Edulane	(335)	Glyceollin II	(364)
Edulenane	(332)	Glyceollin III	(365)
Edulenanol	(326)	Glyceollin IV	(373)
Edulenol	(334)	Glycinol	(350)
Edulin	(308)	Glycitein	(19)
Edunol	(324)	Glycyrin	(483)
Elliptone	(226)	Glycyrol	(475)
Elongatin	(156)	Glyzaglabrin	(26)
Equol	(394)	Glyzarin	(23)
Erosenone	(201)		
Erosnin	(469)	Haginin A	(448)
Erosone	(227)	Haginin B	(445)
Erythrabyssin I	(371)	Haginin C	(446a)
Erythrabyssin II	(344)	Haginin D	(444a)
Erythrinin A	(61)	Hemileiocarpin	(315)
Erythrinin B	(138)	Heminitidulan	(432)
Erythrinin C	(146)	Hirta Substance A	(634)
		Hirta Substance C	(635)
Ferreirin	(205)	Hispaglabridin A	(431)
Ferrugone	(75)	Hispaglabridin B	(430)
Ficifolinol	(343)	Homoedudiol	(314)
Ficinin	(318)	Homoferreirin	(207)
Flemichapparin B	(382)	Homopisatin	(354)
Flemichapparin C	(462)	Homopterocarpin	(280)
Folinin	(340)	Homotectoridin	(605)
Folitenol	(339)	1a-Hydroxyphaseollone	(378)
Formononetin	(5)		
Fujikinetin	(35)	Ichthynone	(76)
Fujikinin	(546)	Inermin	(279)
		Iridin	(609)
Gangetin	(349)	Irifloside	(599)
Gangetinin	(347)	Irigenin	(128)
Genistein	(81)	Irilone	(92)
Genistin	(551)	Irisflorentin	(133)
Genistoside	(551)	Iriskumaonin	(120)
Glabrene	(450)	Irisolidone	(104)

Trivial Name Index *(continued)*

Trivial Name Index *(continued)*

Name[1]	Compound Number	Name[1]	Compound Number
Mundulone	(79)	Orobol	(87)
Munduserone	(223)	Oroboside	(577)
Munetone	(77)	Osajin	(158)
Muningin	(106)	Ougenin	(210)
Myriconol	(234)		
		Pachyrrhizin	(481)
Nallanin	(163)	Pachyrrhizone	(228)
Neobanol	(360)	Paniculatin	(564)
Neobanone	(247)	Parvifuran	(499)
Neobavaisoflavone	(62)	Parvisoflavanone	(209)
Neochanin	(5)	Parvisoflavone A	(143)
Neoduleen	(391)	Parvisoflavone B	(142)
Neodulin	(308)	Pendulone	(439)
Neodunol	(307)	Petalostetin	(54)
Neofolin	(482)	Phaseollidin	(312)
Neoraucarpan	(345)	Phaseollidin Hydrate	(321)
Neoraucarpanol	(338)	Phaseollin	(309)
Neorauflavane	(426)	Phaseollinisoflavan	(418)
Neorauflavene	(451)	Phaseoluteone	(144)
Neoraufurane	(508)	Philenopteran	(296)
Neoraunone	(196)	Piscerythrone	(154)
Neorautane	(325)	Piscidone	(155)
Neorautanin	(337)	Pisatin	(356)
Neorautanol	(333)	Platycarpanetin	(121)
Neorauteen	(390)	Podospicatin	(115)
Neorautenane	(322)	Pomiferin	(166)
Neorautenanol	(331)	Pratensein	(94)
Neorautenol	(310)	Pratensol	(83)
Neorautenone	(197)	Pratol	(5)
Neorautone	(481)	Prunetin	(84)
Neotenone	(197)	Prunetol	(81)
Neovestitol	(399)	Prunitrin	(574)
Nepseudin	(199)	Prunitroside	(574)
Nicouline	(233)	Pseudobaptigenin	(11)
Nissolin	(281)	Pseudobaptisin	(531)
Nitiducarpin	(346)	Pseudotectorigenin	(99)
Nitiducol	(348)	Psoralenol	(67)
Nitidulan	(433)	Psoralidin	(472)
Nitidulin	(434)	Psoralidin Oxide	(474)
Norsantal	(87)	Pterocarpin	(284)
Norwedelolactone	(461)	Pterofuran	(504)
		Puerarin	(518)
Odoratin	(36)	Purpuranin A	(642)
Odoratine	(125)	Purpuranin B	(643)
Onogenin	(191)		
Ononin	(521)	Repensol	(454)
Onospin	(630)	Retusin	(20)

Trivial Name Index *(continued)*

Source Index

Plant Sources[1]	Compound Number
Abrus	(442), (443), (444)
Adenocarpus	(3), (81), (85), (551)
LM *Albizia*	(3), (5), (81), (83), (275)
Aldina	(276), (279)
Amorpha	(5), (42), (46), (233), (234a), (235), (238), (239), (241), (254), (255), (265), (267), (268), (269), (270), (293), (441), (521), (541), (618), (619), (621), (622)
Amphimas	(28)
Andira	(83), (203), (276), (279)
Antheroporum	(233)
Anthyllis	(396), (398), (500)
Apios	(81), (86), (317)
LC *Apuleia*	(323)
Arachis	(82)
Argyrocytisus	(81), (86), (138), (145), (207)
Astragalus	(404), (405), (406), (452), (566)
Baphia	(3), (97), (280), (284), (449)
Baptisia	(3), (5), (11), (15), (17), (21), (28), (81), (83), (87), (88), (98), (293), (511), (516), (521), (523), (531), (533), (534), (535), (536), (538), (541), (542), (551), (557), (566), (567), (576), (578), (581), (590), (626)
* *Belamcanda* (IRID)	(590), (609), (631)
* *Beta* (CHEN)	(100)
Bolusanthus	(87)
Bowdichia	(15), (81), (179), (280)
* *Brassica* (CRU)	(452)
Brya	(383), (385), (386), (387), (389), (392), (393), (412), (507), (510)
Cadia	(15)
Cajanus	(3), (5), (78), (81), (86), (96), (103), (208), (217), (220), (565)
Calicotome	(3), (81), (85)
Calopogonium	(24), (34), (46), (63), (68), (135), (139), (140)
Canavalia	(81), (96), (276), (279), (397)
Caragana	(276), (279), (354), (356), (358)
Carmichaelia	(276), (397), (405)
Castanospermum	(5), (28), (30)
Centrosema	(5), (9), (19), (28), (88), (96), (98), (452)
Chamaecytisus	(3), (5), (81), (85)
Chamaespartium	(3), (5), (81), (85)
Chronanthus	(81), (85)
Cicer	(3), (5), (33), (83), (94), (207), (276), (279), (453), (456), (521), (566)
Cladrastis	(5), (11), (15), (25), (28), (29), (35), (44), (113), (114), (119), (121), (188), (521), (525), (530), (532), (539), (540), (541), (543), (544),

[1] See Table 1 for key to abbreviations and superscript symbols.

Source Index *(continued)*

Plant Sources[1]	Compound Number
	(545), **(546)**, **(547)**, **(548)**, **(549)**, **(600)**, **(601)**, **(602)**, **(606)**, **(607)**, **(608)**
Clitoria	**(243)**, **(261)**
Colutea	**(405)**
Cordyla	**(40)**, **(47)**, **(49)**, **(51)**, **(52)**, **(58)**, **(125)**, **(134)**, **(193)**
Coronilla	**(500)**, **(505)**
* *Cotoneaster* (ROS)	**(83)**, **(566)**
Crotalaria	**(81)**, **(86)**, **(226)**, **(232)**, **(233)**, **(236)**, **(237)**, **(251)**
Cyclolobium	**(15)**, **(16)**, **(397)**, **(427)**, **(436)**, **(438)**, **(440)**
Cytisophyllum	**(632)**
Cytisus	**(3)**, **(5)**, **(81)**, **(85)**, **(87)**, **(94)**, **(551)**, **(560)**
Dalbergia	**(3)**, **(5)**, **(11)**, **(12)**, **(15)**, **(20)**, **(28)**, **(29)**, **(35)**, **(42)**, **(47)**, **(48)**, **(51)**, **(83)**, **(84)**, **(95)**, **(98)**, **(99)**, **(105)**, **(112)**, **(130)**, **(132)**, **(187)**, **(191)**, **(192)**, **(194)**, **(235)**, **(239)**, **(254)**, **(256a)**, **(272)**, **(276)**, **(279)**, **(288)**, **(293)**, **(294)**, **(315)**, **(323)**, **(346)**, **(348)**, **(354)**, **(381)**, **(397)**, **(399)**, **(401)**, **(406)**, **(413)**, **(423)**, **(424)**, **(432)**, **(433)**, **(434)**, **(453)**, **(456)**, **(457)**, **(479)**, **(480)**, **(499)**, **(521)**, **(524)**, **(547a)**, **(550)**, **(563)**, **(564)**, **(566)**, **(568)**, **(571)**, **(575)**, **(579)**, **(580)**, **(584)**, **(585)**, **(589)**, **(590)**, **(591)**, **(595)**, **(596)**, **(597)**, **(598)**, **(610)**, **(611)**, **(612)**, **(613)**, **(614)**, **(615)**, **(617)**, **(618)**, **(620)**, **(623)**, **(623a)**
Derris	**(80)**, **(109)**, **(122)**, **(127)**, **(129)**, **(136)**, **(140)**, **(148)**, **(149)**, **(150)**, **(152)**, **(153)**, **(157)**, **(158)**, **(159)**, **(162)**, **(163)**, **(174)**, **(226)**, **(229)**, **(232)**, **(233)**, **(236)**, **(237)**, **(251)**, **(252)**, **(265)**, **(266)**, **(270)**, **(276)**, **(279)**, **(400)**, **(484)**, **(485)**, **(486)**, **(487)**, **(488)**, **(489)**, **(490)**, **(491)**, **(492)**, **(644)**
Desmodium	**(336)**, **(347)**, **(349)**
* *Dianthus* (CARY)	**(100)**
Diplotropis	**(5)**, **(279)**
Dipteryx	**(20)**, **(29)**, **(36)**, **(37)**, **(38)**, **(119)**, **(125)**
Dolichos	**(81)**, **(86)**, **(204)**, **(206)**, **(213)**, **(312)**, **(319)**, **(320)**, **(452)**, **(472)**
Echinospartum	**(3)**, **(5)**, **(81)**, **(83)**, **(85)**, **(91)**
* *Eclipta* (COMP)	**(461)**, **(465)**, **(629)**
Erinacea	**(3)**, **(81)**, **(85)**
Erythrina	**(3)**, **(61)**, **(81)**, **(135)**, **(138)**, **(146)**, **(158)**, **(275)**, **(309)**, **(312)**, **(316)**, **(344)**, **(350)**, **(368)**, **(371)**, **(396)**, **(398)**, **(511)**, **(551)**
Euchresta	**(279)**, **(456)**, **(626)**
Factorovskya	**(276)**, **(397)**
Ferreirea	**(5)**, **(83)**, **(205)**, **(207)**
Flemingia	**(382)**, **(462)**
Genista	**(3)**, **(5)**, **(81)**, **(84)**, **(85)**, **(521)**, **(551)**, **(574)**
Gliricidia	**(31)**, **(405)**, **(446)**, **(447)**
Glycine	**(3)**, **(6)**, **(9)**, **(19)**, **(350)**, **(361)**, **(364)**, **(365)**, **(366)**, **(367)**, **(369)**, **(370)**, **(373)**, **(374)**, **(375)**, **(379)**, **(452)**, **(471)**, **(511)**, **(512)**, **(529)**, **(551)**, **(552)**

Source Index *(continued)*

Plant Sources[1]	Compound Number
Glycyrrhiza	(2), (4), (5), (22), (23), (26), (64), (73), (84), (141), (144), (211), (405), (418), (419), (421), (425), (430), (431), (435), (450), (475), (476), (478), (483)
Hardenbergia	(81), (86), (144), (145)
Hedysarum	(5), (276)
Hosackia	(396), (397), (398)
* *Iresine* (AMAR)	(110)
* *Iris* (IRID)	(92), (98), (101), (104), (111), (116), (117), (120), (126), (128), (133), (586), (590), (592), (599), (603), (604), (605), (609), (636)
* *Juniperus* (CUPR)	(107), (123), (128), (131), (609)
Lablab	(81), (86), (204), (213), (312), (396), (398), (407), (501)
Laburnum	(3), (81), (85), (86), (135), (138), (145)
Lathyrus	(87), (276), (279), (281), (285), (354), (356), (359a), (577)
Lembotropis	(81), (85)
Lens	(354)
Lespedeza	(3), (81), (192a), (204), (218), (219), (341), (342), (380), (444a), (445), (446a), (448), (498a)
Lonchocarpus	(230), (232), (233), (251), (252), (266), (296), (301), (382), (407), (414), (491), (637)
Lotus	(396), (397), (400), (409), (410)
Lupinus	(81), (83), (85), (86), (90), (138), (145), (169), (551), (554), (563), (579), (582), (583)
Lygos	(3), (81), (85), (551)
Maačkia	(5), (11), (16), (81), (189), (276), (279), (280), (456)
Machaerium	(3), (5), (6), (8), (15), (16), (276), (280), (282), (298), (397), (406), (415), (416), (417), (440), (638), (639), (640)
* *Maclura* (MOR)	(158), (166), (633)
Macroptilium	(213), (312)
Macrotyloma	(204), (213)
Medicago	(3), (5), (81), (83), (185), (187), (190), (276), (397), (400), (401), (452), (453), (455), (456), (458), (459), (460), (463), (464), (511), (521), (624), (628)
Melilotus	(275), (276), (278), (283), (290), (351), (352), (355), (397), (452)
Mildbraedeodendron	(19), (47), (51), (52), (58), (193)
Millettia	(51), (65), (71), (75), (76), (135), (150a), (151), (160), (161), (164), (165), (167), (168), (170), (171), (172), (173), (175), (176), (181), (230), (232), (233), (237), (249), (251), (265), (266), (279), (394), (438), (439), (637)
Moghania	(81), (86), (137)
Monopteryx	(29), (37), (45), (83), (94), (116), (118), (124)
Mucuna	(213) (see also *Stizolobium*)
Mundulea	(77), (79), (223), (224), (232), (233), (251), (252), (266), (641)
* *Myrica* (MYRC)	(234)
Myrocarpus	(28), (34)

Source Index *(continued)*

Plant Sources[1]	Compound Number
Myroxylon	(5), (15), (17), (28), (34), (37), (43), (83), (183), (186), (195), (276), (295), (453), (464), (506)
Neonotonia	(81), (86), (138)
Neorautanenia	(66), (196), (197), (199), (200), (225), (233), (244), (245), (246), (247), (250), (252), (253), (262), (266), (267), (276), (291), (299), (300), (307), (308), (310), (313), (314), (318), (322), (324), (325), (326), (328), (331), (332), (333), (334), (335), (337), (338), (339), (340), (343), (345), (360), (390), (391), (426), (437), (451), (481), (482), (508), (509)
Onobrychis	(5), (28), (185), (276), (397)
Ononis	(3), (5), (98), (191), (521), (530), (616), (624), (626), (630)
Ormocarpum	(233)
Ormosia	(85)
* *Osteophleum* (MYR)	(276), (279)
Ostryoderris	(233)
Ougeinia	(204), (207), (210)
Pachyrrhizus	(66), (197), (201), (225), (227), (228), (233), (242), (244), (246), (248), (252), (263), (275), (307), (469), (481)
Parochetus	(276), (354)
Pericopsis	(5), (28), (29), (81), (83), (104), (182), (276), (279), (280), (494), (495)
Petalostemon	(54)
Phaseolus	(3), (7), (81), (86), (144), (147), (184), (198), (204), (212), (213), (275), (309), (312), (363), (396), (418), (420), (452), (471), (472)
Pickeringia	(5)
Piptanthus	(3), (5), (81), (511), (513), (514), (521), (551), (553), (555)
Piscidia	(70), (76), (154), (155), (180), (230), (231), (232), (233), (237), (251), (252), (264)
Pisum	(3), (5), (11), (27), (28), (81), (279), (286), (292), (303), (356), (382), (452), (453)
Platymiscium	(17), (276), (282)
* *Podocarpus* (POD)	(81), (115)
Poecilanthe	(142), (143), (209)
Poiretia	(233)
LM *Prosopis*	(537)
* *Prunus* (ROS)	(81), (84), (551), (574)
Psophocarpus	(275), (277), (312), (327), (371)
Psoralea	(3), (10), (60), (62), (67), (367), (452), (472), (473), (474), (477), (511)
Pterocarpus	(5), (11), (15), (81), (84), (97), (105), (106), (279), (280), (284), (494), (496), (497), (498), (504)
Pterodon	(3), (12), (28), (35), (36), (39), (49), (51), (53), (56), (57), (58), (59)
Pueraria	(3), (5), (18), (81), (83), (104), (350), (362), (511), (518), (519), (520), (521), (551), (566), (590), (593), (594)

Source Index *(continued)*

Plant Sources[1]	Compound Number
Rothia	(530)
Sophora	(3), (5), (15), (81), (83), (90), (94), (104), (202), (214), (215), (221), (222), (276), (279), (284), (293), (311), (382), (429), (456), (467a), (470), (502), (503), (506a), (553), (559), (566), (569), (570), (593), (625), (626), (627)
Spartium	(5), (81), (85), (86), (287), (521), (551)
Spatholobus	(233)
Sphaerophysa	(407), (428)
* *Spinacia* (CHEN)	(452)
Stauracanthus	(3), (81), (85)
* *Stemona* (STEM)	(259), (260), (261)
Stizolobium	(81), (86), (204), (206), (208), (213), (276), (279)
Swartzia	(276), (279), (280), (284), (289), (291), (293), (295), (297), (300), (304), (305), (306), (381), (384), (388), (457), (466), (468)
Teline	(3), (5), (81), (85)
Tephrosia	(32), (69), (72), (74), (156), (226), (232), (233), (236), (237), (240), (251), (252), (256), (257), (265), (266), (271), (273), (274), (279), (293), (356), (358), (359), (467), (634), (635), (642), (643)
Tetragonolobus	(380), (396), (397), (398), (500)
Thermopsis	(3), (5), (14), (15), (81), (83), (85), (87), (94), (95), (279), (511), (521), (533), (551), (555), (566), (576), (587), (588), (626)
Tipuana	(5), (185), (276), (279), (397)
Trifolium	(3), (5), (11), (13), (15), (81), (83), (92), (94), (185), (275), (276), (279), (280), (289), (290), (293), (302), (351), (353), (354), (355), (356), (357), (397), (398), (400), (401), (402), (452), (453), (454), (456), (458), (511), (521), (522), (526), (527), (528), (530), (551), (561), (562), (566), (572), (573), (574), (586), (587), (624), (626)
Trigonella	(276), (279), (397), (400), (452)
Ulex	(3), (5), (81), (85), (551)
Vatairea	(5)
* *Verbascum* (SCRO)	(233)
Vicia	(276)
Vigna	(3), (213), (276), (309), (312), (422), (452), (501)
* *Virola* (MYR)	(5), (13), (83), (93), (102)
* *Wedelia* (COMP)	(461), (465)
Wisteria	(5), (28), (113), (521), (541), (544)
* *Wyethia* (COMP)	(95), (97)
Xanthocercis	(29), (37), (41), (45), (55)

Source Index *(continued)*

Plant Sources[1]	Compound Number

Microbial and Other Sources

Animal/Insect
   Tissue Homogenates   **(235), (241), (252), (254), (258)**
Animal Urine   **(394), (395), (493)**
Ascochyta cultures   **(353)**
Aspergillus cultures   **(81), (87), (89), (99), (108)**
Botrytis cultures   **(351), (353), (354), (363)**
Cladosporium cultures   **(378)**
Colletotrichum cultures   **(275), (363), (372)**
Fusarium cultures   **(81), (83), (87), (91), (216), (275), (276), (277), (321), (353), (378), (396), (397), (408), (411)**
Gibberella cultures   **(275), (277)**
Micromonospora cultures   **(3), (81)**
Mycobacterium cultures   **(81), (84)**
Nectria cultures   **(185), (189), (351), (353), (376), (377)**
Penicillium cultures   **(1)**
Sclerotinia cultures   **(351), (353)**
Septoria cultures   **(329), (330)**
Stemphylium cultures   **(87), (353), (397), (403), (418)**
Streptomyces cultures   **(3), (50), (81), (87), (116), (118), (177), (178), (515), (517), (556), (558)**

References

1. ABRAMS, C., C. v. D. M. BRINK, and D. H. MEIRING: Pachyrrhizin from *Neorautanenia edulis.* J. South Afr. Chem. Inst. **15**, 78 (1962).

2. ACREE, F., M. JACOBSON, and H. L. HALLER: Amorphin, a Glycoside in *Amorpha fruticosa.* J. Org. Chem. **8**, 572 (1943).

3. — — — *Amorpha fruticosa* Contains no Rotenone. Science, N. Y. **99**, 99 (1944).

4. ADINARAYANA, D., M. RADHAKRISHNIAH, and J. R. RAO: The Occurrence of Caviunin in *Dalbergia paniculata.* Curr. Sci. **40**, 602 (1971).

5. ADINARAYANA, D., M. RADHAKRISHNIAH, J. R. RAO, R. CAMPBELL, and L. CROMBIE: Dalpanol, a New 6'-Hydroxyrotenoid from a *Dalbergia* Species. J. Chem. Soc. C **1971**, 29.

6. ADINARAYANA, D., and J. R. RAO: Isoflavonoid Glycosides of *Dalbergia paniculata.* The Constitutions of Dalpanitin and Dalpatin. Tetrahedron **28**, 5377 (1972).

7. — — Occurrence of Flavones in *Dalbergia paniculata* Flowers. Phytochemistry **12**, 2543 (1973).

8. — — Dalpanin, a *C*-Glycosylisoflavanone from *Dalbergia paniculata.* Proc. Indian Acad. Sci. **81A**, 23 (1975).

9. — — Isoflavonoids of *Dalbergia paniculata.* Indian J. Chem. **13**, 425 (1975).

10. ADITYACHAUDHURY, N., and P. K. GUPTA: A New Pterocarpan and Coumestan in the Roots of *Flemingia chappar.* Phytochemistry **12**, 425 (1973).

11. AHLUWALIA, V. K., M. M. BHASIN, and T. R. SESHADRI: Isoflavones of Soyabeans. Curr. Sci. **22**, 363 (1953).

12. AHLUWALIA, V. K., G. P. SACHDEV, and T. R. SESHADRI: Chemical Components of Immature Green Pods of *Dalbergia sissoo*. Indian J. Chem. **3**, 474 (1965).

13. — — — Chemical Investigation of the Leaves of *Ougeinia dalbergioides*. Indian J. Chem. **4**, 250 (1966).

14. AKISANYA, A., C. W. L. BEVAN, and J. HIRST: Heartwood Constituents of the Genus *Pterocarpus*. J. Chem. Soc. **1959**, 2679.

15. AL-ANI, H. A. M., and P. M. DEWICK: Isoflavone Biosynthesis in *Onobrychis viciifolia:* Formononetin and Texasin as Precursors of Afrormosin. Phytochemistry **19**, 2337 (1980).

16. ALBUQUERQUE, F. B., R. BRAZ FILHO, O. R. GOTTLIEB, M. T. MAGALHÃES, J. G. S. MAIA, A. BRAGA DE OLIVEIRA, G. G. DE OLIVEIRA, and V. C. WILBERG: Isoflavone Evolution in *Monopteryx*. Phytochemistry **20**, 235 (1981).

17. ANDERSON, E. L., and G. F. MARRIAN: The Identification of Equol as 7-Hydroxy-3-(4′-hydroxyphenyl)chroman, and the Synthesis of Racemic Equol Methyl Ether. J. Biol. Chem. **127**, 649 (1939).

18. ANDERSON, T.: Educts from *Baphia nitida* (Barwood). J. Chem. Soc. **30**, 582 (1876).

19. AOYAGI, T., T. HAZATO, M. KUMAGAI, M. HAMADA, T. TAKEUCHI, and H. UMEZAWA: Isoflavone Rhamnosides, Inhibitors of β-Galactosidase Produced by Actinomycetes. J. Antibiot., Tokyo **28**, 1006 (1975).

19a. APPEL DE MATTOS, M.: A Contribution to the Phytochemical Study of *Eupatorium inulaefolium* — Compositae: Flavonoids. Revta. Cent. Cienc. Biomed., Univ. Fed. St. Maria **5**, 85 (1977); Chem. Abstr. **90**, 100124 (1979).

20. ARISAWA, M., H. KIZU, and N. MORITA: The Constituents of *Iris unguicularis* (2). Chem. Pharm. Bull., Tokyo **24**, 1609 (1976).

21. ARISAWA, M., Y. KYOZUKA, T. HAYASHI, M. SHIMIZU, and N. MORITA: Isoflavonoids in the Roots of *Thermopsis fabacea* (Leguminosae). Chem. Pharm. Bull., Tokyo **28**, 3686 (1980).

22. ARISAWA, M., and N. MORITA: The Constituents of *Iris unguicularis* (1). Chem. Pharm. Bull., Tokyo **24**, 815 (1976).

23. ARISAWA, M., N. MORITA, Y. KONDO, and T. TAKEMOTO: The Constituents of *Iris florentina* (2). Chem. Pharm. Bull., Tokyo **21**, 2323 (1973).

24. ARNONE, A., L. CAMARDA, L. MERLINI, G. NASINI, and D. A. H. TAYLOR: Isoflavonoid Constituents of the West African Red Wood *Baphia nitida*. Phytochemistry **20**, 799 (1981).

25. ASSUMPÇÃO, R. M. V., and O. R. GOTTLIEB: Flavonoids from *Poecilanthe parviflora*. Phytochemistry **12**, 1188 (1973).

26. AYABE, S. I., M. KOBAYASHI, M. HIKICHI, K. MATSUMOTO, and T. FURUYA: Flavonoids from the Cultured Cells of *Glycyrrhiza echinata*. Phytochemistry **19**, 2179 (1980).

27. BAILEY, E. T., and C. M. FRANCIS: Isoflavone Concentrations in the Leaves of the Species of the Genus *Trifolium*, Section *Calycomorphum*. Aust. J. Agric. Res. **22**, 731 (1971).

28. BAILEY, J. A.: Pisatin Production by Tissue Cultures of *Pisum sativum*. J. Gen. Microbiol. **61**, 409 (1970).

29. — Production of Antifungal Compounds in Cowpea *(Vigna sinensis)* and Pea *(Pisum sativum)* after Virus Infection. J. Gen. Microbiol. **75**, 119 (1973).

30. BAILEY, J. A., R. S. BURDEN, A. MYNETT, and C. BROWN: Metabolism of Phaseollin by *Septoria nodorum* and other Non-Pathogens of *Phaseolus vulgaris*. Phytochemistry **16**, 1541 (1977).

31. BAILEY, J. A., and J. L. INGHAM: Phaseollin Accumulation in Bean *(Phaseolus vulgaris)* in Response to Infection by Tobacco Necrosis Virus and the Rust *Uromyces appendiculatus*. Physiol. Pl. Path. **1**, 451 (1971).

32. BAILEY, J. A., and J. W. MANSFIELD: Phytoalexins. Glasgow: Blackie and Son. 1982.

33. BAILEY, J. A., and R. A. SKIPP: Toxicity of Phytoalexins. Ann. Appl. Biol. **89**, 354 (1978).
34. BAJWA, B. S., P. L. KHANNA, and T. R. SESHADRI: A New Isoflavone, Neobavaisoflavone, from the Seeds of *Psoralea corylifolia*. Curr. Sci. **41**, 882 (1972).
35. ———— Components of Different Parts of Seeds (Fruits) of *Psoralea corylifolia*. Indian J. Chem. **12**, 15 (1974).
36. BAKER, W.: The Constitution of Irigenin and Iridin. J. Chem. Soc. **1928**, 1022.
37.. BAKER, W., J. B. HARBORNE, and W. D. OLLIS: A New Synthesis of Isoflavones. Part II. 5:7:2'-Trihydroxyisoflavone. J. Chem. Soc. **1953**, 1860.
38.. BALAKRISHNA, S., J. D. RAMANATHAN, T. R. SESHADRI, and B. VENKATARAMANI: Special Chemical Components of the Heartwood of *Ougeinia dalbergioides*. Proc. R. Soc. **268A**, 1 (1962).
39. BANDYUKOVA, V. A., and A. L. KAZAKOV: Advances in the Chemistry of Natural Isoflavonoids. Khim. Prir. Soedin. **1978**, 669.
40. BANERJI, A., V. V. S. MURTI, and T. R. SESHADRI: Occurrence of 7,4'-Dimethyltectorigenin in the Flowers of *Dalbergia sissoo*. Curr. Sci. **34**, 431 (1965).
41. ———— Isolation of Sissotrin, a New Isoflavone Glycoside from the Leaves of *Dalbergia sissoo*. Indian J. Chem. **4**, 70 (1966).
42. BANERJI, A., V. V. S. MURTI, T. R. SESHADRI, and R. S. THAKUR: Chemical Components of the Flowers of *Dalbergia sissoo:* Isolation of 7-Methyltectorigenin, a New Isoflavone. Indian J. Chem. **1**, 25 (1963).
43. BARNES, C. S., J. L. OCCOLOWITZ, N. L. DUTTA, P. M. NAIR, P. S. PHADKE, and K. VENKATARAMAN: The Structure of Munetone. Tetrahedron Lett: **1963**, 281.
44. BARZ, W., C. ADAMEK, and J. BERLIN: The Degradation of Formononetin and Daidzein in *Cicer arietinum* and *Phaseolus aureus*. Phytochemistry **9**, 1735 (1970).
44a. BARZ, W., R. SCHLEPPHORST, and J. LAIMER: Degradation of Polyphenols by Fungi of the Genus *Fusarium*. Phytochemistry **15**, 87 (1976).
45. BATE-SMITH, E. C., T. SWAIN, and G. S. POPE: The Isolation of 7-Hydroxy-4'-methoxyisoflavone (Formononetin) from Red Clover (*Trifolium pratense*) and a Note on the Identity of Pratol. Chemy. Ind. **72**, 1127 (1953).
46. BÁTKAI, L., M. NOGRÁDI, L. FARKAS, L. FEUER, and I. HORVÁTH: Demethylation of 7-Methoxyisoflavone by *Penicillium cyclopium*. Arch. Mikrobiol. **90**, 165 (1973).
47. BATTERHAM, T. J., D. A. SHUTT, N. K. HART, A. W. H. BRADEN, and H. J. TWEEDALE: Metabolism of Intraruminally Administered [4-¹⁴C] Formononetin and [4-¹⁴C] Biochanin A in Sheep. Aust. J. Agric. Res. **22**, 131 (1971).
47a. BAZTAN, J. M., M. REBUELTA, and J. M. VIVAS: Isolation and Identification of Flavonoids from *Ononis spinosa*. An. R. Acad. Farm., Madrid **47**, 303 (1981).
48. BECK, A. B.: The Oestrogenic Isoflavones of Subterranean Clover. Aust. J. Agric. Res. **15**, 223 (1964).
49. BECK, A. B., and J. R. KNOX: The Acylated Isoflavone Glycosides from Subterranean Clover and Red Clover. Aust. J. Chem. **24**, 1509 (1971).
50. BENN, M. H., and C. WATANATADA: On the Occurrence of Ononin in *Thermopsis rhombifolia*. Can. J. Chem. **48**, 1624 (1970).
51. BENNETT, D., F. H. W. MORLEY, and A. AXELSEN: Bioassay Responses of Ewes to Legume Swards. II. Uterine Weight Results from Swards. Aust. J. Agric. Res. **18**, 495 (1967).
52. BENNINGHOFF, H.: Bright Nickel Plating. Chem. Abstr. **53**, 19641 (1959).
53. BERLIN, J., and W. BARZ: Metabolism of Isoflavones and Coumestans in Cell and Callus Suspension Cultures of *Phaseolus aureus*. Planta **98**, 300 (1971).
54. BERRY, R. C., R. A. EADE, and J. J. H. SIMES: The Isolation, Structure and Synthesis of Koparin (7,2',3'-Trihydroxy-4'-methoxyisoflavone). Aust. J. Chem. **30**, 1827 (1977).
55. BEVAN, C. W. L., D. E. U. EKONG, M. E. OBASI, and J. W. POWELL: Extracts from the

Heartwood of *Amphimas pterocarpoides* and *Pterocarpus erinaceus*. J. Chem. Soc. C **1966**, 509.

55a. BEZUIDENHOUDT, B. C. B., E. V. BRANDT, and D. G. ROUX: A Novel α-Hydroxydihydrochalcone from the Heartwood of *Pterocarpus angolensis:* Absolute Configuration, Synthesis, Photochemical Transformations, and Conversion into α-Methyldeoxybenzoins. J. Chem. Soc. Perkin Trans. I **1981**, 263.

56. BEZUIDENHOUDT, B. C. B., E. V. BRANDT, D. G. ROUX, and P. H. VAN ROOYEN: Novel α-Methyldeoxybenzoins from the Heartwood of *Pterocarpus angolensis:* Absolute Configuration and Conformation of the First Sesquiterpenylangolensins, and X-Ray Crystal Structure of 4-*O*-α-Cadinylangolensin. J. Chem. Soc. Perkin Trans. I **1980**, 2179.

57. BHANDARI, P. R., J. L. BOSE, and S. SIDDIQUI: Isoflavones from the Fresh Soya Bean Germ and the Synthesis of 6-Methylformononetin and 6-Methyldaidzein. J. Scient. Ind. Res. **8B**, 217 (1949).

58. BHANUMATI, S., S. C. CHHABRA, and S. R. GUPTA: Cajaisoflavone, a New Prenylated Isoflavone from *Cajanus cajan*. Phytochemistry **18**, 1254 (1979).

59. BHANUMATI, S., S. C. CHHABRA, S. R. GUPTA, and V. KRISHNAMOORTHY: A New Isoflavone Glucoside from *Cajanus cajan*. Phytochemistry **18**, 365 (1979).

60. — — — — 2'-*O*-Methylcajanone: a New Isoflavanone from *Cajanus cajan*. Phytochemistry **18**, 693 (1979).

61. BHARDWAJ, D. K., R. MURARI, T. R. SESHADRI, and R. SINGH: Occurrence of 2-Methylisoflavones in *Glycyrrhiza glabra*. Phytochemistry **15**, 352 (1976).

62. BHARDWAJ, D. K., T. R. SESHADRI, and R. SINGH: Glyzarin, a New Isoflavone from *Glycyrrhiza glabra*. Phytochemistry **16**, 402 (1977).

63. BHARDWAJ, D. K., and R. SINGH: Glyzaglabrin, a New Isoflavone from *Glycyrrhiza glabra*. Curr. Sci. **46**, 753 (1977).

64. BHARGAVA, K. K., N. R. KRISHNASWAMY, and T. R. SESHADRI: Isolation of Demethylwedelolactone and its Glucoside from *Eclipta alba*. Indian J. Chem. **8**, 664 (1970).

65. — — — Demethylwedelolactone Glucoside from *Eclipta alba* Leaves. Indian J. Chem. **10**, 810 (1972).

66. BHRARA, S. C., A. C. JAIN, and T. R. SESHADRI: A New Examination of the Special Components of *Pterocarpus indicus* Heartwood. Curr. Sci. **33**, 303 (1964).

67. BHUTANI, S. P., S. S. CHIBBER, and T. R. SESHADRI: Components of the Roots of *Pueraria tuberosa:* Isolation of a New Isoflavone-*C*-Glycoside (Di-*O*-Acetylpuerarin). Indian J. Chem. **7**, 210 (1969).

68. BICKEL, H., and H. SCHMID: The Constitution of Pachyrrhizone. Helv. Chim. Acta **36**, 664 (1953).

69. BICKOFF, E. M., A. N. BOOTH, R. L. LYMAN, A. L. LIVINGSTON, C. R. THOMPSON, and F. DEEDS: Coumestrol, a New Estrogen Isolated from Forage Crops. Science, N. Y. **126**, 969 (1957).

70. BICKOFF, E. M., A. N. BOOTH, R. L. LYMAN, A. L. LIVINGSTON, C. R. THOMPSON, and G. O. KOHLER: Isolation of a New Estrogen from Ladino Clover. J. Agric. Fd. Chem. **6**, 536 (1958).

71. BICKOFF, E. M., A. L. LIVINGSTON, S. C. WITT, B. E. KNUCKLES, J. GUGGOLZ, and R. R. SPENCER: Isolation of Coumestrol and Other Phenolics from Alfalfa by Countercurrent Distribution. J. Pharm. Sci. **53**, 1496 (1964).

72. BICKOFF, E. M., A. L. LIVINGSTON, S. C. WITT, R. E. LUNDIN, and R. R. SPENCER: Isolation of 4'-*O*-Methylcoumestrol from Alfalfa. J. Agric. Fd. Chem. **13**, 597 (1965).

73. BICKOFF, E. M., G. M. LOPER, C. H. HANSON, J. H. GRAHAM, S. C. WITT, and R. R. SPENCER: Effect of Common Leafspot on Coumestans and Flavones in Alfalfa. Crop. Sci. **7**, 259 (1967).

74. BICKOFF, E. M., R. L. LYMAN, A. L. LIVINGSTON, and A. N. BOOTH: Characterization of Coumestrol, a Naturally Occurring Plant Estrogen. J. Am. Chem. Soc. **80**, 3969 (1958).

75. BICKOFF, E. M., R. R. SPENCER, B. E. KNUCKLES, and R. E. LUNDIN: 3'-Methoxycoumestrol from Alfalfa: Isolation and Characterisation J. Agric. Fd. Chem. **14**, 444 (1966).

76. BICKOFF, E. M., R. R. SPENCER, S. C. WITT, and B. E. KNUCKLES: Studies on the Chemical and Biological Properties of Coumestrol and Related Compounds. U.S.D.A. Technical Bull. No. 1408. Washington D. C.: U. S. Government Printing Office.

77. BIGGS, D. R.: Post-Infectional Compounds from the French Bean *Phaseolus vulgaris;* Isolation and Identification of Genistein and 2',4',5,7-Tetrahydroxyisoflavone. Aust. J. Chem. **28**, 1389 (1975).

78. BIGGS, D. R., and G. A. LANE: Identification of Isoflavones Calycosin and Pseudobaptigenin in *Trifolium pratense.* Phytochemistry **17**, 1683 (1978).

79. BIGGS, D. R., and G. J. SHAW: Wairol, a New Coumestan from *Medicago sativa.* Phytochemistry **19**, 2801 (1980).

80. BILTON, J. N., J. R. DEBNAM, and I. M. SMITH: 6a-Hydroxypterocarpans from Red Clover. Phytochemistry **15**, 1411 (1976).

81. BOAM, J. J., R. S. CAHN, and A. STUART: The Identification of Tephrosin and Deguelin from Different Sources. J. Soc. Chem. Ind., Lond. **56**, 91T (1937).

81a. BOHLMANN, F., C. ZDERO, H. ROBINSON, and R. M. KING: A Diterpene, a Sesquiterpene Quinone and Flavanones from *Wyethia helenioides.* Phytochemistry **20**, 2245 (1981).

82. BONDE, M. R., R. L. MILLAR, and J. L. INGHAM: Induction and Identification of Sativan and Vestitol as Two Phytoalexins from *Lotus corniculatus.* Phytochemistry **12**, 2957 (1973).

83. BOSE, J. L.: A Note on the Possible Identity of Biochanin A and Pratensol. J. Scient. Ind. Res. **15B**, 324 (1956).

84. BOSE, J. L., and S. SIDDIQUI: The Constitution of Biochanin A. J. Scient. Ind. Res. **4**, 231 (1945).

85. — — The Identity of Biochanin B and Formononetin. J. Scient. Ind. Res. **10B**, 291 (1951).

86. BOSE, P. C., C. L. KIRTANIYA, and N. ADITYACHAUDHURY: Occurrence of Dehydrorotenone in *Derris uliginosa.* Indian J. Chem. **14B**, 1012 (1976).

87. BOUWER, D., C. V. D. M. BRINK, J. P. ENGELBRECHT, and G. J. H. RALL: 4-Methoxypterocarpin, a New Pterocarpan from *Neorautanenia ficifolia.* J. South Afr. Chem. Inst. **21**, 159 (1968).

88. BRADBURY, R. B., and D. E. WHITE: The Chemistry of Subterranean Clover. Part I. Isolation of Formononetin and Genistein. J. Chem. Soc. **1951**, 3447.

89. BRADEN, A. W. H., N. K. HART, and J. A. LAMBERTON: The Oestrogenic Activity and Metabolism of Certain Isoflavones in Sheep. Aust. J. Agric. Res. **18**, 335 (1967).

90. BRAGA, A. DA S., V. H. ARNDT, H. M. ALVES, O. R. GOTTLIEB, M. T. MAGALHÃES, and W. D. OLLIS: Chemistry of *Dalbergia barretoana* and *D. villosa,* Two Related Species. Anais Acad. Bras. Cienc. **39**, 249 (1967).

91. BRAGA, A. DA S., O. R. GOTTLIEB, W. B. EYTON, K. KUROSAWA, and W. D. OLLIS: Constituents of *Machaerium villosum.* Part 1. Anais Acad. Bras. Cienc. **40**, 33 (1968).

92. BRAGA DE OLIVEIRA, A., L. G. FONSECA E SILVA, and O. R. GOTTLIEB: Flavonoids and Coumarins from *Platymiscium praecox.* Phytochemistry **11**, 3515 (1972).

93. BRAGA DE OLIVEIRA, A., O. R. GOTTLIEB, T. M. M. GONÇALVES, and W. D. OLLIS: Isoflavonoids of *Cyclolobium clausseni* and *C. vecchii.* Anais Acad. Bras. Cienc. **43**, 129 (1971).

94. BRAGA DE OLIVEIRA, A., O. R. GOTTLIEB, and M. E. LEITE DE ALMEIDA: Extractives of *Tipuana tipu.* Phytochemistry **10**, 2552 (1971).

95. Braga de Oliveira, A., O. R. Gottlieb, and W. D. Ollis: Constituents of *Machaerium villosum*. Part 2. Anais Acad. Bras. Cienc. **40**, 147 (1968).

96. Braga de Oliveira, A., M. I. L. M. Madruga, and O. R. Gottlieb: Isoflavonoids from *Myroxylon balsamum*. Phytochemistry **17**, 593 (1978).

96a. Braz Filho, R., U. S. de Figueiredo, O. R. Gottlieb, and A. P. Mourão: Rotenone in *Lonchocarpus longifolius*. Acta Amazonica **10**, 843 (1980).

97. Braz Filho, R., M. P. L. de Moraes, and O. R. Gottlieb: Pterocarpans from *Swartzia laevicarpa*. Phytochemistry **19**, 2003 (1980).

98. Braz Filho, R., and O. R. Gottlieb: The Flavones of *Apuleia leiocarpa*. Phytochemistry **10**, 2433 (1971).

99. Braz Filho, R., O. R. Gottlieb, and R. M. V. Assumpção: The Isoflavones of *Pterodon pubescens*. Phytochemistry **10**, 2835 (1971).

100. Braz Filho, R., O. R. Gottlieb, A. A. de Moraes, G. Pedreira, S. L. V. Pinho, M. T. Magalhães, and M. N. de S. Ribeiro: Isoflavonoids from Amazonian Myristicaceae. Lloydia **40**, 236 (1977).

101. Braz Filho, R., O. R. Gottlieb, A. P. Mourão, A. I. da Rocha, and F. S. Oliveira: Flavonoids from *Derris* Species. Phytochemistry **14**, 1454 (1975).

102. Braz Filho, R., O. R. Gottlieb, S. L. V. Pinho, F. J. Q. Monte, and A. I. da Rocha: Flavonoids from Amazonian Leguminosae. Phytochemistry **12**, 1184 (1973).

103. Braz Filho, R., M. E. Leite de Almeida, and O. R. Gottlieb: Iso- and Neo-Flavonoids from *Dalbergia riparia*. Phytochemistry **12**, 1187 (1973).

104. Braz Filho, R., G. Pedreira, O. R. Gottlieb, and J. G. S. Maia: Isoflavones from *Virola caducifolia*. Phytochemistry **15**, 1029 (1976).

105. Bredenberg, J. B., and P. K. Hietala: Investigation of the Structure of Trifolirhizin, an Antifungal Compound from *Trifolium pratense*. Acta Chem. Scand. **15**, 696 (1961).

106. —— Confirmation of the Structure of Trifolirhizin. Acta Chem. Scand. **15**, 936 (1961).

107. Bredenberg, J. B., and J. N. Shoolery: A Revised Structure for Pterocarpin. Tetrahedron Lett. **1961**, 285.

108. Breytenbach, J. C., and G. J. H. Rall: Structure and Synthesis of Isoflavonoid Analogues from *Neorautanenia amboensis*. J. Chem. Soc. Perkin Trans. I **1980**, 1804.

109. —— Ambofuranol — the First Natural 3-Methoxybenzofuran. Tetrahedron Lett. **1980**, 4535.

110. Bridel, M., and C. Charaux: Oroboside, a New Glucoside from *Orobus tuberosus* Hydrolysable by Emulsin. Bull. Soc. Chim. Biol. **12**, 615 (1930).

111. —— Hydrolysis Products of Oroboside: Glucose and Orobol. Bull. Soc. Chim. Biol. **12**, 765 (1930).

112. —— Oroboside, a New Glucoside Obtained from *Orobus tuberosus* Capable of Being Hydrolysed by Means of Emulsin. Its Hydrolysis Products: Glucose and Orobol. C. R. Hebd. Séanc. Acad. Sci., Paris **190**, 387 (1930).

113. Briggs, L. H., and B. F. Cain: Constituents of the Heartwood of *Podocarpus spicatus*. Tetrahedron **6**, 143 (1959).

114. Briggs, L. H., R. C. Cambie, and R. K. Montgomery: Constituents of the Wood and Bark of *Sophora microphylla* and *S. tetraptera*. New Zealand J. Sci. **18**, 555 (1975).

115. Briggs, L. H., and T. P. Cebalo: The Isolation of Genistein from *Podocarpus spicatus* and the Constitution of Podospicatin. Tetrahedron **6**, 145 (1959).

116. Brink, A. J., G. J. H. Rall, and J. C. Breytenbach: Pterocarpans from *Neorautanenia edulis* and *N. amboensis*. Phytochemistry **16**, 273 (1977).

117. Brink, A. J., G. J. H. Rall, and J. P. Engelbrecht: The Isolation and Structures of Neorauflavene, (–)-Neorauflavane and Neoraufurane, Three Novel Isoflavonoids from *Neorautanenia edulis*. Tetrahedron **30**, 311 (1974).

118. — — — Structures of Some Minor Pterocarpans of *Neorautanenia edulis*. Phytochemistry **13**, 1581 (1974).

119. BRINK, C. V. D. M., J. J. DEKKER, E. C. HANEKOM, D. H. MEIRING, and G. J. H. RALL: The Interconversion of Neodulin and Dehydroneotenone. J. South Afr. Chem. Inst. **18**, 21 (1965).

120. BRINK, C. V. D. M., J. P. ENGELBRECHT, and D. E. GRAHAM: Ficifolinol, Folitenol and Folinin, Three New Pterocarpans from the Root Bark of *Neorautanenia ficifolia.* J. South Afr. Chem. Inst. **23**, 24 (1970).

121. BRINK, C. V. D. M., W. NEL, G. J. H. RALL, J. C. WEITZ, and K. G. R. PACHLER: Neofolin and Ficinin, Two New Furoisoflavonoids from *Neorautanenia ficifolia.* J. South Afr. Chem. Inst. **19**, 24 (1966).

122. BROOKS, B. T.: The Natural Dyes and Coloring Matters of the Philippines. Philipp. J. Sci. **5A**, 439 (1910).

123. BROWN, M. P., R. H. THOMSON, B. M. HAUSEN, and M. H. SIMATUPANG: Extractives from *Bowdichia nitida:* the First Isoflavonequinone. Justus Liebigs Annln. Chem. **1974**, 1295.

124. BRUM-BOUSQUET, M., Y. LALLEMAND, F. TILLEQUIN, and P. DELAVEAU: Structure of a Novel Isoflavone Glycoside: Sarothamnoside. Tetrahedron Lett. **22**, 1223 (1981).

125. BÜCHI, J., H. SCHMID, and A. L. KAPOOR: Isolation and Chemistry of some Compounds from *Piscidia erythrina* (Jamaica Dogwood). Arch. Pharm. Chem. **68**, 183 (1961).

126. BURDEN, R. S., and J. A. BAILEY: Structure of the Phytoalexin from Soybean. Phytochemistry **14**, 1389 (1975).

127. BURDEN, R. S., J. A. BAILEY, and G. W. DAWSON: Structures of Three New Isoflavonoids from *Phaseolus vulgaris* Infected with Tobacco Necrosis Virus. Tetrahedron Lett. **1972**, 4175.

128. BURDEN, R. S., J. A. BAILEY, and G. G. VINCENT: Metabolism of Phaseollin by *Colletotrichum lindemuthianum.* Phytochemistry **13**, 1789 (1974).

129. BURDEN, R. S., P. M. ROGERS, and R. L. WAIN: Natural Resistance of Plant Roots to Fungal Pathogens. Ann. Appl. Biol. **78**, 59 (1974).

130. BURROWS, B. F., N. FINCH, W. D. OLLIS, and I. O. SUTHERLAND: Mundulone. Proc. Chem. Soc. **1959**, 150.

131. BUTENANDT, A., and G. HILGETAG: Constituents of Varieties of *Derris* and *Tephrosia.* Justus Liebigs Annln. Chem. **495**, 172 (1932).

132. BUTENANDT, A., and W. MCCARTNEY: Rotenone, the Physiologically Active Constituent of *Derris elliptica.* Constitution of Rotenone. Justus Liebigs Annln. Chem. **494**, 17 (1932).

133. CAHN, R. S., and J. J. BOAM: The Constituents of *Derris* Resin. J. Soc. Chem. Ind., Lond. **54**, 42T (1935).

133a. CAMELE, G., F. DELLE MONACHE, G. DELLE MONACHE, and G. B. MARINI-BETTOLO: Three New Flavonoids from *Tephrosia praecana.* Phytochemistry **19**, 707 (1980).

134. CAMPBELL, R. V. M., S. H. HARPER, and A. D. KEMP: Isoflavonoid Constituents of the Heartwood of *Cordyla africana.* J. Chem. Soc. C **1969**, 1787.

135. CAMPBELL, R. V. M., and J. TANNOCK: Isoflavonoid Constituents of the Heartwood of *Cordyla africana.* J. Chem. Soc. Perkin Trans. I **1973**, 2222.

135a. CANNON, P. F.: Systematic Studies in the Genus *Ononis* (Leguminosae — Papilionoideae). Ph. D. Thesis, University of Reading, U. K. 1981.

136. CARLSON, R. E., and D. H. DOLPHIN: Chromatographic Analysis of Isoflavonoid Accumulation in Stressed *Pisum sativum.* Phytochemistry **20**, 2281 (1981).

137. CARTER, G. A., K. CHAMBERLAIN, and R. L. WAIN: The Fungitoxicity of Analogues of the Phytoalexin 2-(2'-Methoxy-4'-hydroxyphenyl)-6-methoxybenzofuran (Vignafuran). Ann. Appl. Biol. **88**, 57 (1978).

138. CAZENEUVE, P., and L. HUGOUNENQ: Pterocarpin and Homopterocarpin from Sandal Wood. C. R. Hebd. Séanc. Acad. Sci., Paris **104**, 1722 (1887).

139. CHAKRAVARTI, D., and C. BHAR: On the Constitution of Prunusetin from the Bark of *Prunus puddum.* J. Indian Chem Soc. **22**, 301 (1945).

140. CHAKRAVARTI, K. K., A. K. BOSE, and S. SIDDIQUI: Chemical Examination of the Seeds of *Psoralea corylifolia*. J. Scient. Ind. Res. **7B**, 24 (1948).

141. CHANG, C. F., A. SUZUKI, S. KUMAI, and S. TAMURA: Chemical Studies on "Clover Sickness". Part II. Biological Functions of Isoflavonoids and their Related Compounds. Agric. Biol. Chem. **33**, 398 (1969).

142. CHARAUX, C., and J. RABATÉ: Sophoricoside, a New Glycoside from the Fruits of *Sophora japonica*. J. Pharm. Chim., Paris **21**, 546 (1935).

143. — — A Biochemical Study of *Sophora japonica* Fruits. I. The Presence of Sophoricoside. Bull. Soc. Chim. Biol. **20**, 454 (1938).

144. — — The Chemical Constitution of Orobol. Bull. Soc. Chim. Biol. **21**, 1330 (1939).

145. — — An Examination of the Flowers of *Genista tinctoria*. J. Pharm. Chim., Paris **1**, 404 (1941).

146. CHAWLA, H. M., and S. S. CHIBBER: Occurrence of 7-Hydroxy-4-methylcoumarin in *Dalbergia volubilis*. Indian J. Chem. **15 B**, 492 (1977).

146a. — — Chemistry of *Dalbergia* Species. J. Scient. Ind. Res. **40**, 313 (1981).

147. CHAWLA, H. M., S. S. CHIBBER, and T. R. SESHADRI: Volubilin, a New Isoflavone-*C*-Glycoside from *Dalbergia volubilis* Flowers. Phytochemistry **13**, 2301 (1974).

148. — — — Isovolubilin, a New Isoflavone-*C*-Rhamnoside from *Dalbergia volubilis* Flowers. Indian J. Chem. **13**, 444 (1975).

149. — — — Volubilinin, a New Isoflavone-*C*-Glycoside from *Dalbergia volubilis* Flowers. Phytochemistry **15**, 235 (1976).

150. CHEN, Y. L., and H. Y. HSU: On the Active Principles of *Tephrosia obovata*. J. Pharm. Soc. Japan **78**, 198 (1958).

151. CHEN, Y. L., and C. S. TSAI: The Paper Chromatography of Rotenone. J. Taiwan Pharm. Assoc. **7**, 31 (1955); Chem. Abstr. **50**, 17305 (1956).

152. CHI, J. J., S. T. HSU, M. HU, and S. WANG: Iso-shehkanin. Part I. J. Chinese Chem. Soc., Peiping **15**, 26 (1947).

152a. CHIBBER, S. S., S. K. DUTT, R. P. SHARMA, and A. SHARMA: Pongachin: a New Pyranoflavanone from Seeds of *Tephrosia candida*. Indian J. Chem. **20B**, 626 (1981).

153. CHIBBER, S. S., and U. KHERA: Dalbinol — a New 12a-Hydroxyrotenoid from *Dalbergia latifolia* Seeds. Phytochemistry **17**, 1442 (1978).

154. — — Dalbin: a 12a-Hydroxy Rotenoid Glycoside from *Dalbergia latifolia*. Phytochemistry **18**, 188 (1979).

155. CHIBBER, S. S., and R. P. SHARMA: Isolation of 6-Hydroxy-2′,7-dimethoxy-4′,5′-methylenedioxyisoflavone from the Pods of *Dalbergia assamica*. Curr. Sci. **47**, 856 (1978).

156. — — Chemical Constituents of the Seeds of *Dalbergia assamica*. Natnl. Acad. Sci. Lett., India **1**, 253 (1978).

157. — — Robustigenin, a New Isoflavone from *Derris robusta* Seed Shells. Phytochemistry **18**, 1082 (1979).

158. — — Derrugenin, a New Isoflavone from *Derris robusta* Seed Shells. Phytochemistry **18**, 1583 (1979).

159. — — 5-Hydroxy-7-methoxyisoflavone from Seeds of *Derris robusta*. Planta Med. **36**, 379 (1979).

160. — — Chemical Constituents of Seeds of *Derris robusta*. Indian J. Chem. **18B**, 471 (1979).

161. — — Robustigenin-5-*O*-Methyl Ether, a New Isoflavone from *Derris robusta*. Indian J. Chem. **17B**, 649 (1979).

162. — — Derrone, a New Pyranoisoflavone from *Derris robusta* Seeds. Phytochemistry **19**, 1857 (1980).

162a. CHIBBER, S. S., R. P. SHARMA, and S. K. DUTT: Derrone-4′-*O*-Methyl Ether from Seeds of *Derris robusta*. Curr. Sci. **50**, 818 (1981).

163. CHIMURA, H., T. SAWA, Y. KUMADA, H. NAGANAWA, M. MATSUZAKI, T. TAKITA, M.

HAMADA, T. TAKEUCHI, and H. UMEZAWA: New Isoflavones, Inhibiting Catechol-*o*-Methyltransferase, Produced by *Streptomyces*. J. Antibiot., Tokyo **28**, 619 (1975).

164. CHIU, S. F.: Effectiveness of Chinese Insecticidal Plants with Reference to the Comparative Toxicity of Botanical and Synthetic Insecticides. J. Sci. Fd. Agric. **1**, 276 (1950).

165. CHOPIN, J., M. BOUILLANT, and P. LEBRETON: 5-Methylgenistein, a New Isoflavone from *Laburnum*. C. R. Hebd. Séanc. Acad. Sci., Paris **256**, 5653 (1963).

166. CLAISSE, J., L. CROMBIE, and R. PEACE: Structure and Stereochemistry of the Vicianoside Amorphin, the First Rotenoid Glycoside. J. Chem. Soc. **1964**, 6023.

167. CLARK, E. P.: Toxicarol. A Constituent of the South American Fish Poison *Cracca (Tephrosia) toxicaria*. J. Am. Chem. Soc. **52**, 2461 (1930).

168. — Tephrosin. I. The Composition of Tephrosin and its Relation to Deguelin. J. Am. Chem. Soc. **53**, 729 (1931).

169. — The Occurrence of Rotenone and Related Compounds in the Roots of *Cracca virginiana*. Science, N. Y. **77**, 311 (1933).

170. — The Occurrence of Rotenone and Related Substances in the Seeds of the Berebera Tree. A Procedure for the Separation of Deguelin and Tephrosin. J. Am. Chem. Soc. **65**, 27 (1943).

171. — Scandenin — a Constituent of the Roots of *Derris scandens*. J. Org. Chem. **8**, 489 (1943).

172. CLARK, E. P., and H. V. CLABORN: Tephrosin. II. Isotephrosin. J. Am. Chem. Soc. **54**, 4454 (1932).

173. CLARK, E. P., and G. L. KEENAN: Note on the Occurrence of Dehydrodeguelin and Dehydrotoxicarol in Some Samples of *Derris* Root. J. Am. Chem. Soc. **55**, 422 (1933).

174. COCKER, W., T. DAHL, C. DEMPSEY, and T. B. H. MCMURRY: Extractives from *Andira inermis*. J. Chem. Soc. **1962**, 4906.

175. COCKER, W., T. B. H. MCMURRY, and P. A. STANILAND: A Synthesis of Demethylhomopterocarpin. J. Chem. Soc. **1965**, 1034.

176. COOK, J. T., W. D. OLLIS, I. O. SUTHERLAND, and O. R. GOTTLIEB: Pterocarpans from *Dalbergia spruceana*. Phytochemistry **17**, 1419 (1978).

177. COOKE, R. G., and R. A. H. FLETCHER: Constituents of *Cotoneaster* Species. Aust. J. Chem. **27**, 1377 (1974).

178. COOKE, R. G., and I. D. RAE: Some New Constituents of *Pterocarpus indicus* Heartwood. Aust. J. Chem. **17**, 379 (1964).

179. COWAN, R. S.: Swartzieae. In: R. M. POLHILL and P. H. RAVEN, Advances in Legume Systematics, p. 209. London: Her Majesty's Stationery Office. 1981.

180. COX, R. I., and A. W. BRADEN: The Metabolism and Physiological Effects of Phyto-Oestrogens in Livestock. Proc. Aust. Soc. Anim. Prod. **10**, 122 (1974).

181. CRABBÉ, P., P. R. LEEMING, and C. DJERASSI: The Structure of the Isoflavone Tlatlancuayin. J. Am. Chem. Soc. **80**, 5258 (1958).

182. CRAVEIRO, A. A., and O. R. GOTTLIEB: Pterocarpans from *Platymiscium trinitatis*. Phytochemistry **13**, 1629 (1974).

183. CROMBIE, L.: Chemistry of the Natural Rotenoids. Fortschr. Chem. org. Naturstoffe **21**, 275 (1963).

184. CROMBIE, L., P. M. DEWICK, and D. A. WHITING: Chalcone, Isoflavone, and Rotenoid Stages in the Formation of Amorphigenin by *Amorpha fruticosa* Seedlings. J. Chem. Soc. Perkin Trans. I **1973**, 1285.

184a. CROMBIE, L., I. HOLDEN, G. W. KILBEE, and D. A. WHITING: Formation and Dehydration of a Prochiral 2-Hydroxyisopropyl Centre During Biosynthesis: the Rot-2'-enonic Acid — Rotenone Transformation in *Amorpha fruticosa*. J. Chem. Soc. Chem. Commun. **1979**, 1143.

184b. CROMBIE, L., I. HOLDEN, G. W. KILBEE, and D. A. WHITTING: Hydroxylation of Rotenone to Amorphigenin in *Amorpha fruticosa* Seedlings. J. Chem. Soc. Chem. Commun. **1979**, 1144.

185. CROMBIE, L., and D. A. WHITTING: The Constitution of Neotenone and Dolichone. Biogenetic Connexions in the Sub-Family Papilionatae. Tetrahedron Lett. **1962**, 801.

186. — — The Extractives of *Neorautanenia pseudopachyrrhiza*: the Isolation and Structure of a New Rotenoid and Two New Isoflavanones. J. Chem. Soc. **1963**, 1569.

187. CRUICKSHANK, I. A. M.: The Antimicrobial Spectrum of Pisatin. Aust. J. Biol. Sci. **15**, 147 (1962).

188. — A Review of the Role of Phytoalexins in Disease Resistance Mechanisms. Pont. Acad. Sci. Script. Var. **41**, 509 (1977).

189. CRUICKSHANK, I. A. M., and D. R. PERRIN: Pisatin Formation by Cultivars of *Pisum sativum* and Several Other *Pisum* Species. Aust. J. Biol. Sci. **18**, 829 (1965).

190. — — The Induction, Antimicrobial Spectrum and Chemical Assay of Phaseollin. Phytopath. Z. **70**, 209 (1971).

191. CUCA SUAREZ, L. E., F. DELLE MONACHE, G. B. MARINI-BETTOLO, and F. MENICHINI: Three New Prenylated Flavanones from *Tephrosia* sp. Farmaco (Ed. Sci.) **35**, 796 (1980).

192. CURNOW, D. H., and R. C. ROSSITER: The Occurrence of Genistein in Subterranean Clover *(T. subterraneum)* and other *Trifolium* Species. Aust. J. Exp. Biol. Med. Sci. **33**, 243 (1955).

193. DARBARWAR, M., V. SUNDARAMURTHY, and N. V. S. RAO: Coumestans. J. Scient. Ind. Res. **35**, 297 (1976).

194. DAYAL, R., and M. R. PARTHASARATHY: Phenolic Constituents of *Dalbergia sericea* Leaves. Planta Med. **31**, 245 (1977).

195. DE ALENCAR, R., R. BRAZ FILHO, and O. R. GOTTLIEB: Pterocarpanoids from *Dalbergia decipularis*. Phytochemistry **11**, 1517 (1972).

196. DEAN, F. M.: Naturally Occurring Oxygen Ring Compounds. London: Butterworth. 1963.

197. DEBNAM, J. R., and I. M. SMITH: Changes in the Isoflavones and Pterocarpans of Red Clover on Infection with *Sclerotinia trifoliorum* and *Botrytis cinerea*. Physiol. Pl. Path. **9**, 9 (1976).

198. DE LAIRE, G., and F. TIEMANN: Iridin, a Glucoside from *Iris* Roots. Ber. Dt. Chem. Ges. **26**, 2010 (1893).

199. DELFEL, N. E., and W. H. TALLENT: Thin Layer Densitometric Determination of Rotenone and Deguelin. J. Assoc. Off. Anal. Chem. **52**, 182 (1969).

200. DELFEL, N. E., W. H. TALLENT, D. G. CARLSON, and I. A. WOLFF: Distribution of Rotenone and Deguelin in *Tephrosia vogelii* and Separation of Rotenoid-Rich Fractions. J. Agric. Fd. Chem. **18**, 385 (1970).

201. DELLE MONACHE, F., L. E. CUCA SUAREZ, and G. B. MARINI-BETTOLO: Flavonoids from the Seeds of Six *Lonchocarpus* Species. Phytochemistry **17**, 1812 (1978).

202. DELLE MONACHE, F., G. C. VALERA, D. SIALER DE ZAPATA, and G. B. MARINI-BETTOLO: 3-Aryl-4-methoxycoumarins and Isoflavones from *Derris glabrescens*. Gazz. Chim. Ital. **107**, 403 (1977).

203. DELLE MONACHE, G., F. DELLE MONACHE, G. B. MARINI-BETTOLO, M. M. F. DE ALBUQUERQUE, J. F. DE MELLO, and O. G. DE LIMA: Isosophoranone, a New Diprenylated Isoflavanone from *Sophora tomentosa*. Gazz. Chim. Ital. **107**, 189 (1977).

204. DEMENT, W. A., and T. J. MABRY: Flavonoids of North American Species of *Thermopsis*. Phytochemistry **11**, 1089 (1972).

205. DENNY, T. P., and H. D. VANETTEN: Metabolism of the Phytoalexins Medicarpin and Maackiain by *Fusarium solani*. Phytochemistry **21**, 1023 (1982).

206. DE PASCUAL, T. J., J. C. H. AUBANELL, and M. GRANDE: Components of *Astragalus lusitanicus*. Flavonoids. An. Quim. **75**, 1005 (1979).

207. DERYUGINA, L. I.: Astroside — a New Isoflavone Glycoside from *Astragalus austriacus*. Khim. Prir. Soedin. **1966**, 315.

208. DESAI, H. K., D. H. GAWAD, B. S. JOSHI, P. C. PARTHASARATHY, K. R. RAVINDRANATH, M. T. SAINDANE, A. R. SIDHAYE, and N. VISWANATHAN: Chemical Investigation of Indian Plants: Part X. Indian J. Chem. **15B**, 291 (1977).

209. DESHPANDE, V. H., A. D. PENDSE, and R. PENDSE: Erythrinins A, B and C, Three New Isoflavones from the Bark of *Erythrina variegata*. Indian J. Chem. **15B**, 205 (1977).

210. DESHPANDE, V. H., and R. K. SHASTRI: Phenolics of *Albizia lebbek, A. amara* and *A. procera*. Indian J. Chem. **15B**, 201 (1977).

210a. DEWICK, P. M.: Unpublished Results.

211. — Pterocarpan Biosynthesis: Chalcone and Isoflavone Precursors of Demethylhomopterocarpin and Maackiain in *Trifolium pratense*. Phytochemistry **14**, 979 (1975).

212. — Biosynthesis of Pterocarpan Phytoalexins in *Trifolium pratense*. Phytochemistry **16**, 93 (1977).

213. — Biosynthesis of the 6-Oxygenated Isoflavone Afrormosin in *Onobrychis viciifolia*. Phytochemistry **17**, 249 (1978).

214. DEWICK, P. M., W. BARZ, and H. GRISEBACH: Biosynthesis of Coumestrol in *Phaseolus vulgaris*. Phytochemistry **9**, 775 (1970).

215. DEWICK, P. M., and J. L. INGHAM: Isopterofuran, a New 2-Arylbenzofuran Phytoalexin from *Coronilla emerus*. Phytochemistry **19**, 289 (1980).

216. DEWICK, P. M., and M. MARTIN: Biosynthesis of Pterocarpan and Isoflavan Phytoalexins in *Medicago sativa*: the Biochemical Interconversion of Pterocarpans and 2'-Hydroxyisoflavans. Phytochemistry **18**, 591 (1979).

217. DHAR, K. L., and A. K. KALLA: Isoflavones of *Iris kumaonensis* and *I. germanica*. Phytochemistry **11**, 3097 (1972).

218. — — A New Isoflavone from *Iris germanica*. Phytochemistry **12**, 734 (1973).

219. — — Irisolidone from *Iris kashmiriana*. J. Indian Chem. Soc. **52**, 784 (1975).

220. DHINGRA, V. K., T. R. SESHADRI, and S. K. MUKERJEE: Isotectorigenin from the Bark of *Dalbergia sissoo*. Indian J. Chem. **12**, 1118 (1974).

221. DIETRICHS, H. H., and M. H. SIMATUPANG: Homopterocarpin from the Heartwood of *Pterocarpus angolensis*. Holzforschung **28**, 186 (1974).

222. DIXON, R. A., and D. S. BENDALL: Changes in Phenolic Compounds Associated with Phaseollin Production in Cell Suspension Cultures of *Phaseolus vulgaris*. Physiol. Pl. Path. **13**, 283 (1978).

223. DONNELLY, B. J., D. M. X. DONNELLY, and A. M. O'SULLIVAN: The Occurrence of Melannein in the Genus *Dalbergia*. Tetrahedron **24**, 2617 (1968).

223a. DONNELLY, D. M. X.: Unpublished Results.

224. DONNELLY, D. M. X., and M. A. FITZGERALD: Pterocarpanoid Constituents of *Swartzia leiocalycina*. Phytochemistry **10**, 3147 (1971).

225. DONNELLY, D. M. X., and P. J. KAVANAGH: Isoflavonoids of *Dalbergia oliveri*. Phytochemistry **13**, 2587 (1974).

226. DONNELLY, D. M. X., P. J. KEENAN, and J. P. PRENDERGAST: Isoflavonoids of *Dalbergia ecastophyllum*. Phytochemistry **12**, 1157 (1973).

227. DONNELLY, D. M. X., J. C. THOMPSON, W. B. WHALLEY, and S. AHMAD: Phytochemical Examination of *Dalbergia stevensonii*. J. Chem. Soc. Perkin Trans. I **1973**, 1737.

228. DUTTA, N. L.: Constitution of Munetone, the Principal Crystalline Product of the Root Bark of *Mundulea suberosa*. J. Indian Chem. Soc. **36**, 165 (1959).

229. DYKE, S. F., W. D. OLLIS, and M. SAINSBURY: Munetone. J. Chem. Soc. C **1966**, 749.

230. EADE, R. A., H. HINTERBERGER, and J. J. H. SIMES: Afromosin (Castanin, 6,4'-Dimethoxy-7-hydroxyisoflavone) from *Castanospermum australe*. Aust. J. Chem. **16**, 188 (1963).

231. EAST, A. J., W. D. OLLIS, and R. E. WHEELER: Phytochemical Examination of *Derris robusta*. J. Chem. Soc. C **1969**, 365.
232. EBEL, J., A. R. AYERS, and P. ALBERSHEIM: Response of Suspension-Cultured Soybean Cells to the Elicitor Isolated from *Phytophthora megasperma* var. *sojae*, a Fungal Pathogen of Soybeans. Pl. Physiol., Lancaster **57**, 775 (1976).
233. EISENBEISS, J., and H. SCHMID: The Structure of Erosnin (Norton & Hansberry's "Compound I"). Helv. Chim. Acta **42**, 61 (1959).
234. EL-EMARY, N. A., Y. KOBAYASHI, and Y. OGIHARA: Two Isoflavonoids from the Fresh Bulbs of *Iris tingitana*. Phytochemistry **19**, 1878 (1980).
235. ELGAMAL, M. H. A., and M. B. E. FAYEZ: Isolation of Formononetin from the Roots of *Glycyrrhiza glabra* Collected Locally. Indian J. Chem. **10**, 128 (1972).
236. EL-MOGHAZY, A. M., A. A. ALI, N. A. EL-EMARY, and F. M. DARWISH: Isoflavones from Rhizomes of *Iris germanica*. Fitoterapia **51**, 237 (1980).
237. ERDTMAN, H., and T. NORIN: Heartwood Constituents of *Laburnum alpinum*. Acta Chem. Scand. **17**, 1781 (1963).
238. FALSHAW, C. P., R. A. HARMER, W. D. OLLIS, R. E. WHEELER, V. R. LALITHA, and N. V. S. RAO: Phytochemical Examination of *Derris scandens*. J. Chem. Soc. C **1969**, 374.
239. FALSHAW, C. P., W. D. OLLIS, J. A. MOORE, and K. MAGNUS: The Constitutions of Lisetin, Piscidone and Piscerythrone. Tetrahedron (Suppl. 7) **1966**, 333.
240. FARKAS, L., and M. NÓGRÁDI: The Structure of Sophorabiose. Tetrahedron Lett. **1964**, 3919.
241. FARKAS, L., M. NÓGRÁDI, S. ANTUS, and A. GOTTSEGEN: About the Existence of Padmakastein and Padmakastin. The Synthesis of 4′,5-Dihydroxy-7-methoxyiso-flavanone and its 4′-Glucoside. Tetrahedron **25**, 1013 (1969).
242. FARKAS, L., M. NÓGRÁDI, H. WAGNER, and L. HÖRHAMMER: Final Structure Determination and Total Synthesis of Sophorabioside, a Glycoside from *Sophora japonica*. Chem. Ber. **101**, 2758 (1968).
243. FARKAS, L., J. VÁRADY, and A. GOTTSEGEN: Synthesis of 7,3′,4′,5′-Tetrahydroxy-isoflavone (Baptigenin ?). Chem. Ber. **96**, 1865 (1963).
244. FELLOWS, L.: Unpublished Results.
245. FELSBERG, A. A., and P. E. ROSENTSVEIG: Glycosides of *Ononis arvensis*. Rast. Resursy **1**, 224 (1965).
246. — — A Comparative Phytochemical Study of the Roots of *Ononis arvensis* and *Ononis spinosa*. Aptechn. Delo **14**, 26 (1965).
246a. FELTWELL, J., and L. R. G. VALADON: Plant Pigments Identified in the Common Blue Butterfly. Nature, Lond. **225**, 969 (1970).
247. FERREIRA, M. A., M. MOIR, and R. H. THOMSON: New Pterocarpenes from *Brya ebenus*. J. Chem. Soc. Perkin Trans I **1974**, 2429.
248. — — Pterocarpenequinones (6*H*-Benzofuro-[3,2-*c*][1]-benzopyranquinones) from *Brya ebenus*. J. Chem. Soc. Perkin Trans. I **1975**, 1113.
249. FINCH, N., and W. D. OLLIS: Munduserone. Proc. Chem. Soc. **1960**, 176.
250. FINNEMORE, H.: Chemical Examination of a Species of *Prunus*. Pharm. J. **31**, 604 (1910).
251. FISCHER, R., and H. EHRLICH: Detection of Baptisin in Radix Baptisiae. Mikrochemie (Hans Molisch Commemorative Issue) **1936**, 99.
252. FITZGERALD, M. A., P. J. M. GUNNING, and D. M. X. DONNELLY: Phytochemical Examination of *Pericopsis* Species. J. Chem. Soc. Perkin Trans. I **1976**, 186.
253. FORMIGA, M. D., O. R. GOTTLIEB, P. H. MENDES, M. KOKETSU, M. E. LEITE DE ALMEIDA, M. O. DA S. PEREIRA, and M. T. MAGALHÃES: Constituents of Brazilian Leguminosae. Phytochemistry **14**, 828 (1975).
254. FOXALL, C. D., and J. W. W. MORGAN: Extractives of *Afrormosia elata*. J. Chem. Soc. **1963**, 5573.

255. FRAISHTAT, P. D., and S. A. POPRAVKO: Identification of 2',7-Dihydroxy-4'-methoxyisoflavan (Vestitol) in the Roots of Alsike Clover. Khim. Prir. Soedin **1979**, 729.

256. FRAISHTAT, P. D., S. A. POPRAVKO, and N. S. WULFSON: Isolation of 5,4'-Dihydroxy-6,7-methylenedioxyisoflavone and its 4'-O-β-D-Glucoside from the Roots of Red Clover *(Trifolium pratense)*. Bio-Org. Khim. **5**, 228 (1979).

257. — — — A New Pterocarpan from the Roots of Red Clover *(Trifolium pratense)*. Bio-Org. Khim. **5**, 1879 (1979).

258. — — — Isoflavones from the Roots of Red Clover *(Trifolium pratense)*. Bio-Org. Khim. **6**, 1722 (1980).

259. — — — Isolation and Identification of Pterocarpans from the Roots of Cultivated Clover Species. Bio-Org. Khim. **7**, 927 (1981).

260. FRANCIS, C. M., and A. J. MILLINGTON: The Oestrogenic Potency of Dry Subterranean Clover Pastures, and of Leaf Blade and Petiole in the Green State. Aust. J. Agric. Res. **16**, 23 (1965).

261. — — The Oestrogenic Activity of Annual Medic Pastures. Aust. J. Agric. Res. **16**, 927 (1965).

262. — — The Presence of Methylated Coumestans in Annual *Medicago* Species: Response to a Fungal Pathogen. Aust. J. Agric. Res. **22**, 75 (1971).

263. FRANCIS, C. M., A. J. MILLINGTON, and E. T. BAILEY: The Distribution of Oestrogenic Isoflavones in the Genus *Trifolium*. Aust. J. Agric. Res. **18**, 47 (1967).

264. FRIEDLANDER, A., and B. SKLARZ: Catecholic Flavonoids from Soybean Flakes. Experientia **27**, 762 (1971).

265. FUCHS, A., F. W. DE VRIES, C. A. LANDHEER, and A. VAN VELDHUIZEN: 3-Hydroxymaackiainisoflavan, a Pisatin Metabolite Produced by *Fusarium oxysporum* f. sp. *pisi*. Phytochemistry **19**, 917 (1980).

266. FUCHS, A., F. W. DE VRIES, and M. PLATERO SANZ: The Mechanism of Pisatin Degradation by *Fusarium oxysporum* f. sp. *pisi*. Physiol. Pl. Path. **16**, 119 (1980).

267. FUJISE, Y., T. TODA, and S. ITÔ: Isolation of Trifolirhizin from *Ononis spinosa*. Chem. Pharm. Bull., Tokyo **13**, 93 (1965).

267a. FUKAMI, H., and M. NAKAJIMA: Rotenone and the Rotenoids. In: M. JACOBSON and D. G. CROSBY, Naturally Occurring Insecticides, p. 71. New York: Marcel Dekker. 1971.

268. FUKAMI, J. I., I. YAMAMOTO, and J. E. CASIDA: Metabolism of Rotenone *In vitro* by Tissue Homogenates from Mammals and Insects. Science, N. Y. **155**, 713 (1967).

269. FUKUI, H., H. EGAWA, K. KOSHIMIZU, and T. MITSUI: A New Isoflavone with Antifungal Activity from Immature Fruits of *Lupinus luteus*. Agric. Biol. Chem. **37**, 417 (1973).

270. FURUYA, T.: Metabolic Products and Their Chemical Regulations in Plant Tissue Cultures. Kitasato Archs. Exp. Med. **41**, 47 (1968).

271. FURUYA, T., and A. IKUTA: The Presence of 1-Maackiain and Pterocarpin in Callus Tissue of *Sophora angustifolia*. Chem. Pharm. Bull., Tokyo **16**, 771 (1968).

272. GAKHOKIDZE, A. M.: A Dye in *Gleditsia triacanthos*. J. Appl. Chem., U.S.S.R. **23**, 789 (1950).

273. GALINA, E., and O. R. GOTTLIEB: Isoflavones from *Pterodon apparicioi*. Phytochemistry **13**, 2593 (1974).

274. GANGULY, A. K., and O. Z. SARRE: Genistein and Daidzein, Metabolites of *Micromonospora halophytica*. Chemy. Ind. **1970**, 201.

275. GAUDIN, O., and R. VACHERAT: Studies on Rotenone and the Fish-Poisoning Properties Associated with some Plants of the French Sudan. Bull. Sci. Pharmac. **45**, 385 (1938).

276. GEIGERT, J., F. R. STERMITZ, G. JOHNSON, D. D. MAAG, and D. K. JOHNSON: Two Phytoalexins from Sugar Beet *(Beta vulgaris)*. Tetrahedron **29**, 2703 (1973).

277. GEOFFROY, E.: A Botanical, Chemical and Physiological Study of *Robinia nicou*. Annls. Inst. Colon. Marseille **2**, 1 (1895).

278. GHANIM, A., and I. JAYARAMAN: Chemical Components of *Tephrosia falciformis*. Indian J. Chem. **17B**, 648 (1979).
279. GHANIM, A., L. PRAKASH, A. ZAMAN, and A. R. KIDWAI: Isolation of Iridin from *Iris kumaonensis*. Indian J. Chem. **1**, 230 (1963).
280. GHIGLIONE, C., D. LEMORDANT, T. PUGNET, and M. GIRAUD: Chemical Composition of *Millettia ferruginea* Seeds. Bull. Soc. Pharm. Marseille **20**, 67 (1971).
281. GHOSE, T. P., and S. KRISHNA: Occurrence of Rotenone in *Millettia pachycarpa*. Curr. Sci. **6**, 57 (1937).
282. GHOSH, A. C., and N. L. DUTTA: Isolation of Mundulone from *Mundulea suberosa*. J. Indian Chem. Soc. **39**, 475 (1962).
283. GILBERT, A. H., A. McGOOKIN, and A. ROBERTSON: Isoshekkangenin and the Synthesis of 4-Hydroxycoumarins. J. Chem. Soc. **1957**, 3740.
284. GIOVAMBATTISTA, N.: The Value as Insecticides of the Roots of Two Species of Argentine *Tephrosia*. Revta. Fac. Cienc. Quim. Farm. Univ. Nac. La Plata **17**, 83 (1942); Chem. Abstr. **38**, 3409 (1944).
285. GNANAMANICKAM, S. S.: Isolation of Isoflavonoid Phytoalexins from Seeds of *Phaseolus vulgaris*. Experientia **35**, 323 (1979).
286. GNANAMANICKAM, S. S., and S. S. PATIL: Accumulation of Antibacterial Isoflavonoids in Hypersensitively Responding Bean Leaf Tissues Inoculated with *Pseudomonas phaseolicola*. Physiol. Pl. Path. **10**, 159 (1977).
287. GNANAMANICKAM, S. S., and D. A. SMITH: Selective Toxicity of Isoflavonoid Phytoalexins to Gram-Positive Bacteria. Phytopathology **70**, 894 (1980).
288. GOEL, R. N., and T. R. SESHADRI: Padmatin, a New Component of the Heartwood of *Prunus puddum*. Tetrahedron **5**, 91 (1959).
289. GOODHUE, L. D., and H. L. HALLER: The Non-Crystalline Constituents of *Tephrosia virginiana* Roots. J. Am. Chem. Soc. **62**, 2520 (1940).
290. GOPINATH, K. W., A. R. KIDWAI, and L. PRAKASH: Structure of Irisolone. Tetrahedron **16**, 201 (1961).
291. GOPINATH, K. W., L. PRAKASH, and A. R. KIDWAI: Isolation of Irigenin from *Iris nepalensis*. Indian J. Chem. **1**, 187 (1963).
292. GORDON, M. A., E. W. LAPA, M. S. FITTER, and M. LINDSAY: Susceptibility of Zoopathogenic Fungi to Phytoalexins. Antimicrob. Agents Chemotherap. **17**, 120 (1980).
293. GORTER, K.: Constituents of *Baptisia tinctoria* Root. Baptisin. Arch. Pharm., Berl. **235**, 301 (1897).
294. — Constituents of *Baptisia tinctoria* Root. Pseudobaptisin. Arch. Pharm., Berl. **235**, 494 (1897).
295. GOTTLIEB, O. R., A. BRAGA DE OLIVEIRA, T. M. M. GONÇALVES, G. G. DE OLIVEIRA, and S. A. PEREIRA: Isoflavonoids from *Cyclolobium* Species. Phytochemistry **14**, 2495 (1975).
296. GOTTLIEB, O. R., and A. I. DA ROCHA: 5-O-Methylgenistein from *Ormosia excelsa*. Phytochemistry **11**, 1183 (1972).
297. GOTTLIEB, O. R., and M. T. MAGALHÃES: Isolation of 3′,4′,7-Trimethoxyisoflavone (Cabreuvin) from *Myroxylon balsamum* and *Myrocarpus fastigiatus*. Anais Assoc. Quim. Bras. **18**, 89 (1959).
298. — — Isolation and Structure of Caviunin. J. Org. Chem. **26**, 2449 (1961).
299. GOVINDACHARI, T. R., K. NAGARAJAN, and B. R. PAI: Wedelolactone from *Eclipta alba*. J. Scient. Ind. Res. **15B**, 664 (1956).
300. — — — Structure of Wedelolactone J. Chem. Soc. **1956**, 629.
301. GOVINDACHARI, T. R., K. NAGARAJAN, B. R. PAI, and P. C. PARTHASARTHY: The Position of the Methoxyl Group in Wedelolactone. J. Chem. Soc. **1957**, 545.
302. GRAYER-BARKMEIJER, R. J., and J. B. HARBORNE: Unpublished Results.
303. GRAYER-BARKMEIJER, R. J., J. L. INGHAM, and P. M. DEWICK: 5-O-Methylbiochanin A, a New Isoflavone from *Echinospartum horridum*. Phytochemistry **17**, 829 (1978).

304. GREGSON, M., W. D. OLLIS, B. T. REDMAN, I. O. SUTHERLAND, H. H. DIETRICHS, and O. R. GOTTLIEB: Obtusastyrene and Obtustyrene, Cinnamylphenols from *Dalbergia retusa*. Phytochemistry 17, 1395 (1978).

305. GREGSON, M., W. D. OLLIS, I. O. SUTHERLAND, O. R. GOTTLIEB, and M. T. MAGALHÃES: Violastyrene and Isoviolastyrene, Cinnamylphenols from *Dalbergia miscolobium*. Phytochemistry 17, 1375 (1978).

306. GROSS, D.: Phytoalexins and Related Plant Substances. Fortschr. Chem. Org. Naturstoffe 34, 187 (1977).

307. GUGGOLZ, J., A. L. LIVINGSTON, and E. M. BICKOFF: Detection of Daidzein, Formononetin, Genistein and Biochanin A in Forages. J. Agric. Fd. Chem. 9, 330 (1961).

308. GUIMARÃES, I. S. DE S., O. R. GOTTLIEB, C. H. ANDRADE, and M. T. MAGALHÃES: Flavonoids from *Dalbergia cearensis*. Phytochemistry 14, 1452 (1975).

309. GUPTA, G. K., K. L. DHAR, and C. K. ATAL: Isolation and Constitution of Corylidin: a New Coumestan from the Fruits of *Psoralea corylifolia*. Phytochemistry 16, 403 (1977).

310. — — — Corylinal: a New Isoflavone from Seeds of *Psoralea corylifolia*. Phytochemistry 17, 164 (1978).

311. GUPTA, B. K., G. K. GUPTA, K. L. DHAR, and C. K. ATAL: A C-Formylated Chalcone from *Psoralea corylifolia*. Phytochemistry 19, 2034 (1980).

312. — — — — Psoralidin Oxide, a Coumestan from the Seeds of *Psoralea corylifolia*. Phytochemistry 19, 2232 (1980).

313. GUPTA, V. N., and T. R. SESHADRI: Special Chemical Components of Commercial Woods and Related Plant Materials: *Pterocarpus indicus*. J. Scient. Ind. Res. 15B, 146 (1956).

313a. GUSTINE, D. L.: Evidence for Sulfhydryl Involvement in Regulation of Phytoalexin Accumulation in *Trifolium repens* Callus Tissue Cultures. Pl. Physiol., Lancaster 68, 1323 (1981).

314. GUSTINE, D. L., R. T. SHERWOOD, and C. P. VANCE: Regulation of Phytoalexin Synthesis in Jackbean Callus Cultures. Stimulation of Phenylalanine Ammonia-Lyase and o-Methyltransferase. Pl. Physiol., Lancaster 61, 226 (1978).

315. GYÖRGY, P., K. MURATA, and H. IKEHATA: Antioxidants Isolated from Fermented Soybeans (Tempeh). Nature, Lond. 203, 870 (1964).

316. HAGEMANN, J. W., M. B. PEARL, J. J. HIGGINS, N. E. DELFEL, and F. R. EARLE: Rotenone and Deguelin in *Tephrosia vogelii* at Several Stages of Maturity. J. Agric. Fd. Chem. 20, 906 (1972).

317. HARBORNE, J. B.: Unpublished Results.

318. — Chemosystematics of the Leguminosae. Flavonoid and Isoflavonoid Patterns in the Tribe Genisteae. Phytochemistry 8, 1449 (1969).

319. HARBORNE, J. B., O. R. GOTTLIEB, and M. T. MAGALHÃES: Occurrence of the Isoflavone Afrormosin in Cabreuva Wood. J. Org. Chem. 28, 881 (1963).

320. HARBORNE, J. B., and J. L. INGHAM: Biochemical Aspects of the Coevolution of Higher Plants with Their Fungal Parasites. In: J. B. HARBORNE, Biochemical Aspects of Plant and Animal Coevolution, p. 343. London: Academic Press. 1978.

321. HARBORNE, J. B., J. L. INGHAM, L. KING, and M. PAYNE: The Isopentenyl Isoflavone Luteone as a Pre-Infectional Antifungal Agent in the Genus *Lupinus*. Phytochemistry 15, 1485 (1976).

322. HARGREAVES, J. A., J. W. MANSFIELD, and D. T. COXON: Identification of Medicarpin as a Phytoalexin in the Broad Bean Plant *(Vicia faba)*. Nature, Lond. 262, 318 (1976).

323. HARGREAVES, J. A., J. W. MANSFIELD, and S. ROSSALL: Changes in Phytoalexin Concentrations in Tissues of the Broad Bean Plant *(Vicia faba)* Following Inoculation with Species of *Botrytis*. Physiol. Pl. Path. 11, 227 (1977).

324. HARPER, S. H.: The Isolation of l-Elliptone from *Derris elliptica*. J. Chem. Soc. 1939, 1099.

325. HARPER, S. H.: The Isolation of Malaccol from *Derris malaccensis.* J. Chem. Soc. **1940**, 309.

326. — The Active Principles of Leguminous Fish-Poison Plants. *Derris malaccensis* and *Tephrosia toxicaria.* J. Chem. Soc. **1940**, 1178.

327. HARPER, S. H., A. D. KEMP, W. G. E. UNDERWOOD, and R. V. M. CAMPBELL: Pterocarpanoid Constituents of the Heartwoods of *Pericopsis angolensis* and *Swartzia madagascariensis.* J. Chem. Soc. C **1969**, 1109.

328. HARPER, S. H., D. B. SHIRLEY, and D. A. TAYLOR: Isoflavones from *Xanthocercis zambesiaca.* Phytochemistry **15**, 1019 (1976).

329. HARPER, S. H., and W. G. E. UNDERWOOD: The Active Principles of Leguminous Fish-Poison Plants. Toxicarol Isoflavone. J. Chem. Soc. **1965**, 4203.

330. HASEGAWA, M.: The Flavonoids in the Wood of *Prunus aequinoctialis, P. nipponica, P. maximowiczii* and *P. avium.* J. Am. Chem. Soc. **79**, 1738 (1957).

331. HASEGAWA, M., and T. SHIRATO: The Flavonoids in the Wood of *Prunus verecunda.* J. Am. Chem. Soc. **79**, 450 (1957).

332. HAYASHI, T., and R. H. THOMSON: Isoflavones from *Dipteryx odorata.* Phytochemistry **13**, 1943 (1974).

333. HAYASHI, Y., T. SHIRATO, K. SAKURAI, and T. TAKAHASHI: Isoflavonoids from the Heartwood of *Millettia pendula.* J. Japan Wood Res. Soc. **24**, 898 (1978).

334. HAZATO, T., H. NAGANAWA, M. KUMAGAI, T. AOYAGI, and H. UMEZAWA: β-Galactosidase-Inhibiting New Isoflavonoids Produced by Actinomycetes. J. Antibiot., Tokyo **32**, 217 (1979).

335. HÁZNAGY, A., G. TÓTH, and J. TAMÁS: Secondary Metabolites from Aqueous Extracts of *Ononis spinosa.* Arch. Pharm., Weinheim **311**, 318 (1978).

336. HERTELENDY, F., and R. H. COMMON: Isolation and Identification of Equol from the Urine of the Domestic Fowl. Poult. Sci. **43**, 954 (1964).

337. HIGGINS, V. J.: Induced Conversion of the Phytoalexin Maackiain to Dihydromaackiain by the Alfalfa Pathogen *Stemphylium botryosum.* Physiol. Pl. Path. **6**, 5 (1975).

338. — The Effect of some Pterocarpanoid Phytoalexins on Germ Tube Elongation of *Stemphylium botryosum.* Phytopathology **68**, 339 (1978).

338a. — Demethylation of the Phytoalexin Pisatin by *Stemphylium botryosum.* Can. J. Bot. **59**, 547 (1981).

339. HIGGINS, V. J., and J. L. INGHAM: Demethylmedicarpin, a Product Formed from Medicarpin by *Colletotrichum coccodes.* Phytopathology **71**, 800 (1981).

340. HIGGINS, V. J., and D. G. SMITH: Separation and Identification of Two Pterocarpanoid Phytoalexins Produced by Red Clover Leaves. Phytopathology **62**, 235 (1972).

341. HIGGINS, V. J., A. STOESSL, and M. C. HEATH: Conversion of Phaseollin to Phaseollinisoflavan by *Stemphylium botryosum.* Phytopathology **64**, 105 (1974).

342. HIGHET, R. J., and P. F. HIGHET: The Structure of Two Isoflavones from the Abyssinian Berebera Tree. J. Org. Chem. **32**, 1055 (1967).

343. HIJWEGEN, T.: Induced Pterocarpan Formation in Two *Melilotus* Species. Neth. J. Pl. Path. **83**, 161 (1977).

344. HLASIWETZ, H.: Constituents of *Ononis spinosa* Root. J. Prakt. Chem. **65**, 419 (1855).

345. HÖRHAMMER, L., and H. WAGNER: Isolation of the Oestrogenic 4',5,7-Trihydroxy-isoflavone 7-Glucoside (Genistin) from Multifoliated Lupin *(Lupinus polyphyllus)* and Broom *(Sarothamnus scoparius).* Arzneimittel-Forsch. **12**, 1002 (1962).

346. HÖRHAMMER, L., H. WAGNER, and H. GRASMAIER: Isolation of an Oestrogenic Isoflavone from *Lupinus polyphyllus.* Naturwissenschaften **45**, 388 (1958).

347. HÖSEL, W., and W. BARZ: Flavonoids of *Cicer arietinum.* Phytochemistry **9**, 2053 (1970).

348. HUDSON, A. T., and R. BENTLEY: The Isolation of Isoflavonoids from Bacteria. J. Chem. Soc. Chem. Commun. **1969**, 830.

348a. IBRAHIM, G.: Investigations on *Fusarium oxysporum* Root Rot of Bean. Ph. D. Thesis, University of Reading, U. K. 1978.

349. IKEHATA, H., M. WAKAIZUMI, and K. MURATA: Antioxidant and Antihemolytic Activity of a New Isoflavone, "Factor 2" Isolated from Tempeh. Agric. Biol. Chem. **32**, 740 (1968).

350. IMAMURA, H., Y. HIBINO, and H. OHASHI: New Isoflavonoids from the Bark of *Cladrastis platycarpa*. J. Japan Wood Res. Soc. **18**, 325 (1972).

351. — — — Structures of Two New Isoflavones, Fujikinetin and Fujikinin from the Bark of *Cladrastis platycarpa*. J. Japan Wood Res. Soc. **19**, 293 (1973).

352. — — —New Isoflavone Glucosides from the Bark of *Cladrastis platycarpa*. Phytochemistry **13**, 757 (1974).

353. IMAMURA, H., Y. HIBINO, H. OHTA, and H. OHASHI: Further New Isoflavone Glycosides from the Bark of *Cladrastis platycarpa*. J. Japan Wood Res. Soc. **21**, 257 (1975).

354. IMAMURA, H., M. ITO, and H. OHASHI: Isoflavonoids of *Erythrina crista-galli* (Leguminosae). Res. Bull. Fac. Agric. Gifu Univ. **1981**, 77.

355. IMAMURA, H., Y. TANNO, and T. TAKAHASHI: Isolation and Identification of Four Isoflavone Derivatives from Eurasian Teak *(Pericopsis* sp.*)*. J. Japan Wood Res. Soc. **14**, 295 (1968).

356. INGHAM, J. L.: Unpublished Results.

357. — Phytoalexins and Other Natural Products as Factors in Plant Disease Resistance. Bot Rev. **38**, 343 (1972).

358. — Disease Resistance in Higher Plants. The Concept of Pre-Infectional and Post-Infectional Resistance. Phytopath. Z. **78**, 314 (1973).

359. — A Comparative Study of Phytoalexins from the Leguminosae. Ph. D. Thesis, University of Reading, U. K. 1976.

360. — Fungal Modification of Pterocarpan Phytoalexins from *Melilotus alba* and *Trifolium pratense*. Phytochemistry **15**, 1489 (1976).

361. — Induced and Constitutive Isoflavonoids from Stems of Chickpeas *(Cicer arietinum)* Inoculated with Spores of *Helminthosporium carbonum*. Phytopath. Z. **87**, 353 (1976).

362. — Isosativan: an Isoflavan Phytoalexin from *Trifolium hybridum* and Other *Trifolium* Species. Z. Naturforsch. **31c**, 331 (1976).

363. — Induced Isoflavonoids from Fungus-Infected Stems of Pigeon Pea *(Cajanus cajan)*. Z. Naturforsch. **31c**, 504 (1976).

364. — Isoflavan Phytoalexins from *Anthyllis, Lotus* and *Tetragonolobus*. Phytochemistry **16**, 1279 (1977).

365. — An Isoflavan Phytoalexin from Leaves of *Glycyrrhiza glabra*. Phytochemistry **16**, 1457 (1977).

366. — Medicarpin as a Phytoalexin of the Genus *Melilotus*. Z. Naturforsch. **32c**, 449 (1977).

367. — Phytoalexins of Hyacinth Bean *(Lablab niger)*. Z. Naturforsch. **32c**, 1018 (1977).

368. — Phaseollidin, a Phytoalexin of *Psophocarpus tetragonolobus*. Phytochemistry **17**, 165 (1978).

369. — Isoflavonoid and Stilbene Phytoalexins of the Genus *Trifolium*. Biochem. Syst. Ecol. **6**, 217 (1978).

370. — Flavonoid and Isoflavonoid Compounds from Leaves of Sainfoin *(Onobrychis viciifolia)*. Z. Naturforsch. **33c**, 146 (1978).

371. — Isoflavonoid Phytoalexins of the Genus *Medicago*. Biochem. Syst. Ecol. **7**, 29 (1979).

372. — A Revised Structure for the Phytoalexin Cajanol. Z. Naturforsch. **34c**, 159 (1979).

373. — Isoflavonoid Phytoalexins of *Parochetus communis* and *Factorovskya aschersoniana*. Z. Naturforsch. **34c**, 290 (1979).

374. — Phytoalexin Production by Species of the Genus *Caragana*. Z. Naturforsch. **34c**, 293 (1979).

375. INGHAM, J. L.: Phytoalexin Production by Flowers of Garden Pea *(Pisum sativum)*. Z. Naturforsch. **34c**, 296 (1979).
376. — Isoflavonoid Phytoalexins from Leaflets of *Dalbergia sericea*. Z. Naturforsch. **34c**, 630 (1979).
377. — Isoflavonoid Phytoalexins of Yam Bean *(Pachyrrhizus erosus)*. Z. Naturforsch. **34c**, 683 (1979).
378. — Induced Isoflavonoids of *Erythrina sandwicensis*. Z. Naturforsch. **35c**, 384 (1980).
379. — Phytoalexin Induction and its Taxonomic Significance in the Leguminosae (Subfamily Papilionoideae). In: R. M. POLHILL and P. H. RAVEN, Advances in Legume Systematics, p. 599. London: Her Majesty's Stationery Office. 1981.
380. — Phytoalexins from the Leguminosae. In: J. A. BAILEY and J. W. MANSFIELD, Phytoalexins, p. 21. Glasgow: Blackie and Son. 1982.
381. — A New Isoflavanone Phytoalexin from *Medicago rugosa*. Plant. Med.**45**, 46 (1982).
382. — Isolation and Identification of *Cicer* Isoflavonoids. Biochem. Syst. Ecol. **9**, 125 (1981).
383. — Phytoalexin Induction and Its Chemosystematic Significance in the Genus *Trigonella* (Leguminosae). Biochem. Syst. Ecol. **9**, 275 (1981).
384. INGHAM, J. L., and P. M. DEWICK: Isoflavonoid Phytoalexins from Leaves of *Trifolium arvense*. Z. Naturforsch. **32c**, 446 (1977).
385. — — 6-Demethylvignafuran as a Phytoalexin of *Tetragonolobus maritimus*. Phytochemistry **17**, 535 (1978).
386. — — A New Isoflavan Phytoalexin from Leaflets of *Lotus hispidus*. Phytochemistry **18**, 1711 (1979).
387. — — Astraciceran: a New Isoflavan Phytoalexin from *Astragalus cicer*. Phytochemistry **19**, 1767 (1980).
388. — — Isolation of a New Isoflavan Phytoalexin from Two *Lotus* Species. Phytochemistry **19**, 2799 (1980).
389. — — Sparticarpin: a Pterocarpan Phytoalexin from *Spartium junceum*. Z. Naturforsch. **35c**, 197 (1980).
390. INGHAM, J. L., and J. B. HARBORNE: Phytoalexin Induction as a New Dynamic Approach to the Study of Systematic Relationships Among Higher Plants. Nature, Lond. **260**, 241 (1976).
391. INGHAM, J. L., N. T. KEEN, and T. HYMOWITZ: A New Isoflavone Phytoalexin from Fungus-Inoculated Stems of *Glycine wightii*. Phytochemistry **16**, 1943 (1977).
392. INGHAM, J. L., N. T. KEEN, K. R. MARKHAM, and L. J. MULHEIRN: Dolichins A and B, Two New Pterocarpans from Bacteria-Treated Leaves of *Dolichos biflorus*. Phytochemistry **20**, 807 (1981).
393. INGHAM, J. L., N. T. KEEN, L. J. MULHEIRN, and R. L. LYNE: Inducibly-Formed Isoflavonoids from Leaves of Soybean *(Glycine max)*. Phytochemistry **20**, 795 (1981).
394. INGHAM, J. L., and K. R. MARKHAM: Identification of the *Erythrina* Phytoalexin Cristacarpin and a Note on the Chirality of other 6a-Hydroxypterocarpans. Phytochemistry **19**, 1203 (1980).
395. INGHAM, J. L., and R. L. MILLAR: Sativin: an Induced Isoflavan from the Leaves of *Medicago sativa*. Nature, Lond. **242**, 125 (1973).
396. INGHAM, J. L., and L. J. MULHEIRN: Isoflavonoid Phytoalexins from Fungus-Inoculated Leaves of *Apios tuberosa*. Phytochemistry. In press.
397. IRVINE, J. E., and R. H. FREYRE: Occurrence of Rotenoids in some Species of the Genus *Tephrosia*. J. Agric Fd. Chem. **7**, 106 (1959).
398. ISOGAI, Y., Y. KOMODA, and T. OKAMOTO: Plant Growth Regulators in the Pea Plant *(Pisum sativum)*. Chem. Pharm. Bull., Tokyo **18**, 1872 (1970).
399. JACKSON, B., P. J. OWEN, and F. SCHEINMANN: The Structure of Alpinumisoflavone, a New Pyranoisoflavone from *Laburnum alpinum*. J. Chem. Soc. C **1971**, 3389.

400. JAIN, A. C., G. K. GUPTA, and P. R. RAO: Isolation and Constitution of Corylin: a New Isoflavone from the Fruits of *Psoralea corylifolia*. Indian J. Chem. **12**, 659 (1974).

401. JAIN, A. C., and S. KOUL: Isolation of Biochanin A from the Heartwood of *Sophora mollis*. Curr. Sci. **41**, 414 (1972).

402. JAIN, A. C., A. KUMAR, and A. K. KOHLI: Recent Developments in the Chemistry of Rotenoids (Derronoids). J. Scient. Ind. Res. **37**, 606 (1978).

403. JAIN, A. C., and D. K. TULI: Pterocarpanoids — Recent Developments in their Chemistry. J. Scient. Ind. Res. **37**, 287 (1978).

404. JAIN, S. C.: Unpublished Results.

405. JOHNSON, A. P., and A. PELTER: The Structure of Robustic Acid, a New 4-Hydroxy-3-phenylcoumarin. J. Chem. Soc. C **1966**, 606.

406. JOHNSON, A. P., A. PELTER, and P. STAINTON: The Structures of Scandenin and Lonchocarpic Acid. J. Chem. Soc. C **1966**, 192.

407. JONES, H. A.: Rotenone in a Species of *Spatholobus*. J. Am. Chem. Soc. **55**, 1737 (1933).

408. — Notes on the Occurrence of Rotenone in Species of *Derris* and *Lonchocarpus*. J. Wash. Acad. Sci. **23**, 493 (1933).

409. — Lonchocarpic Acid, a New Compound from a Species of *Lonchocarpus*. J. Am. Chem. Soc. **56**, 1247 (1934).

410. — A List of Plants Reported to Contain Rotenone or Rotenoids. Bur. Entomol. Pl. Quarant., U. S. Dept. Agric., Mon. E-571 (1942).

411. JOSHI, B. S., and V. N. KAMAT: Tuberosin, a New Pterocarpan from *Pueraria tuberosa*. J. Chem. Soc. Perkin Trans. I **1973**, 907.

412. JURD, L.: A Phenolic Isoflav-3-ene from *Gliricidia sepium*. Tetrahedron Lett. **1976**, 1741.

413. JURD, L., and G. D. MANNERS: Isoflavene, Isoflavan and Flavonoid Constituents of *Gliricidia sepium*. J. Agric. Fd. Chem. **25**, 723 (1977).

414. JURD, L., K. STEVENS, and G. MANNERS: Isoflavones of the Heartwood of *Dalbergia retusa*. Phytochemistry **11**, 2535 (1972).

415. KALLA, A. K., M. K. BHAN, and K. L. DHAR: A New Isoflavone from *Iris kumaonensis*. Phytochemistry **17**, 1441 (1978).

416. KALRA, A. J., M. KRISHNAMURTI, and M. NATH: Chemical Investigation of Indian Yam Beans *(Pachyrrhizus erosus)*: Isolation and Structures of Two New Rotenoids and a New Isoflavanone, Erosenone. Indian J. Chem. **15B**, 1084 (1977).

417. KAMAL, R., and S. C. JAIN: *Tephrosia falciformis* — a New Source of Rotenoids. Planta Med. **33**, 418 (1978).

418. — — Occurrence of Rotenoids in *Tephrosia strigosa*. Agric. Biol. Chem. **44**, 2985 (1980).

419. KAMAT, V. S., F. Y. CHUO, I. KUBO, and K. NAKANISHI: Antimicrobial Agents from an East African Medicinal Plant, *Erythrina abyssinica*. Heterocycles **15**, 1163 (1981).

420. KANEDA, M., T. SAITOH, Y. IITAKA, and S. SHIBATA: Structure of Licoricone, a New Isoflavone from Licorice Root. Chem. Pharm. Bull., Tokyo **21**, 1338 (1973).

421. KAPLAN, D. T., N. T. KEEN, and I. J. THOMASON: Studies on the Mode of Action of Glyceollin in Soybean Incompatibility to the Root Knot Nematode, *Meloidogyne incognita*. Physiol. Pl. Path. **16**, 319 (1980).

422. KAPOOR, A. L., A. AEBI, and J. BÜCHI: Constituents of *Piscidia erythrina*. Helv. Chim. Acta **40**, 1574 (1957).

423. KARIYONE, T., K. ATSUMI, and M. SHIMADA: The Constituents of *Millettia taiwaniana*. J. Pharm. Soc. Japan **1923**, 739; Chem. Abstr. **18**, 408 (1924).

424. KASYMOV, A. U., E. S. KONDRATENKO, and N. K. ABUBAKIROV: Amorphigenin β-D-Glucoside from *Amorpha*. Khim. Prir. Soedin. **1968**, 326.

425. — — — Dihydroamorphigenin from the Seeds of *Amorpha fruticosa*. Khim. Prir. Soedin. **1972**, 115.

426. — — — Structure of Amorphol, a Rotenoid Bioside from Plants of the Genus *Amorpha*. Khim. Prir. Soedin. **1974**, 464.

427. KASYMOV, A. U., E. S. KONDRATENKO, Y. V. RASHKES, and N. K. ABUBAKIROV: Amorphigenol β-D-Glucopyranoside from *Amorpha*. Khim. Prir. Soedin. **1970**, 197.
428. KATTAEV, N. S., I. A. KHARLAMOV, N. M. AKHMEDKHODZHAEVA, G. K. NIKONOV, and K. K. KHALMATOV: Trifoside, an Isoflavone from *Trifolium pratense*. Khim. Prir. Soedin. **1972**, 806.
429. KATTAEV, N. S., and G. K. NIKONOV: Flavonoids of *Thermopsis alterniflora*. Khim. Prir. Soedin. **1972**, 648.
430. — — Flavonoids of *Glycyrrhiza glabra*. Khim. Prir. Soedin. **1974**, 93.
431. — — Flavonoids of *Thermopsis alterniflora*. Khim. Prir. Soedin. **1975**, 140.
432. KATTAEV, N. S., G. K. NIKONOV, and Y. V. RASHKES: Sphaerosin and Sphaerosinin, New Isoflavans from *Sphaerophysa salsula*. Khim. Prir. Soedin. **1975**, 147.
433. KAWASE, A., N. OHTA, and K. YAGISHITA: On the Chemical Structure of a New Isoflavone Glucoside, Homotectoridin, Isolated together with Tectoridin from the Rhizomes of *Iris germanica*. Agric. Biol. Chem. **37**, 145 (1973).
434. KAZAKOV, A. L.: Biochanin A and its Glucoside in some Species of Clover. Khim. Prir. Soedin. **1973**, 274.
435. — Prunitrin from *Trifolium medium* and *T. pratense*. Khim. Prir. Soedin. **1976**, 538.
436. KAZAKOV, A. L., E. A. KECHATOV, and V. M. CHEMERKO: Isoflavones of the Oil Cake of the Seeds of *Glycine hispida*. Khim. Prir. Soedin. **1975**, 256.
437. KAZAKOV, A. L., V. A. KOMPANTSEV, and T. P. LEONT'EVA: Flavonoids of *Cicer arietinum*. Khim. Prir. Soedin. **1980**, 721.
438. KAZAKOV, A. L., A. L. SHINKARENKO, and E. T. OGANESYAN: Ononin and Formononetin from Representatives of the Genus *Trifolium*. Khim. Prir. Soedin. **1972**, 804.
439. KEEN, N. T.: Unpublished Results.
440. — Accumulation of Wyerone in Broad Bean and Demethylhomopterocarpin in Jack Bean after Inoculation with *Phytophthora megasperma* var. *sojae*. Phytopathology **62**, 1365 (1972).
441. — The Isolation of Phytoalexins from Germinating Seeds of *Cicer arietinum*, *Vigna sinensis*, *Arachis hypogaea*, and other Plants. Phytopathology **65**, 91 (1975).
442. KEEN, N. T., and R. HORSCH: Hydroxyphaseollin Production by Various Soybean Tissues: a Warning Against Use of "Unnatural" Host-Parasite Systems. Phytopathology **62**, 439 (1972).
443. KEEN, N. T., and J. L. INGHAM: Phytoalexins from *Dolichos biflorus*. Z. Naturforsch. **35c**, 923 (1980).
444. KEEN, N. T., and B. W. KENNEDY: Hydroxyphaseollin and Related Isoflavonoids in the Hypersensitive Resistance Reaction of Soybeans to *Pseudomonas glycinea*. Physiol. Pl. Path. **4**, 173 (1974).
445. KEEN, N. T., A. I. ZAKI, and J. J. SIMS: Biosynthesis of Hydroxyphaseollin and Related Isoflavonoids in Disease-Resistant Soybean Hypocotyls. Phytochemistry **11**, 1031 (1972).
446. KENNY, T. S., A. ROBERTSON, and S. W. GEORGE: The Synthesis of Dehydrotetrahydrosumatrol. J. Chem. Soc. **1939**, 1601.
447. KHANNA, P.: Unpublished Results.
448. KHASTGIR, H. N., P. C. DUTTAGUPTA, and P. SENGUPTA: The Structure of Psoralidin. Tetrahedron **14**, 275 (1961).
449. KHERA, U., and S. S. CHIBBER: Tectorigenin 7-Gentiobioside from *Dalbergia volubilis* Stem Bark. Phytochemistry **17**, 596 (1978).
450. — — Chemical Constituents of *Dalbergia volubilis:* Isolation of Cearoin and (+)-Medicarpin. Indian J. Chem. **16B**, 78 (1978).
451. — — Isolation of the 4'-*O*-Gentiobioside and 4'-*O*-Glucoside of 7-*O*-Methyltectorigenin from *Dalbergia volubilis*. Indian J. Chem. **16B**, 641 (1978).

452. KING, F. E., C. B. COTTERILL, D. H. GODSON, L. JURD, and T. J. KING: Colourless Constituents of *Pterocarpus* Species. J. Chem. Soc. **1953**, 3693.
453. KING, F. E., M. F. GRUNDON, and K. G. NEILL: Constituents of the Heartwood of *Ferreirea spectabilis.* J. Chem. Soc. **1952**, 4580.
454. KING, F. E., and L. JURD: The Isolation of 5:4′-Dihydroxy-7-methoxyisoflavone (Prunetin) from the Heartwood of *Pterocarpus angolensis* and a Synthesis of 7:4′-Dihydroxy-5-methoxyisoflavone Hitherto Known as Prunusetin. J. Chem. Soc. **1952**, 3211.
455. KING, F. E., T. J. KING, and A. J. WARWICK: Constituents of Muninga, the Heartwood of *Pterocarpus angolensis.* A 6:4′-Dihydroxy-5:7-dimethoxyisoflavone (Muningin). J. Chem. Soc. **1952**, 96.
456. — — — Constituents of Muninga, the Heartwood of *Pterocarpus angolensis.* B: 2:4-Dihydroxyphenyl 1-*p*-Methoxyphenylethyl Ketone (Angolensin). J. Chem. Soc. **1952**, 1920.
457. KING, F. E., and K. G. NEILL: The Constitution of Ferreirin and of Homoferreirin. J. Chem. Soc. **1952**, 4752.
458. KINOSHITA, T., T. SAITOH, and S. SHIBATA: The Occurrence of an Isoflavene and the Corresponding Isoflavone in Licorice Root. Chem. Pharm. Bull., Tokyo **24**, 991 (1976).
459. — — — A New 3-Arylcoumarin from Licorice Root. Chem. Pharm. Bull., Tokyo **26**, 135 (1978).
460. — — — A New Isoflavone from Licorice Root. Chem. Pharm. Bull., Tokyo **26**, 141 (1978).
461. KIRKIACHARIAN, B. S., and A. RAVISÉ: Synthesis and Biological Properties of (±)-*O*-Methylsativan. Phytochemistry **15**, 907 (1976).
462. KLYNE, W., and A. A. WRIGHT: Steroids and other Lipids of Pregnant Goat's Urine. Biochem. J. **66**, 92 (1957).
463. — — Steroids and other Lipids of Pregnant Cow's Urine. J. Endocr. **18**, 32 (1959).
464. KNUCKLES, B. E., D. DEFREMERY, and G. O. KOHLER: Coumestrol Content of Fractions Obtained During Wet Processing of Alfalfa. J. Agric. Fd. Chem. **24**, 1177 (1976).
465. KODAMA, T., T. YAMAKAWA, and Y. MINODA: Rotenoid Biosynthesis by Tissue Cultures of *Derris elliptica.* Agric. Biol. Chem. **44**, 2387 (1980).
466. KOIKE, H.: Pharmacological Investigation on the Flavone Compounds, Particularly on their Diuretic Activity. Folia Pharmac. Jap. **12**, 89 (1931).
467. KOJIMA, R., S. FUKUSHIMA, A. UENO, and Y. SAIKI: Antitumor Activity of Constituents of *Sophora subprostrata.* Chem. Pharm. Bull., Tokyo **18**, 2555 (1970).
467a. KOMATSU, M., Y. ICHIRO, and Y. SHIRATAKI: Constituents of the Root of *Sophora franchetiana* (2). Chem. Pharm. Bull., Tokyo **29**, 2069 (1981).
468. KOMATSU, M., T. TOMIMORI, K. HATAYAMA, and Y. MAKIGUCHI: Constituents of the Root of *Sophora subprostrata* (3). J. Pharm. Soc. Japan **90**, 459 (1970).
469. KOMATSU, M., I. YOKOE, and Y. SHIRATAKI: Constituents of the Root of *Sophora japonica* (1). J. Pharm. Soc. Japan **96**, 254 (1976).
470. — — — Constituents of the Aerial Parts of *Sophora tomentosa* (1). Chem. Pharm. Bull., Tokyo **26**, 1274 (1978).
471. — — — Constituents of the Aerial Parts of *Sophora tomentosa* (2). Chem. Pharm. Bull., Tokyo **26**, 3863 (1978).
472. — — — Constituents of the Root of *Sophora franchetiana* (1). Chem. Pharm. Bull., Tokyo **29**, 532 (1981).
473. KOMATSU, M., I. YOKOE, Y. SHIRATAKI, and J. CHEN: Trifolirhizin 6′-Monoacetate, a New Glycoside from the Roots of *Sophora subprostrata.* Phytochemistry **15**, 1089 (1976).
474. KÖNIG, W. A., C. KRAUSS, and H. ZÄHNER: Metabolites from Micro-Organisms. 6-Chlorogenistein and 6,3′-Dichlorogenistein. Helv. Chim. Acta **60**, 2071 (1977).

475. KOVALEV, V. M.: Flavonoids of *Ononis arvensis*. Farmat. Zh., Kiev **30**, 93 (1975); Chem. Abstr. **83**, 128662 (1975).

476. KOVALEV, V. N., M. I. BORISOV, and V. N. SPIRIDONOV: Phenolic Compounds of *Ononis arvensis*. I. Khim. Prir. Soedin. **1974**, 795.

477. KOVALEV, V. N., M. I. BORISOV, V. N. SPIRIDONOV, I. P. KOVALEV, and V. G. GORDIENKO: Phenolic Compounds of *Ononis arvensis*. III. Khim. Prir. Soedin. **1976**, 104.

478. KOVALEV, V. N., V. N. SPIRIDONOV, M. I. BORISOV, I. P. KOVALEV, V. G. GORDIENKO, and D. D. KOLESNIKOV: Phenolic Compounds of *Ononis arvensis*. The Structure of Onogenin. Khim. Prir. Soedin. **1975**, 354.

479. KRISHNA, S., and T. P. GHOSE: Indian *Tephrosia* sp. as a Source of Rotenone. Curr. Sci. **6**, 454 (1938).

480. KRISHNAMOORTHY, V., N. R. KRISHNASWAMY, and T. R. SESHADRI: Myriconol from the Stem Bark of *Myrica nagi*. Curr. Sci. **32**, 16 (1963).

481. KRISHNAMURTI, M., Y. R. SAMBHY, and T. R. SESHADRI: Chemical Study of Indian Yam Beans *(Pachyrrhizus erosus)*. Isolation of Two New Rotenoids: 12a-Hydroxydolineone and 12a-Hydroxypachyrrhizone. Tetrahedron **26**, 3023 (1970).

482. KRISHNAMURTY, H. G., and J. S. PRASAD: Isoflavones of *Moghania macrophylla*. Phytochemistry **19**, 2797 (1980).

483. KRUKOFF, B. A., and A. C. SMITH: Rotenone-Yielding Plants of South America. Am J. Bot. **24**, 573 (1937).

484. KRUPADANAM, G. L. D., P. N. SARMA, G. SRIMANNARAYANA, and N. V. S. RAO: New C-6 Oxygenated Rotenoids from *Tephrosia villosa* — Villosin, Villosone, Villol and Villinol. Tetrahedron Lett. **1977**, 2125.

485. KUBO, M.: Unpublished Results.

486. KUBO, M., K. FUJITA, H. NISHIMURA, S. NARUTO, and K. NAMBA: A New Irisolidone-7-O-glucoside and Tectoridin from *Pueraria* Species. Phytochemistry **12**, 2547 (1973).

487. KUBO, M., T. ODANI, S. HOTTA, S. ARICHI, and K. NAMBA: Isolation of an Antibacterial Compound from Honggi *(Hedysarum polybotrys)*. Syôyakugaku Zasshi **31**, 82 (1977).

488. KUBO, M., M. SASAKI, K. NAMBA, S. NARUTO, and H. NISHIMURA: Isolation of a New Isoflavone from Chinese *Pueraria* Flowers. Chem. Pharm. Bull., Tokyo **23**, 2449 (1975).

489. KUHN, P. J., D. A. SMITH, and D. F. EWING: 5,7,2′,4′-Tetrahydroxy-8-(3″-hydroxy-3″-methyl-butyl)isoflavanone, a Metabolite of Kievitone Produced by *Fusarium solani* f. sp. *phaseoli*. Phytochemistry **16**, 296 (1977).

490. KUKLA, A. S., and T. R. SESHADRI: Constitution and Synthesis of Maxima Isoflavones-A and -B. Tetrahedron **18**, 1443 (1962).

491. KURIHARA, T., and M. KIKUCHI: On the Components of Flowers of *Pueraria thunbergiana* 1. J. Pharm. Soc. Japan **93**, 1201 (1973).

492. — — On the Components of Flowers of *Pueraria thunbergiana* 2. Isolation of a New Isoflavone Glycoside. J. Pharm. Soc. Japan **95**, 1283 (1975).

493. — — On the Components of Flowers of *Pueraria thunbergiana* 3. J. Pharm. Soc. Japan **96**, 1486 (1976).

494. — — On the Components of Flowers of *Cytisus scoparius*. J. Pharm. Soc. Japan **100**, 1054 (1980).

495. KUROSAWA, K., W. D. OLLIS, B. T. REDMAN, I. O. SUTHERLAND, A. BRAGA DE OLIVEIRA, O. R. GOTTLIEB, and H. M. ALVES: The Natural Occurrence of Isoflavans and an Isoflavanquinone. J. Chem. Soc. Chem. Commun. **1968**, 1263.

496. KUROSAWA, K., W. D. OLLIS, B. T. REDMAN, I. O. SUTHERLAND, and O. R. GOTTLIEB: Vestitol and Vesticarpan, Isoflavonoids from *Machaerium vestitum*. Phytochemistry **17**, 1413 (1978).

497. KUROSAWA, K., W. D. OLLIS, I. O. SUTHERLAND, and O. R. GOTTLIEB: Variabilin, a 6a-Hydroxypterocarpan from *Dalbergia variabilis*. Phytochemistry **17**, 1417 (1978).

498. KUROSAWA, K., W. D. OLLIS, I. O. SUTHERLAND, O. R. GOTTLIEB, and A. BRAGA DE OLIVEIRA: Mucronustyrene, Mucronulastyrene and Villostyrene, Cinnamylphenols from *Machaerium mucronulatum* and *M. villosum*. Phytochemistry **17**, 1389 (1978).

499. — — — — — Mucronulatol, Mucroquinone and Mucronucarpan, Isoflavonoids from *Machaerium mucronulatum* and *M. villosum*. Phytochemistry **17**, 1405 (1978).

500. KUTANI, N., and K. HAYASHI: Glycosides of the Rhizomes of *Belamcanda chinensis*. J. Pharm. Soc. Japan **64**, 16 (1944).

501. KYOGOKU, K., K. HATAYAMA, and M. KOMATSU: Constituents of the Chinese Crude Drug "Kushen" (the Root of *Sophora flavescens*). Isolation of Five New Flavonoids and Formononetin. Chem. Pharm. Bull., Tokyo **21**, 2733 (1973).

502. KYOGOKU, K., K. HATAYAMA, K. SUZUKI, S. YOKOMORI, K. MAEJIMA, and M. KOMATSU: Studies on the Constituents of Guang-Dou-Gen (the Root of *Sophora subprostrata*). Isolation of Two New Flavanones and Daidzein. Chem. Pharm. Bull., Tokyo **21**, 1436 (1973).

503. LAFORGE, F. B., and H. L. HALLER: The Nature of the Alkali Soluble Hydrogenation Products of Rotenone and its Derivatives and their Bearing on the Structure of Rotenone. J. Am. Chem. Soc. **54**, 810 (1932).

504. LAMAN, N. A.: Physiological Activity of a Total Preparation of Isoflavone Glycosides Isolated from Yellow Lupin *(Lupinus luteus)*. Soviet Pl. Physiol. **21**, 244 (1974).

505. — Flavonoid Aglycones of *Lupinus luteus*. Khim. Prir. Soedin. **1975**, 252.

506. LAMAN, N. A., and A. P. VOLYNETS: Isoflavones of the Root of *Lupinus luteus*. Khim. Prir. Soedin. **1974**, 162.

507. — — Flavonoids in Ontogenesis of Yellow Lupin *(Lupinus luteus)*. Soviet Pl. Physiol. **21**, 604 (1974).

508. LAMPARD, J. F.: Demethylhomopterocarpin: an Antifungal Compound in *Canavalia ensiformis* and *Vigna unguiculata* Following Infection. Phytochemistry **13**, 291 (1974).

509. LAPPE, U., and W. BARZ: Degradation of Pisatin by Fungi of the Genus *Fusarium*. Z. Naturforsch. **33c**, 301 (1978).

510. LEBRETON, P., K. R. MARKHAM, W. T. SWIFT, OUNG-BORAN, and T. J. MABRY: Flavonoids of *Baptisia australis* (Leguminosae). Phytochemistry **6**, 1675 (1967).

511. LEITE DE ALMEIDA, M. E., and O. R. GOTTLIEB: Iso- and Neo-Flavonoids from *Dalbergia inundata*. Phytochemistry **13**, 751 (1974).

512. — — Further Isoflavones from *Pterodon apparicioi*. Phytochemistry **14**, 2716 (1975).

513. LEMESHEV, N. N., G. P. KUDRYAVTSEV, and A. P. VOLYNETS: Flavonoid Glycosides of some *Lupinus* Species. Khim. Prir. Soedin. **1980**, 569.

514. LETCHER, R. M., and I. M. SHIRLEY: Phenolic Compounds from the Heartwood of *Dalbergia nitidula*. Phytochemistry **15**, 353 (1976).

515. LI, K. N., and S. WANG: Isoshehkanin. Part II. The Linkage Between Hexose and Isoshehkangenin. J. Chinese Chem. Soc., Peiping **17**, 132 (1950); Chem. Abstr. **47**, 3306 (1953).

516. LINDAHL, P. E., and K. E. ÖBERG: The Effect of Rotenone on Respiration and its Point of Attack. Expl. Cell Res. **23**, 228 (1961).

517. LIVINGSTON, A. L.: Forage Plant Estrogens. J. Toxicol. Environ. Health **4**, 301 (1978).

518. LIVINGSTON, A. L., E. M. BICKOFF, R. E. LUNDIN, and L. JURD: Trifoliol, a New Coumestan from Ladino Clover. Tetrahedron **20**, 1963 (1964).

519. LIVINGSTON, A. L., S. C. WITT, R. E. LUNDIN, and E. M. BICKOFF: Medicagol, a New Coumestan from Alfalfa. J. Org. Chem. **30**, 2353 (1965).

520. LLOYD, D.: Inhibition of Electron Transport in *Prototheca zopfii*. Phytochemistry **5**, 527 (1966).

520a. LOOKHART, G. L., B. L. JONES, and K. F. FINNEY: Determination of Coumestrol in Soybeans by High-Performance Liquid and Thin-Layer Chromatography. Cereal Sci. **55**, 967 (1978).

521. LOPER, G. M.: Effect of Aphid Infestation on the Coumestrol Content of Alfalfa Varieties Differing in Aphid Resistance. Crop Sci. **8**, 104 (1968).

522. — Accumulation of Coumestrol in Barrel Medic *(Medicago littoralis)*. Crop Sci. **8**, 317 (1968).

523. LOPER, G. M., and C. H. HANSON: Influence of Controlled Environmental Factors and Two Foliar Pathogens on Coumestrol in Alfalfa. Crop Sci. **4**, 480 (1964).

524. LOPER, G. M., C. H. HANSON, and J. H. GRAHAM: Coumestrol Content of Alfalfa as Affected by Selection for Resistance to Foliar Diseases. Crop Sci. **7**, 189 (1967).

525. LUPI, A., F. D. DELLE MONACHE, G. B. MARINI-BETTOLO, D. L. B. COSTA, and I. L. DE ALBUQUERQUE: Abruquinones: New Natural Isoflavanquinones. Gazz. Chim. Ital. **109**, 9 (1979).

526. LYMAN, R. L., E. M. BICKOFF, A. N. BOOTH, and A. L. LIVINGSTON: Detection of Coumestrol in Leguminous Plants. Archs. Biochem. Biophys. **80**, 61 (1959).

527. LYNE, R. L., and L. J. MULHEIRN: Minor Pterocarpinoids of Soybean. Tetrahedron Lett. **1978**, 3127.

528. LYNE, R. L., L. J. MULHEIRN, and N. T. KEEN: Novel Pterocarpinoids from *Glycine* Species. Tetrahedron Lett. **1981**, 2483.

529. LYNE, R. L., L. J. MULHEIRN, and D. P. LEWORTHY: New Pterocarpinoid Phytoalexins of Soybean. J. Chem. Soc. Chem. Commun. **1976**, 497.

530. LYON, F. M., and R. K. S. WOOD: Production of Phaseollin, Coumestrol and Related Compounds in Bean Leaves Inoculated with *Pseudomonas* spp. Physiol. Pl. Path. **6**, 117 (1975).

531. MAEKAWA, E., and K. KITAO: Isolation of Pterocarpanoid Compounds as Heartwood Constituents of *Maackia amurensis* var. *buergeri*. Wood Res., Kyoto **1970**, 29.

532. MALHOTRA, A., V. V. S. MURTI, and T. R. SESHADRI: Lanceolarin, a New Isoflavone Glycoside of *Dalbergia lanceolaria*. Tetrahedron **23**, 405 (1967).

533. — — Chemical Components of *Dalbergia lanceolaria* (Flowers and Leaves). Curr. Sci. **36**, 484 (1967).

534. MANNERS, G. D., and L. JURD: Additional Flavonoids in *Gliricidia sepium*. Phytochemistry **18**, 1037 (1979).

535. MANNICH, C., P. SCHUMANN, and W. H. LIN: Shekanin (Tectoridin), a Glucoside from *Belamcanda chinensis (Pardanthus chinensis)*. Arch. Pharm., Berl. **275**, 317 (1937).

536. MARANDUBA, A., A. BRAGA DE OLIVEIRA, G. G. DE OLIVEIRA, J. E. DE P. REIS, and O. R. GOTTLIEB: Isoflavonoids from *Myroxylon peruiferum*. Phytochemistry **18**, 815 (1979).

537. MARKHAM, K. R., and J. L. INGHAM: Tectorigenin, a Phytoalexin of *Centrosema haitiense* and other *Centrosema* Species. Z. Naturforsch. **35c**, 919 (1980).

538. MARKHAM, K. R., and T. J. MABRY: The Identification of Twenty-Three 5-Deoxy and Ten 5-Hydroxy-Flavonoids from *Baptisia lecontei* (Leguminosae). Phytochemistry **7**, 791 (1968).

539. MARKHAM, K. R., T. J. MABRY, and T. W. SWIFT: New Isoflavones from the Genus *Baptisia* (Leguminosae). Phytochemistry **7**, 803 (1968).

540. — — — Distribution of Flavonoids in the Genus *Baptisia* (Leguminosae). Phytochemistry **9**, 2359 (1970).

541. MARKHAM, K. R., W. T. SWIFT, and T. J. MABRY: A New Isoflavone Glycoside from *Baptisia australis*. J. Org. Chem. **33**, 462 (1968).

542. MARRIAN, G. F., and D. BEALL: The Constitution of Equol. Biochem. J. **29**, 1586 (1935).

543. MARRIAN, G. F., and G. A. D. HASLEWOOD: Equol, a New Inactive Phenol Isolated from the Ketohydroxyoestrin Fraction of Mares' Urine. Biochem. J. **26**, 1227 (1932).

544. MARTIN, J. T.: Occurrence of Rotenone in *Tephrosia macropoda*. Nature, Lond. **137**, 1075 (1936).

545. MARTIN, M., and P. M. DEWICK: Biosynthesis of the 2-Arylbenzofuran Phytoalexin Vignafuran in *Vigna unguiculata*. Phytochemistry **18**, 1309 (1979).

546. Martinod, P., J. Hidalgo, C. Guevara, and M. Muñoz: Investigation of the Active Principles of Lonchocarpus utilis (Barbasco). Politecnica 3, 137 (1973).
547. Matos, F. J. de A., O. R. Gottlieb, and C. H. S. Andrade: Flavonoids from Dalbergia ecastophyllum. Phytochemistry 14, 825 (1975).
548. McGookin, A., A. Robertson, and W. B. Whalley: The Chemistry of the "Insoluble Red" Woods. Pterocarpin and Homopterocarpin. J. Chem. Soc. 1940, 787.
549. McMurry, T. B. H., and C. Y. Theng: The Constitution and Synthesis of Afromosin. J. Chem. Soc. 1960, 1491.
550. Meegan, M. J., and D. M. X. Donnelly: Isoflavonoids of Mildbraedeodendron excelsa. Phytochemistry 14, 2283 (1975).
551. Mehta, R.: Unpublished Results.
552. Meijer, T. M.: The Insecticidal Constituents of Pachyrrhizus erosus. I. Recl. Trav. Chim. Pays-Bas Belg. 65, 835 (1946).
553. — Chemical Constituents of Mundulea suberosa. I. Recl. Trav. Chim. Pays-Bas Belg. 66, 177 (1947).
553a. Menichini, F.: Unpublished Results.
554. Merz, K. W.: Toxic Constituents of the Seed of Tephrosia vogelii. Arch. Pharm., Berl. 270, 362 (1932).
555. Merz, K. W., and G. Schmidt: Toxic Principles in the Seed of Tephrosia vogelii. Arch. Pharm., Berl. 273, 1 (1935).
556. Meunier, A.: The Nature and Distribution of some Carbohydrates in Various Native Vicia Species. Bull. Sci. Pharmac. 43, 270 (1936).
557. Minhaj, N., H. Khan, S. K. Kapoor, and A. Zaman: Extractives of Millettia auriculata. Tetrahedron 32, 749 (1976).
558. Minhaj, N., H. Khan, A. Zaman, and F. M. Dean: Unanisoflavan, a New Isoflavan from Sophora secundiflora. Tetrahedron Lett. 1976, 2391.
559. Minhaj, N., K. Tasneem, K. Z. Khan, and A. Zaman: Secondifloran, a Novel Isoflavanone from Sophora secundiflora. Tetrahedron Lett. 1977, 1145.
560. Mitscher, L. A., A. Al-Shamma, T. Haas, P. B. Hudson, and Y. H. Park: A New Rotenoid, 11-Hydroxytephrosin from Amorpha fruticosa. Heterocycles 12, 1033 (1979).
561. Mitscher, L. A., Y. H. Park, and D. Clark: Antimicrobial Isoflavonoids and Related Substances from Glycyrrhiza glabra var. typica. J. Nat. Products (Lloydia) 43, 259 (1980).
562. Miyase, T., A. Ueno, T. Noro, and S. Fukushima: Studies on the Constituents of Lespedeza cyrtobotrya. The Structures of a New Chalcone and Two New Isoflav-3-enes. Chem. Pharm. Bull., Tokyo 28, 1172 (1980).
562a. — — — — The Structures of Haginin C, Haginin D and Lespedeol C. Chem. Pharm. Bull., Tokyo 29, 2205 (1981).
563. Moore, J. A., and S. Eng: Some New Constituents of Piscidia erythrina. J. Am. Chem. Soc. 78, 395 (1956).
564. Morgan, J. W. W., and R. J. Orsler: Isolation of 7-Methyltectorigenin from the Heartwood of Muninga (Pterocarpus angolensis). Chemy. Ind. 1967, 1173.
565. Morita, N., M. Arisawa, Y. Kondo, and T. Takemoto: The Constituents of Iris florentina (1). Chem. Pharm. Bull., Tokyo 21, 600 (1973).
566. Morita, N., M. Shimokoriyama, M. Shimizu, and M. Arisawa: The Components of the Rhizome of Iris tectorum (Iridaceae). 1. Chem. Pharm. Bull., Tokyo 20, 730 (1972).
567. — — — — The Components of the Rhizome of Iris tectorum (Iridaceae). 2. J. Pharm. Soc. Japan 92, 1052 (1972).
568. Muangnoicharoen, N., and A. W. Frahm: Arylbenzofurans from Dalbergia parviflora. Phytochemistry 20, 291 (1981).
569. Murakami, T., Y. Nishikawa, and T. Ando: On the Constituents of Pueraria Root (2). Chem. Pharm. Bull., Tokyo 8, 688 (1960).

570. Naim, M., B. Gestetner, A. Bondi, and Y. Birk: Antioxidative and Antihemolytic Activities of Soybean Isoflavones. J. Agric. Fd. Chem. **24**, 1174 (1976).

571. Naim, M., B. Gestetner, I. Kirson, Y. Birk, and A. Bondi: A New Isoflavone from Soya Beans. Phytochemistry **12**, 169 (1973).

572. Naim, M., B. Gestetner, S. Zilkah, Y. Birk, and A. Bondi: Soybean Isoflavones. Characterisation, Determination and Antifungal Activity. J. Agric. Fd. Chem. **22**, 806 (1974).

573. Nair, A. G. R., and S. S. Subramanian: Rothindin — a New Isoflavone Glucoside from *Rothia indica*. Indian J. Chem. **14B**, 801 (1976).

574. Nakano, T., J. Alonso, R. Grillet, and A. Martin: Isoflavonoids of the Bark of *Dipteryx odorata*. J. Chem. Soc. Perkin Trans. I **1979**, 2107.

575. Nakov, N., C. Achtardzhiev, M. Elnain-Voitashek, and Z. Kovalevski: Flavonoid Composition of *Genista ovata*. Probl. Farm. **6**, 65 (1978); Chem. Abstr. **89**, 193855 (1978).

576. Nakov, N., C. Achtardzhiev, and O. N. Tolkatschev: Flavonoid Composition of *Genista rumelica*. Part II. Farmatsiya, Sofia **28**, 27 (1978).

577. Nakov, N., T. Sarmova, and C. Achtardzhiev: Flavonoid Composition of *Genista rumelica*. Part I. Pharmazie **28**, 680 (1973).

578. Nakov, N., O. N. Tolkatschev, and C. Achtardzhiev: Flavonoid Composition of *Genista carinalis*. Pharmazie **33**, 463 (1978).

579. Narasimhachari, N., and T. R. Seshadri: A Note on the Components of the Bark of *Prunus puddum*. Proc. Indian Acad. Sci. **30A**, 271 (1949).

580. — — Components of the Bark of *Prunus puddum*. Padmakastin and Padmakastein. Proc. Indian Acad. Sci. **35A**, 202 (1952).

581. Narasimhachari, N., T. R. Seshadri, and S. Sethuraman: Methyl Ethers of Genistein. Proc. Indian Acad. Sci. **36A**, 194 (1952).

582. Narayana, C. S., and S. Rangaswami: Chemical Examination of Plant Insecticides: Seeds of *Mundulea suberosa*. J. Scient. Ind. Res. **14B**, 105 (1955).

583. Narayanan, V., and T. R. Seshadri: Paniculatin, a New Isoflavone-Di-C-Glucoside of *Dalbergia paniculata* Bark. Indian J. Chem. **9**, 14 (1971).

583a. Narayanaswamy, P., and A. Mahadevan: Phytoalexin Production by Germinating Seeds of *Mucuna utilis*. Curr. Sci. **50**, 905 (1981).

584. Nico, R., and C. M. Pérez Cambet: Rotenone or Rotenoids in Argentine Plants. Revta. Fac. Cienc. Quim. Farm. Univ. Nac. La Plata **20**, 7 (1945); Chem. Abstr. **42**, 2390 (1948).

585. Norton, L. B.: Rotenone in the Yam Bean *(Pachyrrhizus erosus)*. J. Am. Chem. Soc. **65**, 2259 (1943).

586. Norton, L. B., and R. Hansberry: Constituents of the Insecticidal Resin of the Yam Bean *(Pachyrrhizus erosus)*. J. Am. Chem. Soc. **67**, 1609 (1945).

586a. Obara, Y., and H. Matsubara: Isolation and Identification of (−)-Maackiain from *Derris* Roots. Meijo Daigaku Nogakubu Gakujutsu Hokoku **17**, 40 (1981); Chem. Abstr. **95**, 200536 (1981).

587. Obdulio, F., and M. P. Lobete: Presence of Rotenone in *Verbascum thapsus* (Mullin). Farm. Nueva **8**, 129 (1943); Chem. Abstr. **39**, 3036 (1945).

588. — — Distribution of Rotenone in *Verbascum thapsus*. Farm. Nueva **8**, 204 (1943); Chem. Abstr. **39**, 3036 (1945).

589. Oberholzer, M. E., G. J. H. Rall, and D. G. Roux: The Concurrence of 12a-Hydroxy and 12a-O-Methylrotenoids. Isolation of the First Natural 12a-O-Methylrotenoids. Tetrahedron Lett. **1974**, 2211.

590. — — — New Natural Rotenoid and Pterocarpanoid Analogues from *Neorautanenia amboensis*. Phytochemistry **15**, 1283 (1976).

591. — — — Structure and Synthesis of Ambanol, the First Natural Occurring Isoflavan-4-ol. Tetrahedron Lett. **1977**, 1165.

592. OCKENDON, D. J., R. E. ALSTON, and K. NAIFEH: The Flavonoids of *Psoralea* (Leguminosae). Phytochemistry **5**, 601 (1965).

593. OGIYAMA, K., and M. YASUE: Constituents of *Machaerium pedicellatum* Heartwood. Phytochemistry **12**, 2544 (1973).

594. OGNYANOV, I., and T. SOMLEVA: Rotenoids and 7,2′,4′,5′-Tetramethoxyisoflavone in *Amorpha fruticosa* Fruits. Planta Med. **38**, 279 (1980).

595. OHASHI, H.: Unpublished Results.

596. OHASHI, H., T. FUJIYAMA, and H. IMAMURA: Isoflavonoids of *Wisteria floribunda* Affected with *Milletia*-Gall Disease. Res. Bull. Fac. Agric. Gifu Univ. **42**, 123 (1979).

597. OHASHI, H., M. GOTO, and H. IMAMURA: Flavonoids from the Wood of *Cladrastis platycarpa*. Phytochemistry **15**, 354 (1976).

598. OHASHI, H., and H. IMAMURA: Syntheses of Formononetin-7-*O*-β-Laminarabioside and Four Isoflavone Glucosides. Res. Bull. Fac. Agric. Gifu Univ. **39**, 79 (1976).

599. — — Chemotaxonomical Comparison of Flavonoid Constituents Between *Cladrastis platycarpa* and *C. shikokiana*. J. Japan Wood Res. Soc. **24**, 750 (1978).

600. OHASHI, H., Y. ISHIDA, T. GIGA, and H. IMAMURA: Chemical Constituents from the Leaves of *Cladrastis platycarpa*. Res. Bull. Fac. Agric. Gifu Univ. **38**, 97 (1975).

601. OHASHI, H., K. NOZAKI, Y. HIBINO, and H. IMAMURA: Structures of Two New Isoflavones, Platycarpanetin and 5-Methoxyafrormosin, from the Wood of *Cladrastis platycarpa*. J. Japan Wood Res. Soc. **20**, 336 (1974).

602. OHTA, N., G. KUWATA, H. AKAHORI, and T. WATANABE: Isoflavonoid Constituents of Soybeans and Isolation of a New Acetyl Daidzin. Agric. Biol. Chem. **43**, 1415 (1979).

603. — — — — Isolation of a New Isoflavone Acetyl Glucoside, 6″-*O*-Acetyl Genistin, from Soybeans. Agric. Biol. Chem. **44**, 469 (1980).

604. OHTA, N., K. MIKUMO, R. IKEDA, and T. WATANABE: Effect of Soybean Isoflavones on Lipase Activity. Kumamoto Joshi Daigaku Gakujutsu Kiyo **33**, 56 (1981); Chem. Abstr. **95**, 110824 (1981).

605. OKANO, K., and I. BEPPU: Isolation of Four Kinds of Isoflavone from Soybean. J. Agric. Chem. Soc. Japan **15**, 645 (1939); Chem. Abstr. **34**, 429 (1940).

606. OKU, H., S. OUCHI, T. SHIRAISHI, K. UTSUMI, and S. SENO: Toxicity of a Phytoalexin, Pisatin, to Mammalian Cells. Proc. Japan Acad. **52**, 33 (1976).

607. OLAH, A. F., and R. T. SHERWOOD: Flavones, Isoflavones, and Coumestans in Alfalfa Infected by *Ascochyta imperfecta*. Phytopathology **61**, 65 (1971).

607a. OLIVARIES, E. M.: Unpublished Results.

608. OLLIS, W. D.: The Isoflavonoids. In: T. A. GEISSMAN, The Chemistry of Flavonoid Compounds, p. 353. Oxford: Pergamon Press. 1962.

609. — Structural Relationships Involving the Rotenoids. In: H. R. ARTHUR, Symposium on Phytochemistry, Hong Kong 1961, p. 128. Hong Kong University Press. 1964.

610. — New Structural Variants Among the Isoflavonoid and Neoflavonoid Classes. In: T. J. MABRY, R. E. ALSTON, and V. C. RUNECKLES, Recent Advances in Phytochemistry, Vol. 1, p. 329. New York: Appleton-Century-Crofts. 1968.

611. OLLIS, W. D., B. T. REDMAN, R. J. ROBERTS, I. O. SUTHERLAND, O. R. GOTTLIEB, and M. T. MAGALHÃES: Neoflavonoids and the Cinnamylphenol Kuhlmannistyrene from *Machaerium kuhlmanii* and *M. nictitans*. Phytochemistry **17**, 1383 (1978).

612. OLLIS, W. D., B. T. REDMAN, I. O. SUTHERLAND, and O. R. GOTTLIEB: Petrostyrene, a Cinnamylphenol from *Machaerium acutifolium*. Phytochemistry **17**, 1379 (1978).

613. OLLIS, W. D., C. A. RHODES, and I. O. SUTHERLAND: The Constitutions of Durlettone, Durmillone, Milldurone, Millettone and Millettosin. Tetrahedron **23**, 4741 (1967).

614. OLLIS, W. D., I. O. SUTHERLAND, H. M. ALVES, and O. R. GOTTLIEB: Duartin, an Isoflavan from *Machaerium opacum*. Phytochemistry **17**, 1401 (1978).

615. O'NEILL, P., and A. G. PERKIN: The Colouring Matters of Camwood, Barwood, and Sanderswood. J. Chem. Soc. **113**, 125 (1918).

616. ORTH, H., and P. FORSCHNER: Formononetin in *Pterocarpus vidalianus*. Investigation of the Extractives of *Pterocarpus vidalianus*. Holzforschung **19**, 111 (1965).

617. OUNG-BORAN, P. LEBRETON, and G. NETIEN: A Biochemical and Pharmacological Study of *Baptisia australis*. Planta Med. **17**, 301 (1969).

618. OZIMINA, I. I.: Flavonoid Compounds of *Genista patula*. Khim. Prir. Soedin. **1981**, 242.

619. OZIMINA, I. I., V. A. BANDYUKOVA, and A. L. KAZAKOV: Flavonoids of *Spartium junceum*. Isoflavones. Khim. Prir. Soedin. **1979**, 858.

620. PACHÉCO, H.: Flavonoid Compounds of *Prunus mahaleb*. Bull. Soc. Chim. Biol. **41**, 111 (1959).

621. PAILER, M., and F. FRANKE: Constituents of *Iris germanica*. Mh. Chem. **104**, 1394 (1973).

622. PANKHURST, C. E., and D. R. BIGGS: Sensitivity of *Rhizobium* to Selected Isoflavonoids. Can. J. Microbiol. **26**, 542 (1980).

623. PARIS, R. R., and G. FAUGERAS: Flavonoids of the Flowers of *Retama raetam*. Isolation and Identification of Genistin (Genistoside). C. R. Hebd. Séanc. Acad. Sci., Paris **257**, 1728 (1963).

624. — — Isolation of Genistin (Genistoside) from *Ulex nanus* and *Adenocarpus complicatus*, and 5-*O*-Methylgenistein from *Genista hispanica*. C. R. Hebd. Séanc. Acad. Sci., Paris **261**, 1761 (1965).

625. PARIS, R. R., G. FAUGERAS, and J. F. DOBREMEZ: Isoflavones of the Twigs of *Piptanthus nepalensis*. Planta Med. **29**, 32 (1976).

626. PARIS, R. R., and A. STAMBOULI: Isolation of a New Glycoside, Cytifolioside, from the Leaves and Fruits of *Cytisus sessilifolius*. C. R. Hebd. Séanc. Acad. Sci., Paris **254**, 3031 (1962).

627. PARTHASARATHY, M. R., R. N. PURI, and T. R. SESHADRI: New Components of *Pterocarpus dalbergioides* Heartwood. Indian J. Chem. **7**, 118 (1969).

628. PARTHASARATHY, M. R., T. R. SESHADRI, and R. S. VARMA: Minor Isoflavonoid Glycosides of the Stem Bark of *Dalbergia paniculata* — Isolation of a New *C*-Glycoside. Curr. Sci. **43**, 74 (1974).

629. — — — 7-*O*-Rutinosides of Biochanin A and Formononetin: Two New Rhamnoglucosides of the Bark of *Dalbergia paniculata*. Indian J. Chem. **12**, 518 (1974).

630. — — — Triterpenoids and Flavonoids of *Dalbergia sericea* Bark. Phytochemistry **15**, 226 (1976).

631. — — — New Isoflavonoid Glycosides from *Dalbergia paniculata*. Phytochemistry **15**, 1025 (1976).

632. PARTHASARATHY, M. R., P. SHARMA, and S. B. KALIDHAR: Isocaviunin-7-*O*-glucoside, a New Isoflavone Glucoside from the Bark of *Dalbergia paniculata*. Indian J. Chem. **19B**, 429 (1980).

633. PARTRIDGE, J. E.: Fungus Disease Defense Responses in Legumes. Ph. D. Thesis, University of California (Riverside), U.S.A. 1973.

634. PARTRIDGE, J. E., and N. T. KEEN: Association of the Phytoalexin Kievitone with Single-Gene Resistance of Cowpeas to *Phytophthora vignae*. Phytopathology **66**, 426 (1976).

635. PELTER, A., and P. I. AMENECHI: Isoflavonoid and Pterocarpinoid Extractives of *Lonchocarpus laxiflorus*. J. Chem. Soc. C **1969**, 887.

636. PELTER, A., and P. STAINTON: The Extractives from *Derris scandens*. The Isolation of Osajin and Two New Isoflavones, Scandenone and Scandinone. J. Chem. Soc. C **1966**, 701.

637. PERKIN, A. G., and L. H. HORSFALL: Genistein. Part II. J. Chem. Soc. **77**, 1310 (1900).

638. PERKIN, A. G., and F. G. NEWBURY: The Colouring Matters Contained in Dyer's Broom *(Genista tinctoria)* and Heather *(Calluna vulgaris)*. J. Chem. Soc. **75**, 830 (1899).

639. PERRIN, D. R.: The Structure of Phaseolin. Tetrahedron Lett. **1964**, 29.

640. PERRIN, D. R., D. R. BIGGS, and I. A. M. CRUICKSHANK: Phaseollidin, a Phytoalexin from *Phaseolus vulgaris*: Isolation, Physicochemical Properties and Antifungal Activity. Aust. J. Chem. **27**, 1607 (1974).

641. PERRIN, D. R., and W. BOTTOMLEY: The Structure of Pisatin from *Pisum sativum*. J. Am. Chem. Soc. **84**, 1919 (1962).

642. PERRIN, D. R., and I. A. M. CRUICKSHANK: The Antifungal Activity of Pterocarpans Towards *Monilinia fructicola*. Phytochemistry **8**, 971 (1969).

643. PERRIN, D. R., C. P. WHITTLE, and T. J. BATTERHAM: The Structure of Phaseollidin. Tetrahedron Lett. **1972**, 1673.

644. PÉTARD, P.: Ichthyotoxic Plants of Polynesia. Méd. Trop. **11**, 498 (1951).

645. PIATAK, D. M., G. A. FLYNN, and P. D. SORENSEN: Rotenoids from *Amorpha canescens* Roots. Phytochemistry **14**, 1391 (1975).

646. PLOUVIER, V.: Some Constituents of the Bark of *Prunus mahaleb* (Rosaceae). C. R. Hebd. Séanc. Acad. Sci., Paris **250**, 594 (1960).

647. POPRAVKO, S. A., S. A. SOKOLOVA, P. D. FRAISHTAT, and G. P. KONONENKO: Changes in Concentration of Growth Inhibitors of Red Clover Roots in Autumn. Bio-Org. Khim. **5**, 1654 (1979).

648. POPRAVKO, S. A., S. A. SOKOLOVA, and G. P. KONONENKO: Isoflavones and Pterocarpans in Red Clover Roots Inoculated with Nodule Bacteria. Bio-Org. Khim. **6**, 1255 (1980).

649. POWER, F. B., and A. H. SALWAY: The Constituents of Red Clover Flowers. J. Chem. Soc. **97**, 231 (1910).

650. PRAKASH, L., A. ZAMAN, and A. R. KIDWAI: Isolation and Structure of Irisolidone. J. Org. Chem. **30**, 3561 (1965).

651. PRASAD, J. S., and R. S. VARMA: 5,7,2′,4′-Tetrahydroxyisoflavone in *Moghania macrophylla*. Phytochemistry **16**, 1120 (1977).

652. PRASUNAMBA, K. L., G. SRIMANNARAYANA, and N. V. S. RAO: Flavonoid Components of *Mundulea suberosa* Stem Bark. Curr. Sci. **46**, 726 (1977).

653. PRESTON, N. W.: 2′-O-Methylphaseollidinisoflavan from Infected Tissue of *Vigna unguiculata*. Phytochemistry **14**, 1131 (1975).

654. — Cajanone: an Antifungal Isoflavanone from *Cajanus cajan*. Phytochemistry **16**, 143 (1977).

655. — Induced Pterocarpans of *Psophocarpus tetragonolobus*. Phytochemistry **16**, 2044 (1977).

656. PRESTON, N. W., K. CHAMBERLAIN, and R. A. SKIPP: A 2-Arylbenzofuran Phytoalexin from Cowpea *(Vigna unguiculata)*. Phytochemistry **14**, 1843 (1975).

657. PUEPPKE, S. G., and H. D. VANETTEN: Identification of Three New Pterocarpans (6a,11a-Dihydro-6*H*-benzofuro-[3,2-*c*][1]-benzopyrans) from *Pisum sativum* Infected with *Fusarium solani* f. sp. *pisi*. J. Chem. Soc. Perkin Trans. I **1975**, 946.

658. — — Accumulation of Pisatin and Three Additional Antifungal Pterocarpans in *Fusarium solani*-Infected Tissues of *Pisum sativum*. Physiol. Pl. Path. **8**, 51 (1976).

659. PURUSHOTHAMAN, K. K., S. CHANDRASEKHARAN, K. BALAKRISHNA, and J. D. CONNOLLY: Gangetinin and Desmodin, Two Minor Pterocarpanoids of *Desmodium gangeticum*. Phytochemistry **14**, 1129 (1975).

660. PURUSHOTHAMAN, K. K., V. M. KISHORE, V. NARAYANASWAMI, and J. D. CONNOLLY: The Structure and Stereochemistry of Gangetin, a New Pterocarpan from *Desmodium gangeticum* (Leguminosae). J. Chem. Soc. C **1971**, 2420.

661. QUINLIVAN, B. J., W. J. COLLINS, and M. L. BROWN: Varietal Identification of Seedling Subterranean Clover Using Thin Layer Chromatography. Seed Sci. Technol. **4**, 185 (1976).

662. RADHAKRISHNIAH, M.: Isoflavonoids of *Dalbergia paniculata* Seeds. Phytochemistry **12**, 3003 (1973).

663. — Chemical Constituents of *Dalbergia paniculata* Roots. J. Indian Chem. Soc. **56**, 81 (1979).

664. RAJU, K. V. S., and G. SRIMANNARAYANA: Aurmillone, a New Isoflavone from the Seeds of *Millettia auriculata*. Phytochemistry **17**, 1065 (1978).

665. RAJU, K. V. S., G. SRIMANNARAYANA, B. TERNAI, R. STANLEY, and K. R. MARKHAM: Structure of Millettin, a Novel Isoflavone Isolated from *Millettia auriculata*. Tetrahedron **37**, 957 (1981).
666. RAJULU, K. G., and J. R. RAO: Caviunin-7-*O*-rhamnoglucoside from *Dalbergia paniculata* Root. Phytochemistry **19**, 1563 (1980).
667. RALL, G. J. H., A. J. BRINK, and J. P. ENGELBRECHT: Pterocarpanoid Constituents of the Root Bark of *Neorautanenia edulis*. J. South Afr. Chem. Inst. **25**, 25 (1972).
668. — — — The Isolation and Structure of Edulenol, a New Pterocarpan from *Neorautanenia edulis*. J. South Afr. Chem. Inst. **25**, 131 (1972).
669. RALL, G. J. H., J. P. ENGELBRECHT, and A. J. BRINK: The Isolation, Structure and Absolute Configuration of (—)-2-Hydroxypterocarpin, a New Pterocarpan from *Neorautanenia edulis*. Tetrahedron **26**, 5007 (1970).
670. — — — The Constitution of (—)-2-Isopentenyl-3-hydroxy-8,9-methylenedioxy-pterocarpan, a New Pterocarpan from the Root Bark of *Neorautanenia edulis*. J. South Afr. Chem. Inst. **24**, 56 (1971).
671. RAMANUJAM, S., and T. R. SESHADRI: Synthesis of Padmakastein and its Derivatives. Proc. Indian Acad. Sci. **48A**, 175 (1958).
672. RANGASWAMI, S., and B. V. R. SASTRY: Crystalline Components of the Roots of *Tephrosia maxima*. Curr. Sci. **23**, 397 (1954).
673. — — Chemical Examination of the Root Bark of *Tephrosia hirta*. Indian J. Pharm. **18**, 43 (1956).
674. — — Chemical Components of the Seeds of *Tephrosia candida*. Indian J. Pharm. **18**, 333 (1956).
675. — — Chemical Examination of the Seeds of *Tephrosia vogelii*. Indian J. Pharm. **18**, 339 (1956).
676. — — The Constitution of Maxima Substance C. Arch. Pharm., Berl. **292**, 170 (1959).
677. — — Constitution of Maxima Substance B. Proc. Indian Acad. Sci. **57A**, 135 (1963).
678. RANGASWAMI, S., B. V. R. SASTRY, and E. V. RAO: Chemical Examination of the Roots of *Tephrosia maxima*. Indian J. Pharm. **23**, 31 (1961).
679. RAO, N. V. S.: Chemical Components of Leguminous Plant Insecticides Occurring in India. J. Scient. Ind. Res. **13A**, 31 (1954).
680. — Chemical Components of the Pods of *Tephrosia purpurea* var. *maxima*. Curr. Sci. **25**, 396 (1956).
681. RAO, N. V. S., and T. R. SESHADRI: Chemical Components of *Derris ferruginea*. Proc. Indian Acad. Sci. **24A**, 344 (1946).
682. — — Chemical Components of *Derris scandens*. Proc. Indian Acad. Sci. **24A**, 365 (1946).
683. — — Chemical Components of *Derris robusta*. Proc. Indian Acad. Sci. **24A**, 465 (1946).
684. RAO, P. P., and G. SRIMANNARAYANA: Tephrosol, a New Coumestone from the Roots of *Tephrosia villosa*. Phytochemistry **19**, 1272 (1980).
685. RATHMELL, W. G., and D. S. BENDALL: Phenolic Compounds in Relation to Phytoalexin Biosynthesis in Hypocotyls of *Phaseolus vulgaris*. Physiol. Pl. Path. **1**, 351 (1971).
686. RATHMELL, W. G., and D. A. SMITH: Lack of Activity of Selected Isoflavonoid Phytoalexins as Protectant Fungicides. Pestic. Sci. **11**, 568 (1980).
687. RAUDNITZ, H., and G. PERLMANN: Santal, Pterocarpin and Homopterocarpin, the Colourless Co-Constituents of Santalin. Ber. Dt. Chem. Ges. **68B**, 1862 (1935).
688. RAVISÉ, A., and B. S. KIRKIACHARIAN: Effects of the Structure of Phenolic Compounds on the Inhibition of *Phytophthora parasitica* and on Lytic Enzymes. I. Isoflavonoids and Coumestans. Phytopath. Z. **85**, 74 (1976).
689. REINERS, W.: 7-Hydroxy-4′-methoxyisoflavone (Formononetin) from Licorice Root. Experientia **22**, 359 (1966).
690. REISCH, J., M. GOMBOS, K. SZENDREI, and I. NOVÁK: 6a,12a-Dehydro-α-toxicarol, a New Rotenoid from *Amorpha fruticosa*. Phytochemistry **15**, 234 (1976).

691. RICH, J. R., N. T. KEEN, and I. J. THOMASON: Association of Coumestans with the Hypersensitivity of Lima Bean Roots to *Pratylenchus scribneri*. Physiol. Pl. Path. **10,** 105 (1977).

691a. RICHARDSON, P. M.: Unpublished Results.

692. — Phytoalexin Induction in *Beta* and *Spinacia*. Biochem. Syst. Ecol. **9,** 105 (1981).

693. RIZK, A. F., and G. E. WOOD: Phytoalexins of Leguminous Plants. C. R. C. Crit. Rev. Fd. Sci. Nutr. **13,** 245 (1980).

694. ROARK, R. C.: The Chemical Relationship Between Certain Insecticidal Species of Fabaceous Plants. J. Econ. Entomol. **26,** 587 (1933).

695. — *Lonchocarpus* Species (Barbasco, Cubé, Haiari, Nekoe and Timbo) Used as Insecticides. Bur. Entomol. Pl. Quarant., U. S. Dept. Agric., Mon. E-367 (1936).

696. — *Tephrosia* as an Insecticide — a Review of the Literature. Bur. Entomol. Pl. Quarant., U. S. Dept. Agric., Mon. E-402 (1937).

697. ROBERTSON, A.: The Synthesis of Rissic Acid and of Derric Acid, and the Constitution of Rotenone, Deguelin and Tephrosin. J. Chem. Soc. **1932,** 1380.

698. ROBERTSON, A., C. W. SUCKLING, and W. B. WHALLEY: The Structure of Santal and a Note on Orobol. J. Chem. Soc. **1949,** 1571.

699. ROBESON, D. J.: A Comparative Study of Phytoalexin Induction in the Tribe Vicieae. Ph. D. Thesis, University of Reading, U. K. 1978.

700. — Furanoacetylene and Isoflavonoid Phytoalexins in *Lens culinaris*. Phytochemistry **17,** 807 (1978).

701. ROBESON, D. J., and J. B. HARBORNE: A Chemical Dichotomy in Phytoalexin Induction Within the Tribe Vicieae of the Leguminosae Phytochemistry **19,** 2359 (1980).

702. ROBESON, D. J., and J. L. INGHAM: New Pterocarpan Phytoalexins from *Lathyrus nissolia*. Phytochemistry **18,** 1715 (1979).

703. ROBESON, D. J., J. L. INGHAM, and J. B. HARBORNE: Identification of Two Chromone Phytoalexins in the Sweet Pea, *Lathyrus odoratus*. Phytochemistry **19,** 2171 (1980).

704. ROGERSON, H.: The Constituents of the Flowers of *Trifolium incarnatum*. J. Chem. Soc. **97,** 1004 (1910).

705. RÖSLER, H., T. J. MABRY, and J. KAGAN: Sphaerobioside, an Isoflavone Glycoside from *Baptisia sphaerocarpa*. Chem. Ber. **98,** 2193 (1965).

706. RUSSELL, A., and E. A. KACZKA: Fish Poisons from *Ichthyomethia piscipula*. J. Am. Chem. Soc. **66,** 548 (1944).

707. RUSSELL, G. B., O. R. W. SUTHERLAND, R. F. N. HUTCHINS, and P. E. CHRISTMAS: Vestitol: a Phytoalexin with Insect Feeding-Deterrent Activity. J. Chem. Ecol. **4,** 571 (1978).

707a. RUSSELL, G. B., O. R. W. SUTHERLAND, G. A. LANE, and D. R. BIGGS: Is There a Common Factor for Insect and Disease Resistance in Pasture Legumes? Proc. 2nd. Australasian Conf. Grassl. Invertebr. Ecol., p. 95. 1980.

708. SAITOH, T., T. KINOSHITA, and S. SHIBATA: New Isoflavan and Flavanone from Licorice Root. Chem. Pharm. Bull., Tokyo **24,** 752 (1976).

709. SAITOH, S., H. NOGUCHI, and S. SHIBATA: A New Isoflavone and the Corresponding Isoflavanone of Licorice Root. Chem. Pharm. Bull., Tokyo **26,** 144 (1978).

710. SAITOH, T., and S. SHIBATA: Some New Constituents of Licorice Root. (2). Glycyrol, 5-*O*-Methylglycyrol and Isoglycyrol. Chem. Pharm. Bull., Tokyo **17,** 729 (1969).

711. SAKAGAMI, Y., S. KUMAI, and A. SUZUKI: Isolation and Structure of Medicarpin-β-D-glucoside in Alfalfa. Agric. Biol. Chem. **38,** 1031 (1974).

712. SARMA, P. N., G. SRIMANNARAYANA, and N. V. S. RAO: Constitution of Villosol and Villosinol, Two New Rotenoids from *Tephrosia villosa* Pods. Indian J. Chem. **14B,** 152 (1976).

713. SAVAGE, N., and P. W. G. GROENEWOUD: Preliminary Chemical Investigation of the Tuber of *Neorautanenia ficifolia*. J. South Afr. Chem. Inst. **7,** 1 (1954).

714. SAWHNEY, P. L., and T. R. SESHADRI: The Neutral Components from Heartwoods and Sapwoods of *Pterocarpus dalbergioides* (Andaman Padauk) and *P. macrocarpus* (Burma Padauk). J. Scient. Ind. Res. **13B**, 5 (1954).

715. — — Phenolic Components of some *Pterocarpus* Species. J. Scient. Ind. Res. **15C**, 154 (1956).

716. SAYAGAVER, B. M., N. B. MULCHANDANI, and S. NARAYANASWAMY: Isolation of Formononetin from Tissue Cultures of *Cicer arietinum*. Lloydia **32**, 108 (1969).

717. SAYED, S. A., and M. I. BORISOV: Flavonoids of *Pueraria lobata* Rhizomes. Farmat. Zh., Kiev **1979**, 76; Chem. Abstr. **91**, 71727 (1979).

718. SCHULTZ, G.: Isoflavone Glycosides in *Trifolium* Species Supplied with Different Mineral Salts. Ber. Dt. Bot. Ges. **79**, 108 (1966).

719. SCHULTZ, G., and M. I. ELGHAMRY: Isolation of Biochanin A from *Lupinus termis* and Estimation of its Estrogenic Activity. Naturwissenschaften **58**, 98 (1971).

720. SCHWARZ, J. S. P., A. I. COHEN, W. D. OLLIS, E. A. KACZKA, and L. M. JACKMAN: The Constitution of Ichthynone. Tetrahedron **20**, 1317 (1964).

721. SESHADRI, T. R.: Advances in the Phytochemistry of Isoflavonoids. In: H. R. ARTHUR, Symposium on Phytochemistry, Hong Kong 1961, p. 145. Hong Kong University Press. 1964.

722. SETHI, M. L., S. C. TANEJA, S. G. AGARWAL, K. L. DHAR, and C. K. ATAL: Isoflavones and Stilbenes from *Juniperus macropoda*. Phytochemistry **19**, 1831 (1980).

723. SETHI, M. L., S. C. TANEJA, K. L. DHAR, and C. K. ATAL: Two Isoflavones from *Juniperus macropoda*. Phytochemistry **20**, 341 (1981).

724. SHABBIR, M., and A. ZAMAN: Structure of Isoauriculatin and Auriculin, Extractives of *Millettia auriculata*. Tetrahedron **26**, 5041 (1970).

725. SHABBIR, M., A. ZAMAN, L. CROMBIE, B. TUCK, and D. A. WHITING: Structure of Auriculatin, an Extractive of *Millettia auriculata*. J. Chem. Soc. C **1968**, 1899.

726. SHAMMA, M., and L. D. STIVER: Two New Isoflavones from the Heartwood of *Cladrastis lutea*. Tetrahedron **25**, 3887 (1969).

727. SHARMA, A., S. S. CHIBBER, and H. M. CHAWLA: Isocaviunin from Mature Pods of *Dalbergia sissoo*. Indian J. Chem. **18B**, 472 (1979).

728. — — — Caviunin 7-*O*-Gentiobioside from *Dalbergia sissoo* Pods. Phytochemistry **18**, 1253 (1979).

729. — — — Isocaviudin, a New Isoflavone Glucoside Isolated from *Dalbergia sissoo*. Indian J. Chem. **19B**, 237 (1980).

730. — — — Isocaviunin 7-Gentiobioside, a New Isoflavone Glycoside from *Dalbergia sissoo*. Phytochemistry **19**, 715 (1980).

731. SHARMA, R., and P. KHANNA: Production of Rotenoids from *Tephrosia* spp. *In vivo* and *In vitro* Tissue Cultures. Indian J. Exp. Biol. **13**, 84 (1975).

732. SHARMA, R. D.: Isoflavones and Hypercholesterolemia in Rats. Lipids **14**, 535 (1979).

733. — Effect of Various Isoflavones on Lipid Levels in Triton-Treated Rats. Atherosclerosis **33**, 371 (1979).

734. SHEMESH, M., N. AYALON, and H. R. LINDNER: Identification of Phyto-Estrogens in Berseem Clover *(Trifolium alexandrinum)*. J. Agric. Sci., Camb. **87**, 467 (1976).

735. SHIBATA, B.: Constituents of *Iris tectorum*. J. Pharm. Soc. Japan **47**, 380 (1927); Chem. Abstr. **21**, 3050 (1927).

736. SHIBATA, H., and S. SHIMIZU: Amorphaquinone, a New Isoflavanquinone from *Amorpha fruticosa*. Heterocycles **10**, 85 (1978).

737. SHIBATA, S.: Unpublished Results.

738. SHIBATA, S., T. MURAKAMI, and Y. NISHIKAWA: On the Constituents of *Pueraria* Root. J. Pharm. Soc. Japan **79**, 757 (1959).

739. SHIBATA, S., T. MURAKAMI, Y. NISHIKAWA, and M. HARADA: The Constituents of *Pueraria* Root. Chem. Pharm. Bull., Tokyo **7**, 134 (1959).

740. SHIBATA, S., T. MURATA, and M. FUJITA: Wistin, a New Isoflavone Glucoside of *Wisteria* spp. Chem. Pharm. Bull., Tokyo **11**, 382 (1963).

741. SHIBATA, S., and Y. NISHIKAWA: On the Constituents of the Roots of *Sophora subprostrata* (2) and *Sophora japonica* (1). Chem. Pharm. Bull., Tokyo **11**, 167 (1963).

742. SHIBATA, S., and T. SAITOH: Some New Constituents of Licorice Root. (1). The Structure of Licoricidin. Chem. Pharm. Bull., Tokyo **16**, 1932 (1968).

743. SHIENGTHONG, D., T. DONAVANIK, V. UAPRASERT, S. ROENGSUMRAN, and R. A. MASSY-WESTROPP: Constituents of Thai Medicinal Plants. New Rotenoid Compounds — Stemonacetal, Stemonal and Stemonone. Tetrahedron Lett. **1974**, 2015.

744. SHIRAISHI, T., H. OKU, M. ISONO, and S. OUCHI: The Injurious Effect of Pisatin on the Plasma Membrane of Pea. Pl. Cell Physiol., Tokyo **16**, 939 (1975).

745. SHIRAISHI, T., H. OKU, S. OUCHI, and Y. TSUJI: Local Accumulation of Pisatin in Tissues of Pea Seedlings Infected by Powdery Mildew Fungi. Phytopath. Z. **88**, 131 (1977).

745a. SHIRATAKI, Y., M. KOMATSU, I. YOKOE, and A. MANAKA: Constituents of the Root of *Euchresta japonica* (1). Chem. Pharm. Bull., Tokyo **29**, 3033 (1981).

746. SHRINER, R. L., and C. J. HULL: A Synthesis of Methylgenistein. J. Org. Chem. **10**, 228 (1945).

747. SHUKLA, R. V., and K. MISRA: Two Flavonoid Glycosides from the Bark of *Prosopis juliflora*. Phytochemistry **20**, 339 (1981).

748. SHUTT, D. A.: The Effects of Plant Oestrogens on Animal Reproduction. Endeavour **35**, 110 (1976).

749. SIALER DE ZAPATA, D., F. DELLE MONACHE, G. C. VALERA, and G. B. MARINI-BETTOLO: Flavonoids and Rotenoids in *Lonchocarpus* Genus: Rotenoids from *Lonchocarpus urucu* and *Lonchocarpus* sp. (Uaicà). Atti Accad. Naz. Lincei Rc. **62**, 829 (1977); Chem. Abstr. **89**, 103756 (1978).

750. SIDDIQUI, M. T., and M. SIDDIQI: Hypolipidemic Principles of *Cicer arietinum:* Biochanin A and Formononetin. Lipids **11**, 243 (1976).

751. SIMON, J. P., and D. W. GOODALL: Relationship in Annual Species of *Medicago*. Two-Dimensional Chromatography of the Phenolics and Analysis of the Results by Probabilistic Similarity Methods. Aust. J. Bot. **16**, 89 (1968).

752. SIMONITSCH, E., H. FREI, and H. SCHMID: The Constitution of Pachyrrhizin. Mh. Chem. **88**, 541 (1957).

753. SIMS, J. J., N. T. KEEN, and V. K. HONWAD: Hydroxyphaseollin, an Induced Antifungal Compound from Soybeans. Phytochemistry **11**, 827 (1972).

754. SINGHAL, A. K., R. P. SHARMA, K. P. MADHUSUDANAN, G. THYAGARAJAN, W. HERZ, and S. V. GOVINDAN: New Prenylated Isoflavones from *Millettia pachycarpa*. Phytochemistry **20**, 803 (1981).

755. SINGHAL, A. K., R. P. SHARMA, G. THYAGARAJAN, W. HERZ, and S. V. GOVINDAN: New Prenylated Isoflavones and a Prenylated Dihydroflavonol from *Millettia pachycarpa*. Phytochemistry **19**, 929 (1980).

756. SKIPP, R. A., and J. A. BAILEY: The Fungitoxicity of Isoflavonoid Phytoalexins Measured Using Different Types of Bioassay. Physiol. Pl. Path. **11**, 101 (1977).

757. SKIPP, R. A., C. SELBY, and J. A. BAILEY: Toxic Effects of Phaseollin on Plant Cells. Physiol. Pl. Path. **10**, 221 (1977).

758. SMALBERGER, T. M., R. VLEGGAAR, and J. C. WEBER: The Structure of Elongatin, an Isoflavone from *Tephrosia elongata*. Tetrahedron **31**, 2297 (1975).

759. SMITH, D. A.: Some Effects of the Phytoalexin, Kievitone, on the Vegetative Growth of *Aphanomyces euteiches, Rhizoctonia solani* and *Fusarium solani* f. sp. *phaseoli*. Physiol. Pl. Path. **9**, 45 (1976).

760. SMITH, D. A., P. J. KUHN, J. A. BAILEY, and R. S. BURDEN: Detoxification of Phaseollidin by *Fusarium solani* f. sp. *phaseoli*. Phytochemistry **19**, 1673 (1980).

761. Smith, D. A., H. D. Vanetten, J. W. Serum, T. M. Jones, D. F. Bateman, T. H. Williams, and D. L. Coffen: Confirmation of the Structure of Kievitone, an Antifungal Isoflavanone Isolated from *Rhizoctonia*-Infected Bean Tissues. Physiol. Pl. Path. **3**, 293 (1973).

762. Smith, D. G., A. G. McInnes, V. J. Higgins, and R. L. Millar: Nature of the Phytoalexin Produced by Alfalfa in Response to Fungal Infection. Physiol. Pl. Path. **1**, 41 (1971).

762a. Soladoye, M. O.: Systematic Studies in the Genus *Baphia*. Ph. D. Thesis, University of Reading, U. K. 1981.

763. Späth, E., and J. Schläger: The Constitution of Homopterocarpin. Ber. Dt. Chem. Ges. **73B**, 1 (1940).

764. Späth, E., and O. Schmidt: The Constitution of Pseudobaptisin. Mh. Chem. **53/54**, 454 (1929); Chem. Abstr. **24**, 371 (1930).

765. Spencer, R. R., E. M. Bickoff, R. E. Lundin, and B. E. Knuckles: Lucernol and Sativol, Two New Coumestans from Alfalfa *(Medicago sativa)*. J. Agric. Fd. Chem. **14**, 162 (1966).

766. Spencer, R. R., B. E. Knuckles, and E. M. Bickoff: 7-Hydroxy-11,12-dimethoxycoumestan. Characterization and Synthesis. J. Org. Chem. **31**, 988 (1966).

767. Stamm, O. A., H. Schmid, and J. Büchi: The Constitution of Jamaicin. Helv. Chim. Acta **41**, 2006 (1958).

768. Steiner, P. W., and R. L. Millar: Degradation of Medicarpin and Sativan by *Stemphylium botryosum*. Phytopathology **64**, 586 (1974).

769. Stoessl, A.: Inermin Associated with Pisatin in Peas Inoculated with the Fungus *Monilinia fructicola*. Can. J. Biochem. **50**, 107 (1972).

770. — Phytoalexins — a Biogenetic Perspective. Phytopath. Z. **99**, 251 (1980).

771. Suginome, H.: A New Isoflavanone from *Sophora japonica*. J. Org. Chem. **24**, 1655 (1959).

772. — Maackiain, a New Naturally Occurring Chromanocoumaran. Experientia **18**, 161 (1962).

773. Suginome, H., and T. Kio: The Co-Occurrence of Isoflavonoids at Different Oxidation Levels. Bull. Chem. Soc. Japan **39**, 1541 (1966).

773a. Sukumaran, K., and S. S. Gnanamanickam: Isolation of Antifungal Compounds from Indian Fodder Legume Plants. Indian J. Microbiol. **20**, 204 (1980).

774. Suri, J. L., G. K. Gupta, K. L. Dhar, and C. K. Atal: Psoralenol: a New Isoflavone from the Seeds of *Psoralea corylifolia*. Phytochemistry **17**, 2046 (1978).

775. Sutherland, O. R. W., G. B. Russell, D. R. Biggs, and G. A. Lane: Insect Feeding Deterrent Activity of Phytoalexin Isoflavonoids. Biochem. Syst. Ecol. **8**, 73 (1980).

776. Szabo, V., R. Bognar, E. Farkas, and G. Litkei: The Glycosides of the Fruit of *Sophora japonica*. Acta Univ. Debrecen, Ser. Phys. Chem. **13**, 129 (1967); Chem. Abstr. **69**, 16776 (1968).

777. Taguchi, H., P. Kanchanapee, and T. Amatayakul: The Constituents of *Clitoria macrophylla*, a Thai Medicinal Plant. The Structure of a New Rotenoid, Clitoriacetal. Chem. Pharm. Bull., Tokyo **25**, 1026 (1977).

777a. Tahara, S.: Unpublished Results.

778. Takai, M., H. Yamaguchi, T. Saitoh, and S. Shibata: The Chemical Constituents of the Heartwood of *Maackia amurensis* var. *buergeri*. Chem. Pharm. Bull., Tokyo **20**, 2488 (1972).

779. Takeda, T., I. Ishiguro, M. Masegi, and Y. Ogihara: New Isoflavone Glycosides from the Wood of *Sophora japonica*. Phytochemistry **16**, 619 (1977).

780. Tamura, S., C. F. Chang, A. Suzuki, and S. Kumai: Chemical Studies on "Clover Sickness". Part I. Isolation and Structural Elucidation of Two New Isoflavonoids in Red Clover. Agric. Biol. Chem. **33**, 391 (1969).

781. TANAKA, I., K. OHSAKI, and K. TAKAHASHI: The Constituents of the Bark and Wood of *Wisteria floribunda*. J. Pharm. Soc. Japan **95**, 1388 (1975).
782. TOBE, H., H. NAGANAWA, T. TAKITA, T. TAKEUCHI, and H. UMEZAWA: Structure of a New Isoflavone from Fungi and *Streptomyces* Inhibiting DOPA Decarboxylase. J. Antibiot., Tokyo **29**, 623 (1976).
783. TORRANCE, S. J., R. M. WIEDHOPF, J. J. HOFFMANN, and J. R. COLE: Petalostetin, a New Isoflavone from *Petalostemon candidum*. Phytochemistry **18**, 366 (1979).
784. TSUKAYAMA, M., T. HORIE, Y. YAMASHITA, M. MASUMURA, and M. NAKAYAMA: The Synthesis of 5,5'-Dihydroxy-7,2',4'-trimethoxyisoflavone and its Isomer: a Revised Structure of Derrugenin. Heterocycles **14**, 1283 (1980).
785. TSUKIDA, K., K. SAIKI, and M. ITO: New Isoflavone Glycosides from *Iris florentina*. Phytochemistry **12**, 2318 (1973).
786. TURNER, R. B., D. L. LINDSEY, D. D. DAVIS, and R. D. BISHOP: Isolation and Identification of 5,7-Dimethoxyisoflavone, an Inhibitor of *Aspergillus flavus* from Peanuts. Mycopathologia **57**, 39 (1975).
787. UDDIN, A., and P. KHANNA: Rotenoids in Tissue Cultures of *Crotalaria burhia*. Planta Med. **36**, 181 (1979).
788. UENO, A., M. ICHIKAWA, S. FUKUSHIMA, Y. SAIKI, and K. MORINAGA: The Structure of Lespedeol B. Chem. Pharm. Bull., Tokyo **21**, 2712 (1973).
789. UENO, A., M. ICHIKAWA, S. FUKUSHIMA, Y. SAIKI, T. NORO, K. MORINAGA, and H. KUWANO: The Structures of Lespein and Lespedezin. Chem. Pharm. Bull., Tokyo **21**, 2715 (1973).
790. UENO, A., M. ICHIKAWA, T. MIYASE, S. FUKUSHIMA, Y. SAIKI, and K. MORINAGA: The Structure of Lespedeol A. Chem. Pharm. Bull., Tokyo **21**, 1734 (1973).
791. ULUBELEN, A., and T. DOGUC: Flavonoid Compounds from the Flowers of *Genista lydia*. Planta Med. **25**, 39 (1974).
792. UMEZAWA, H., H. TOBE, N. SHIBAMOTO, F. NAKAMURA, K. NAKAMURA, M. MATSUZAKI, and T. TAKEUCHI: Isolation of Isoflavones Inhibiting DOPA Decarboxylase from Fungi and *Streptomyces*. J. Antibiot., Tokyo **28**, 947 (1975).
793. UTKIN, L. M., and A. P. SEREBRYAKOVA: The Isoflavone Glycoside of *Piptanthus nanus*. Khim. Prir. Soedin. **1965**, 70.
793a. VAN DEN HEUVEL, J., and J. A. GLAZENER: Comparative Abilities of Fungi Pathogenic and Nonpathogenic to Bean (*Phaseolus vulgaris*) to Metabolise Phaseollin. Neth. J. Pl. Path. **81**, 125 (1975).
794. VAN DEN HEUVEL, J., H. D. VANETTEN, J. W. SERUM, D. L. COFFEN, and T. H. WILLIAMS: Identification of 1a-Hydroxyphaseollone, a Phaseollin Metabolite Produced by *Fusarium solani*. Phytochemistry **13**, 1129 (1974).
795. VAN DUUREN, B. L.: Chemistry of Edulin, Neorautone, and Related Compounds from *Neorautanenia edulis*. J. Org. Chem. **26**, 5013 (1961).
796. VANETTEN, H. D.: Antifungal and Hemolytic Activities of Four Pterocarpan Phytoalexins. Phytopathology **62**, 795 (1972).
797. — Identification of a Second Antifungal Isoflavan from Diseased *Phaseolus vulgaris* Tissue. Phytochemistry **12**, 1791 (1973).
798. VANETTEN, H. D., and D. F. BATEMAN: Studies on the Mode of Action of the Phytoalexin Phaseollin. Phytopathology **61**, 1363 (1971).
799. VANETTEN, H. D., and S. G. PUEPPKE: Isoflavonoid Phytoalexins. In: J. FRIEND and D. R. THRELFALL, Biochemical Aspects of Plant-Parasite Relationships, p. 239. London: Academic Press. 1976.
800. VANETTEN, H. D., S. G. PUEPPKE, and T. C. KELSEY: 3,6a-Dihydroxy-8,9-methylenedioxypterocarpan as a Metabolite of Pisatin Produced by *Fusarium solani* f. sp. *pisi*. Phytochemistry **14**, 1103 (1975).
801. VAN HEERDEN, F. R., E. V. BRANDT, and D. G. ROUX: Structure and Synthesis of some

Complex Pyranoisoflavonoids from the Bark of *Dalbergia nitidula*. J. Chem. Soc. Perkin Trans. I **1978**, 137.

802. VAN HEERDEN, F. R., E. V. BRANDT, and D. G. ROUX: Synthesis of the Pyrano-isoflavonoid, Heminitidulan. Isoflavonoid and Rotenoid Glycosides from the Bark of *Dalbergia nitidula*. J. Chem. Soc. Perkin Trans. I **1980**, 2463.

803. VAN 'T LAND, B. G., E. D. WIERSMA-VAN DUIN, and A. FUCHS: *In vitro* and *In vivo* Conversion of Pisatin by *Ascochyta pisi*. Acta Bot. Neerl. **24**, 251 (1975).

804. VASCONCELOS, M. N. L., and J. G. S. MAIA: A Chemical Investigation of *Derris negrensis*. Acta Amazonica **6**, 59 (1976).

805. VENKATARAMAN, K.: Flavones and Isoflavones. Fortschr. Chem. Org. Naturstoffe **17**, 1 (1959).

806. VILAIN, C.: Barbigerone, a New Pyranoisoflavone from Seeds of *Tephrosia barbigera*. Phytochemistry **19**, 988 (1980).

807. VILAIN, C., and J. JADOT: Isoflavones of *Calopogonium mucunoides*. Structure of a New 6'',6''-Dimethyl-Pyrano-Isoflavone. Bull. Soc. R. Sci. Liège **44**, 306 (1975).

808. — — Isoflavones of *Calopogonium mucunoides*. Bull. Soc. R. Sci. Liège **45**, 468 (1976).

809. WADA, H.: Estrogenic Activity in Fresh and Dried Forages. Jap. J. Zootech. Sci. **34**, 248 (1963).

810. WADDELL, T. G., M. H. THOMASSON, M. W. MOORE, H. W. WHITE, D. SWANSON-BEAN, M. E. GREEN, G. S. VAN HORN, and H. M. FALES: Constituents of the Medicinal Herb *Wyethia mollis*. Phytochemistry. In press.

811. WALTER, E. D.: Genistin (an Isoflavone Glucoside) and its Aglucone, Genistein, from Soybeans. J. Am. Chem. Soc. **63**, 3273 (1941).

812. WALZ, E.: Isoflavone and Saponin Glucosides in *Soja hispida*. Justus Liebigs Annln. Chem. **489**, 118 (1931).

813. WANG, S., and M. HU: The Constitution of Belamcangenin and Belamcandin. J. Chem. Soc. **1944**, 307.

814. WARBURTON, W. K.: The Isoflavones. Q. Rev. Chem. Soc. **8**, 67 (1954).

815. WARSI, S. A., and A. KAMAL: Constituents of the Germs of *Cicer arietinum*. Pakistan J. Scient. Res. **3**, 85 (1951).

816. WEINSTEIN, L. I., M. G. HAHN, and P. ALBERSHEIM: Isolation and Biological Activity of Glycinol, a Pterocarpan Phytoalexin Synthesized by Soybeans. Pl. Physiol., Lancaster **68**, 358 (1981).

817. WELTRING, K. M., and W. BARZ: Degradation of 3,9-Dimethoxypterocarpan and Medicarpan by *Fusarium proliferatum*. Z. Naturforsch. **35c**, 399 (1980).

817a. WELTRING, K. M., W. BARZ, and P. M. DEWICK: Degradation of 3,9-Dimethoxypterocarpan and Medicarpin by *Fusarium* Fungi. Arch. Microbiol. **130**, 381 (1981).

818. WESSELY, F., F. LECHNER, and K. DINJAŠKI: Ononin. Mh. Chem. **63**, 201 (1933); Chem. Abstr. **28**, 2715 (1934).

819. WHALLEY, W. B.: Some Isoflavones Derived from Genistein. J. Am. Chem. Soc. **75**, 1059 (1953).

820. — 5:4'-Dihydroxy-8-methylisoflavone, and a Note on Lotoflavin. J. Chem. Soc. **1957**, 1833.

821. WILBAUX, R.: A Preliminary Examination of the Stems of *Ostryoderris lucida*. Bull. Agric. Congo Belge **37**, 434 (1946).

822. WOLFROM, M. L., W. D. HARRIS, G. F. JOHNSON, J. E. MAHAN, S. M. MOFFETT, and B. WILDI: Complete Structures of Osajin and Pomiferin. J. Am. Chem. Soc. **68**, 406 (1946).

823. WONG, E.: Pratensein. 5,7,3'-Trihydroxy-4'-methoxyisoflavone. J. Org. Chem. **28**, 2336 (1963).

824. — Structural and Biogenetic Relationships of Isoflavonoids. Fortschr. Chem. Org. Naturstoffe **28**, 1 (1970).

825. — The Isoflavonoids. In: J. B. HARBORNE, T. J. MABRY, and H. MABRY, The Flavonoids, p. 743. London: Chapman and Hall. 1975.

826. WONG, E., and C. M. FRANCIS: Flavonoids in Genotypes of *Trifolium subterraneum*. Mutants of the Geraldton Variety. Phytochemistry **7**, 2131 (1968).

827. WONG, E., and G. C. M. LATCH: Coumestans in Diseased White Clover. Phytochemistry **10**, 466 (1971).

828. WONG, E., P. I. MORTIMER, and T. A. GEISSMAN: Flavonoid Constituents of *Cicer arietinum*. Phytochemistry **4**, 89 (1965).

829. WOODWARD, M. D.: Phaseoluteone and other 5-Hydroxyisoflavonoids from *Phaseolus vulgaris*. Phytochemistry **18**, 363 (1979).

830. — New Isoflavonoids Related to Kievitone from *Phaseolus vulgaris*. Phytochemistry **18**, 2007 (1979).

831. — Phaseollin Formation and Metabolism in *Phaseolus vulgaris*. Phytochemistry **19**, 921 (1980).

832. — Identification of the Biosynthetic Precursors of Medicarpin in Inoculation Droplets on White Clover. Physiol. Pl. Path. **18**, 33 (1981).

833. WORSLEY, R. R. LE G.: The Insecticidal Properties of some East African Plants. *Tephrosia vogelii*. Ann. Appl. Biol. **21**, 649 (1934).

834. — The Insecticidal Properties of some East African Plants. *Mundulea suberosa*. Part 2. Chemical Constituents. Ann. Appl. Biol. **24**, 651 (1937).

835. — Report on Biochemistry. A. Rep. Biochem. Sect., East Afr. Agric. Res. Sta., Amani, **1938**, 27; Chem. Abstr. **34**, 4111 (1940).

835a. — The Histology and Physiology of Rotenoids in some Papilionaceae. Ann. Appl. Biol. **26**, 649 (1939).

836. WYMAN, J. G., and H. D. VANETTEN: Antibacterial Activity of Selected Isoflavonoids. Phytopathology **68**, 583 (1978).

837. ZÄHRINGER, U., E. SCHALLER, and H. GRISEBACH: Induction of Phytoalexin Synthesis in Soybean. Structure and Reactions of Naturally Occurring and Enzymatically Prepared Prenylated Pterocarpans from Elicitor-Treated Cotyledons and Cell Cultures of Soybean. Z. Naturforsch. **36c**, 234 (1981).

838. ZAPESOCHNAYA, G. G., and N. A. LAMAN: The Structures of Isoflavone C-Glycosides from *Lupinus luteus*. Khim. Prir. Soedin. **1977**, 862.

839. ZAPESOCHNAYA, G. G., and I. A. SAMYLINA: Daidzin from *Psoralea acaulis*. Khim. Prir. Soedin. **1974**, 671.

840. ZEMPLÉN, G., and R. BOGNÁR: Sophorabioside, a New Glucoside from *Sophora japonica*. Ber. Dt. Chem. Ges. **75B**, 482 (1942).

841. ZEMPLÉN, G., R. BOGNÁR, and L. FARKAS: Determination of the Structure of Sophoricoside, an Isoflavone Glycoside of *Sophora japonica*. Ber. Dt. Chem. Ges. **76B**, 267 (1943).

842. ZEMPLÉN, G., L. FARKAS, and A. BIEN: Synthesis of Ononin. Ber. Dt. Chem. Ges. **77B**, 452 (1944).

843. ZILG, H., and H. GRISEBACH: Identification and Biosynthesis of Coumestans in *Soja hispida*. Phytochemistry **7**, 1765 (1968).

844. — — Coumestans in *Cicer arietinum*. Phytochemistry **8**, 2261 (1969).

845. ZWEEKHORST-VAN LAER, A. M. H., and T. H. A. NELEN: Flavonoids of *Cadia purpurea*. Pharm. Weekbl. Ned. **111**, 1289 (1976).

(Received February 18, 1982)

Note Added in Proof

Apart from daidzein (3), daidzin (511) and puerarin (518), the roots of *Pueraria lobata* (= *thunbergiana*) have been found to contain daidzein-7,4'-di-*O*-glucoside (514). Isoflavones (3), (511) and (518) are reported to relieve headaches and other symptoms of hypertension (Chem. Abstr. **82**, 47672). *Pueraria* roots may also contain 3'-hydroxypuerarin and 3'-methoxypuerarin (in: Experimental Methods of Organic Natural Products, Natori, S. *et al.,* eds., p. 470, Kodansha Publishers, Tokyo 1977). 6-Hydroxy-dehydrotoxicarol and the anti-bacterial 2-arylbenzofuran, licobenzofuran, have been isolated from fruits and roots of *Amorpha fruticosa* and *Glycyrrhiza glabra* respectively (Chem. Abstr. **97**, 107015, and **97**, 20701). Daidzein (3) and glycitein (19) both exhibit weak anti-oxidant activity (Chem. Abstr. **92**, 57009).

The Sarpagine-Ajmaline Group
of Indole Alkaloids

By A. KOSKINEN and M. LOUNASMAA, Laboratory for Organic Chemistry, Department of Chemistry, Technical University of Helsinki, Finland

With 26 Figures

Contents

1. Introduction*

The indole alkaloids of the sarpagine-ajmaline type comprise one of the largest groups of structurally related indolic natural products: 45 compounds of the sarpagine type, 53 of the ajmaline type, and four bisindole alkaloids containing one or the other. The compounds are widely dispersed

---

* Throughout the text, the "biogenetic numbering" of J. Le Men and W. I. Taylor [*Experientia* **21** (1965) 508] is used.

in 25 plant genera, mainly in *Apocynaceae*, the most important genus being *Rauwolfia* (*1*). Some 100 plant species are currently known to contain members of this vast group of alkaloids.

The sarpagine-ajmaline type alkaloids have only twice been the subject of comprehensive reviews, both by W. I. TAYLOR in the 1960's (*2*a, *2*b). Since the last review (*2*b) in 1968 the number of known structures has doubled from 49 to the present 102.

Synthetic work in the field has met with little success: only three total syntheses leading to ajmaline are known and the other members of the group have so far eluded the synthetic organic chemist.

Ajmaline, pharmacologically the most important representative of the group, was isolated as early as 1931 by SIDDIQUI and SIDDIQUI (*5*) from *Rauwolfia serpentina*. The compound was named after Hakim Ajmal Khan, founder of the Tibbi College in Delhi. Independently VAN ITALLIE and STEENHAUSER (*6*) isolated the same plant base in 1932 and named it rauwolfine. This latter name has been discarded. The structure elucidation of ajmaline with its hexacyclic ring system proved exceedingly difficult. The structure was studied (*7*) in ROBINSON's laboratory in Oxford and the important structural elements were found in 1949. Strychnine or yohimbine skeleta were proposed. Five years later, the skeleton had been worked out and in 1956 WOODWARD was able to deduce the stereochemistry of the compound (*8*).

The biogenesis of the indole alkaloids has recently attracted much interest. In this review it is our aim to bring together in an organized way accepted views of the metabolic processes leading to the sarpagine-ajmaline type alkaloids. The structural complexity suggests that the biogenesis and chemistry of these alkaloids displays many interesting features. However, it cannot be overemphasized that the compounds form an overwhelmingly homogeneous group of natural substances.

In the context of the biogenetic speculations presented later in section 4, it would not seem inappropriate to be reminded of EINSTEIN's words: "I want to know how God created this world. I am not interested in this or that phenomenon; in the spectrum of this or that element. I want to know His thoughts, the rest are details." Bearing these words in mind, we have tried to apply some form of teleological principles in scheduling the biosynthetic pathways leading to the various sarpagine-ajmaline type alkaloids. Thus, it seems that Nature works with a minimum number of transformations to bring about a maximum number of related substances. We hope that this review will be of aid to chemists working in the field and that the ideas put forth will add to our understanding of the chemistry of living Nature.

2. Skeletal Types

We shall begin by considering the different structural features of the sarpagine type and ajmaline type alkaloids. The bisindole alkaloids, which contain either sarpagine or ajmaline type monomeric building blocks, are discussed separately in section 2.3.

The sarpagine alkaloids contain the common structural element of the parent pentacyclic sarpagan (**1**) (*9*) ring system. The ajmaline alkaloids on the other hand contain the hexacyclic ajmalan (**2**) (*9*) ring system. Additional cyclic structures are allowed within each of these groupings, but alkaloids in which the basic ring systems are modified or destroyed, are excluded. Koumine, for example, which it has been proposed is bio-genetically derived from sarpagine type alkaloids, (*10*) will not be considered in detail in this review.

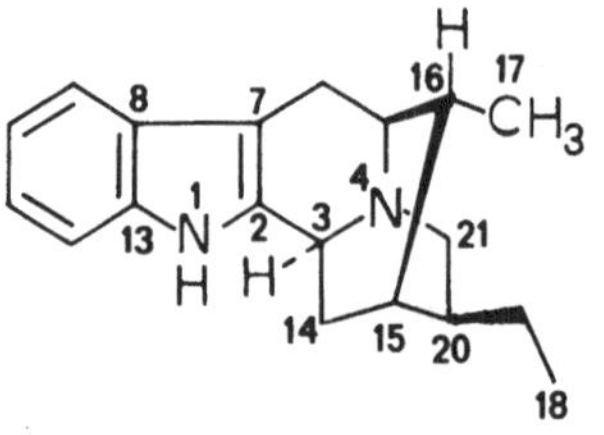

(**1**) Sarpagan

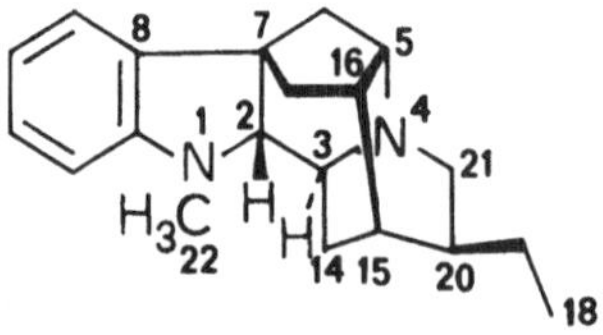

(**2**) Ajmalan

Fig. 1

3-Oxygenated sarpagine type alkaloids, which usually exist as the corresponding 2-acylindole alkaloids, will also be excluded since reviews on these compounds have recently been published (*11, 12*).

2.1. Sarpagine Type Alkaloids

As already mentioned, these alkaloids share the common structural feature of the sarpagan ring system. Usually they contain a double bond at $C_{19}-C_{20}$ and C_{21} lacks other functionality. Many of them have oxygen substituents on the aromatic ring. In all cases known, C_3 has S-stereochemistry, i. e. H_3 is α. The rigid ring system also requires C_5 to be S and C_{15} has the natural α-configuration. Thus, three of the stereochemical centres have a fixed configuration because of the rigid framework. In all compounds known, the $C_{19}-C_{20}$ double bond has E stereochemistry. The subgroups presented below are arranged according to the biogenetic relationships presented later in the review.

2.1.1. Polyneuridine Subgroup

All the indole alkaloids belonging to this subgroup (**3—14**) are derived from polyneuridine (**4**) having the R configuration at C_{16}.

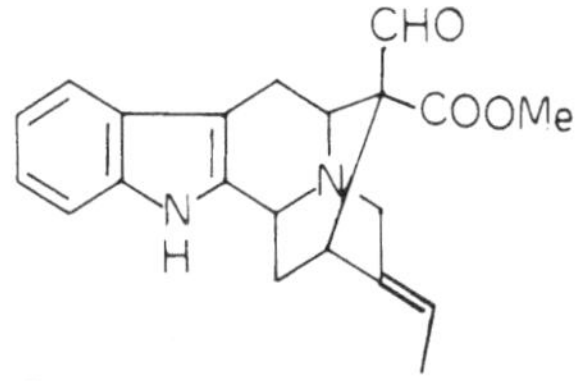

(**3**) Polyneuridine aldehyde

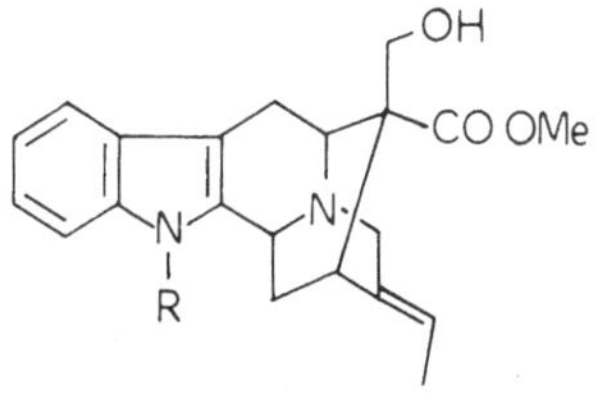

(**4**) R = H   Polyneuridine
(**5**) R = Me   Voachalotine

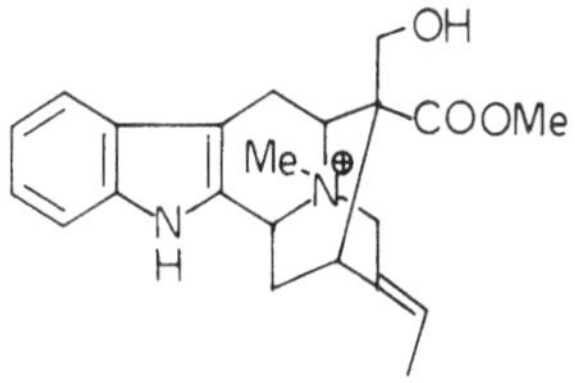

(**6**) Macusine A

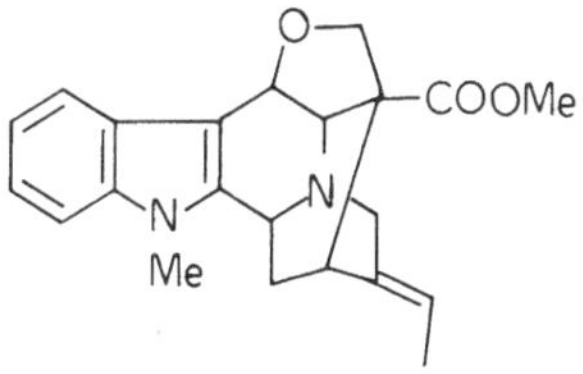

(**7**) Dehydrovoachalotine

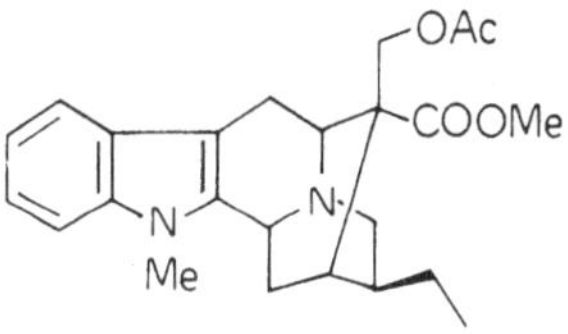

(**8**) 17-O-Acetyl-19,20-dihydro-
voachalotine

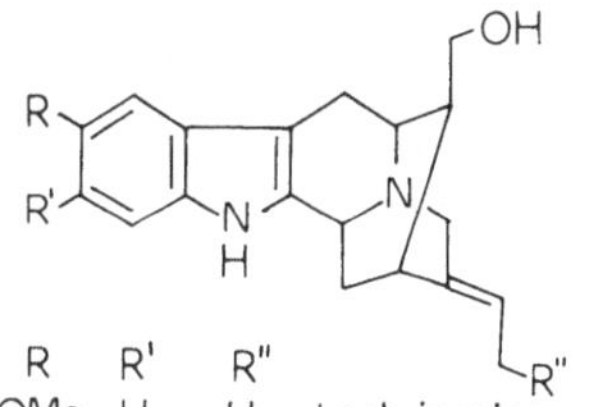

	R	R'	R"	
(**9**)	OMe	H	H	Lochvinerine
(**10**)	H	OMe	H	Gardnerine
(**11**)	H	OMe	OH	Hydroxygardnerine

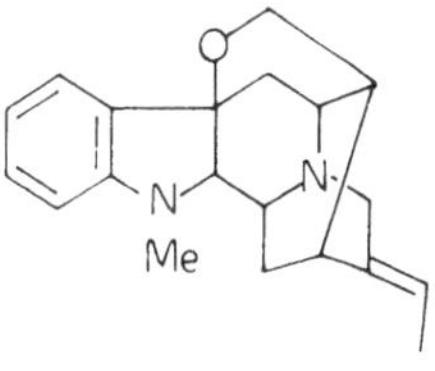

(**12**) Rauwolfinine

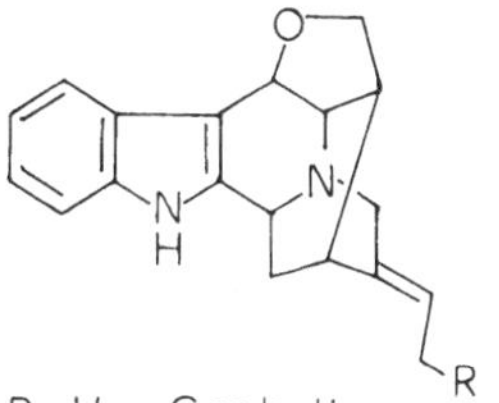

(**13**) R= H   Gardnutine
(**14**) R= OH   Hydroxygardnutine

Fig. 2

2.1.2. Akuammidine Subgroup

The members of this subgroup (**15—41**) are derived from the hypothetic 16S isomer of polyneuridine aldehyde (**3**). The subgroup consists of 27 alkaloids.

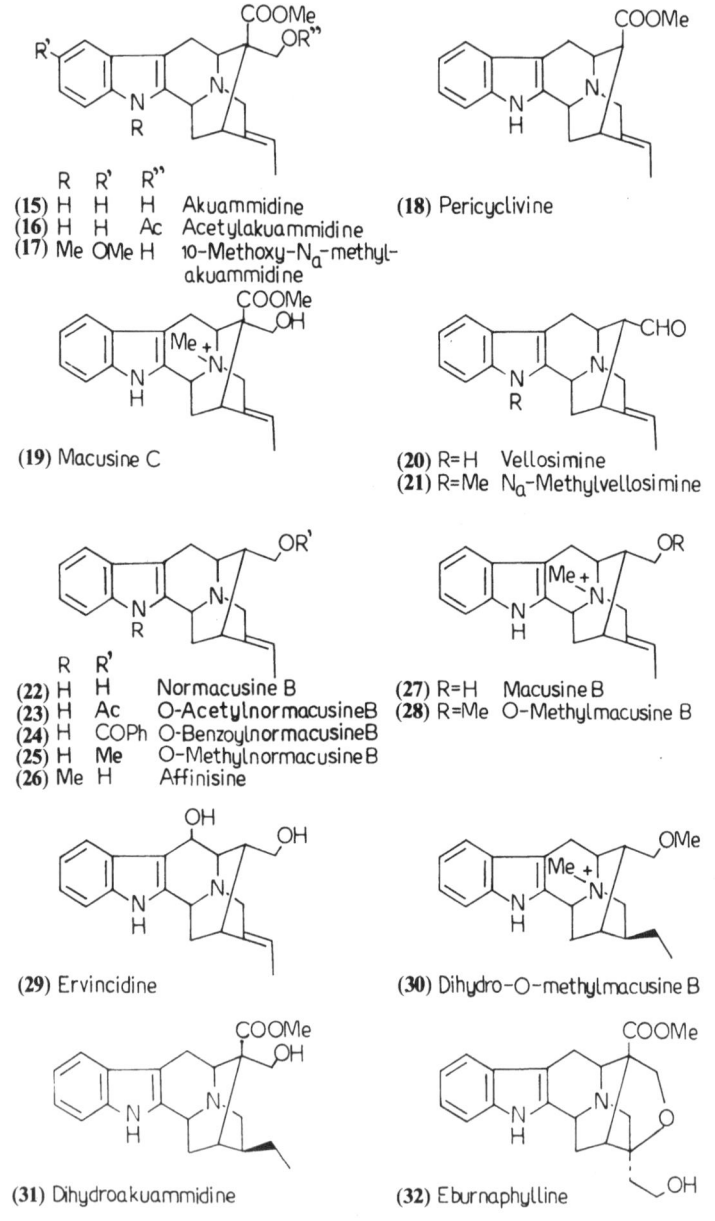

	R	R'	R"	
(15)	H	H	H	Akuammidine
(16)	H	H	Ac	Acetylakuammidine
(17)	Me	OMe	H	10-Methoxy-N_a-methyl-akuammidine

(18) Pericyclivine

(19) Macusine C

(20) R=H Vellosimine
(21) R=Me N_a-Methylvellosimine

	R	R'	
(22)	H	H	Normacusine B
(23)	H	Ac	O-Acetylnormacusine B
(24)	H	COPh	O-Benzoylnormacusine B
(25)	H	Me	O-Methylnormacusine B
(26)	Me	H	Affinisine

(27) R=H Macusine B
(28) R=Me O-Methylmacusine B

(29) Ervincidine

(30) Dihydro-O-methylmacusine B

(31) Dihydroakuammidine

(32) Eburnaphylline

Fig. 3

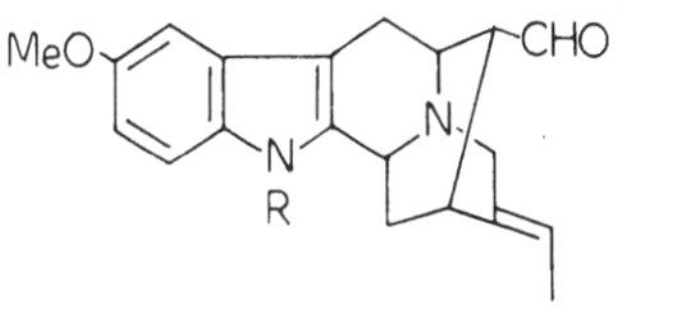

(33) R= H   10-Methoxyvellosimine
(34) R =Me   Majvinine

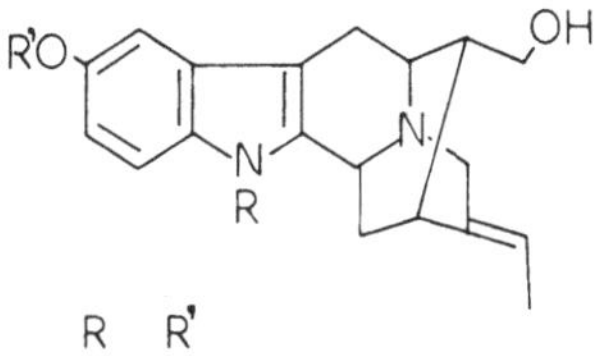

	R	R'	
(35)	H	H	Sarpagine
(36)	H	Me	Lochnerine
(37)	Me	H	N_α-Methylsarpagine

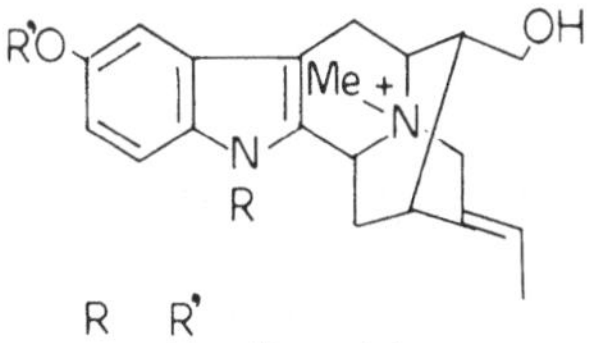

	R	R'	
(38)	H	H	Spegatrine
(39)	H	Me	Lochneram
(40)	Me	H	N_α-Methylsarpagine metho salt

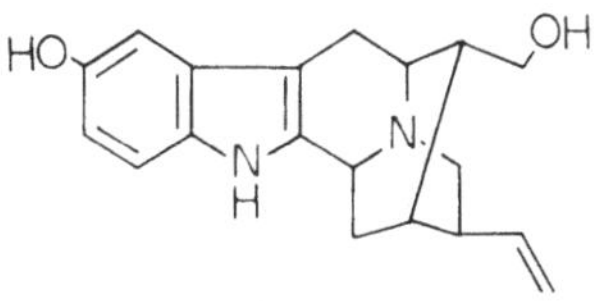

(41) Neosarpagine

Fig. 3 (continued)

2.1.3. Voamonine-Voacoline

Voamonine (42) und voacoline (43) are biogenetically closely related alkaloids and characterized by a hemiacetal structure between the C_{19}-carbonyl and C_{17}-hydroxyl groups.

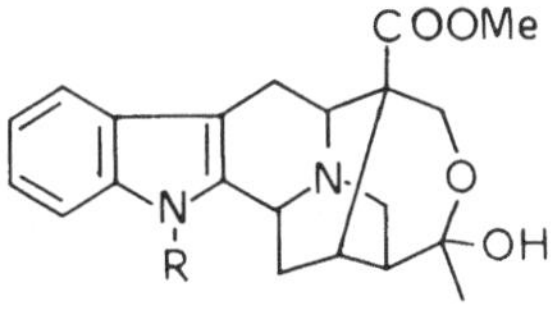

(42) R= H   Voamonine
(43) R = Me   Voacoline

Fig. 4

2.1.4. Talpinine

Talpinine (44) is unique among the known sarpagine type alkaloids in possessing a cyclic ether ring system.

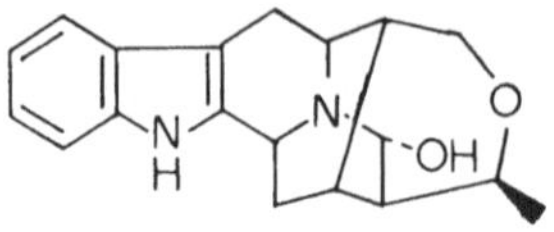

(44) Talpinine

Fig. 5

2.1.5. Vomifoline Subgroup

This subgroup is also small, having only three representatives (45), (46), and (47). Characteristic to all of them is the rearranged carbon skeleton having a C-methyl group α to the non-indolic nitrogen and a cyclic hemiacetal system [(45) and (46)].

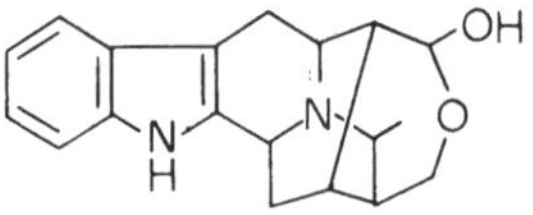

(45) Vomifoline

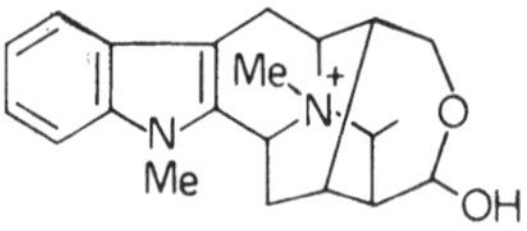

(46) Macrosalhine

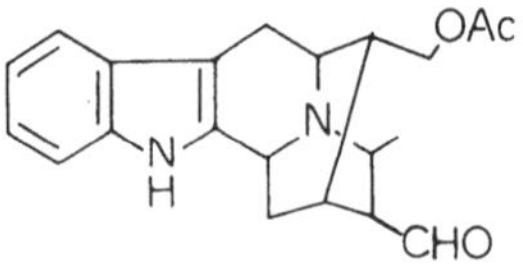

(47) O-Acetylpreperakine

Fig. 6

2.2. Ajmaline Type Alkaloids

Common to the ajmaline type indole alkaloids is the rigid hexacyclic ajmalan ring structure (2). As with the sarpagine type alkaloids, the aromatic ring may bear an oxygen substituent. However, the presence of a hydroxyl group at C_{21} and saturation of the $C_{19} - C_{20}$ double bond are far more common in these alkaloids than in the previous group.

The rigid molecular framework again requires C_3 and C_5 to be of S stereochemistry, and C_{15} has the natural α-configuration. In cases where stereochemical diversity is possible at C_{17}, the R configuration is usually seen. This seems reasonable considering the mechanism of the cyclisation, maximum overlap of the π-system making eq 1 more favourable than eq 2.

In fact, the 17R configuration would appear to be the "natural" one produced by direct cyclisation, the epimeric 17S configuration being derived from the ajmalan ring *via* oxidation to the corresponding 17-keto alkaloid and re-reduction from the less hindered side [as in hydride reduction of purpeline (**90**), for example (*13*)].

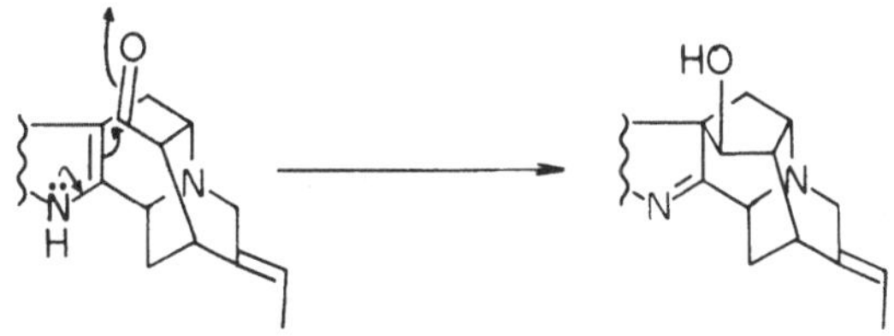

Equation 1

Equation 2

The alkaloids of this group, like the sarpagine type alkaloids, are divided into subgroups according to their proposed biogenetic relationships.

2.2.1. Perakine Subgroup

The perakine type alkaloids (**48**) and (**49**) contain a rearranged carbon skeleton derivable from vomilenine (**62**).

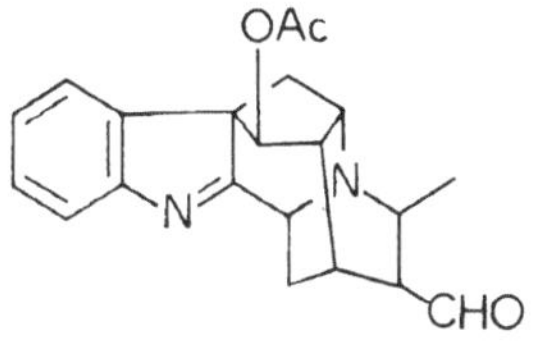

(**48**) Perakine

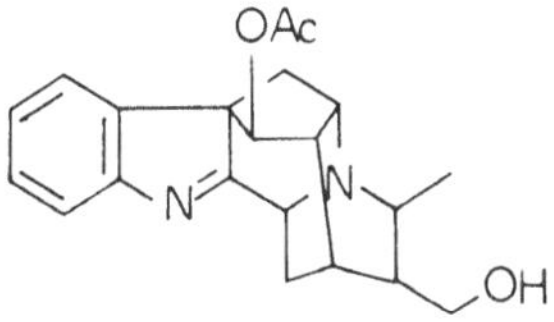

(**49**) Raucaffrinoline

Fig. 7

2.2.2. Quebrachidine Subgroup

Common to these alkaloids (**50**—**60**) is the 2S stereochemistry. Also, C_{16} bears a carbomethoxy substituent and a $C_{19}-C_{20}$ double bond is present. No further functionality is observed at C_{21}. It has been suggested that vincarine (**50′**) is a stereoisomer of quebrachidine (**50**) (*14*).

	R	R'	
(**50**)	H	H	Quebrachidine
(**51**)	H	TMB	17–O–Trimethoxy-benzoylquebrachidine
(**52**)	Me	H	Vincamajine
(**53**)	Me	Ac	Vincamedine
(**54**)	Me	COPh	Benzoylvincamajine
(**55**)	Me	TMC	17–O–Trimethoxy-cinnamoylvincamajine

	R	R'	
(**56**)	H	TMB	10-Hydroxy-17-O-trimethoxy-benzoylvincamajine
(**57**)	H	TMC	10-Hydroxy-17-O-trimethoxy-cinnamoylvincamajine
(**58**)	Me	H	Methoxyvincamajine
(**59**)	Me	TMB	10-Methoxy-17-O-trimethoxy-benzoylvincamajine
(**60**)	Me	TMC	10-Methoxy-17-O-trimethoxy-cinnamoylvincamajine

Fig. 8

2.2.3. Indolenine Subgroup

These natural products (**61**—**66**) contain an indolenine structure as their common feature. The hydroxyl group at C_{17} is acetylated in all cases known.

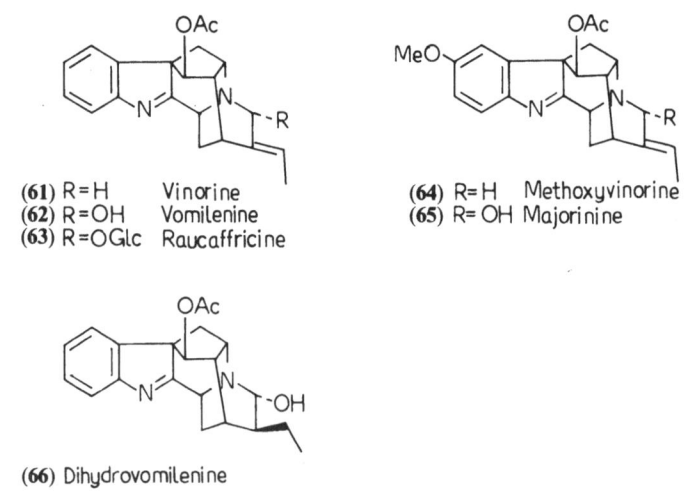

(**61**) R = H   Vinorine
(**62**) R = OH   Vomilenine
(**63**) R = OGlc   Raucaffricine

(**64**) R = H   Methoxyvinorine
(**65**) R = OH   Majorinine

(**66**) Dihydrovomilenine

Fig. 9

2.2.4. Tetraphyllicine Subgroup

The tetraphyllicine alkaloids have the 2β configuration (2R) and bear an oxygen substituent at C_{17}. The $C_{19} - C_{20}$ double bond is intact.

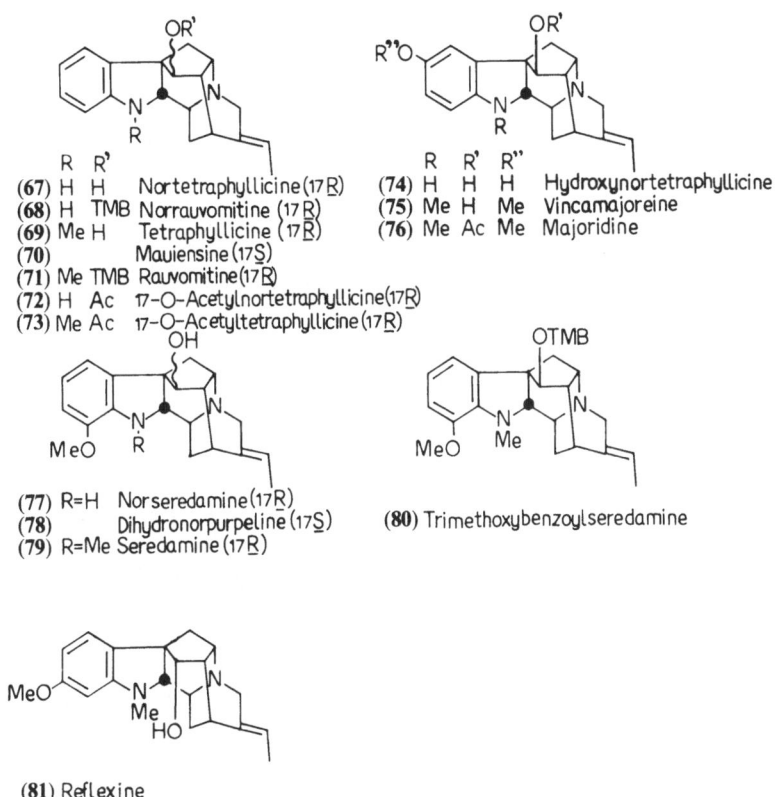

	R	R'	
(67)	H	H	Nortetraphyllicine (17R)
(68)	H	TMB	Norrauvomitine (17R)
(69)	Me	H	Tetraphyllicine (17R)
(70)			Mauiensine (17S)
(71)	Me	TMB	Rauvomitine (17R)
(72)	H	Ac	17-O-Acetylnortetraphyllicine (17R)
(73)	Me	Ac	17-O-Acetyltetraphyllicine (17R)

	R	R'	R"	
(74)	H	H	H	Hydroxynortetraphyllicine
(75)	Me	H	Me	Vincamajoreine
(76)	Me	Ac	Me	Majoridine

(77)	R=H	Norseredamine (17R)
(78)		Dihydronorpurpeline (17S)
(79)	R=Me	Seredamine (17R)

(80) Trimethoxybenzoylseredamine

(81) Reflexine

Fig. 10

2.2.5. 17-Keto Alkaloids

Being derived from nortetraphyllicine (67), these alkaloids have (2R) configuration. The 17-hydroxyl group is oxidized to the level of a carbonyl group (Fig. 11).

2.2.6. Ajmaline Subgroup

The members of the ajmaline subgroup bear an additional characteristic oxygen substituent on C_{21}. The $C_{19} - C_{20}$ double bond is hydrogenated and the configuration at C_{20} is usually S (Fig. 12).

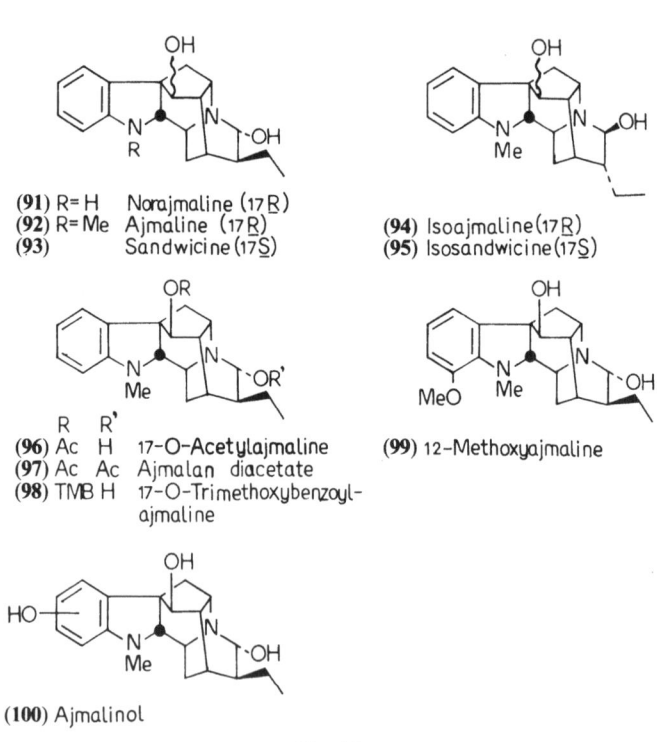

(82) R=H   Rauflorine
(83) R=OH   Endolobine

(84) Rauflexine

(85) R=H   Ajmalidine
(86) R=OMe   Vomalidine

	R	R'	
(87)	H	H	Normitoridine
(88)	H	Me	Norpurpeline
(89)	Me	H	Mitoridine
(90)	Me	Me	Purpeline

Fig. 11

(91) R=H   Norajmaline (17$\underline{R}$)
(92) R=Me   Ajmaline (17$\underline{R}$)
(93)   Sandwicine (17$\underline{S}$)

(94) Isoajmaline(17$\underline{R}$)
(95) Isosandwicine(17$\underline{S}$)

	R	R'	
(96)	Ac	H	17-O-Acetylajmaline
(97)	Ac	Ac	Ajmalan diacetate
(98)	TMB	H	17-O-Trimethoxybenzoyl-ajmaline

(99) 12-Methoxyajmaline

(100) Ajmalinol

Fig. 12

2.3. Bisindole Alkaloids

Four bisindole alkaloids of sarpagine-ajmaline ring skeleton have been reported. Alkaloids in which C_3 is oxygenated will not be considered in this review [e. g. accedinine (**102**)].

Geissolosimine (**103**) (*15*) was reported in 1958, the first alkaloid of bisindole type to be isolated. The other members are accedinisine (**101**) (*16*), macralstonidine (**104**) (*17*) and alstonisidine (**105**) (*18*).

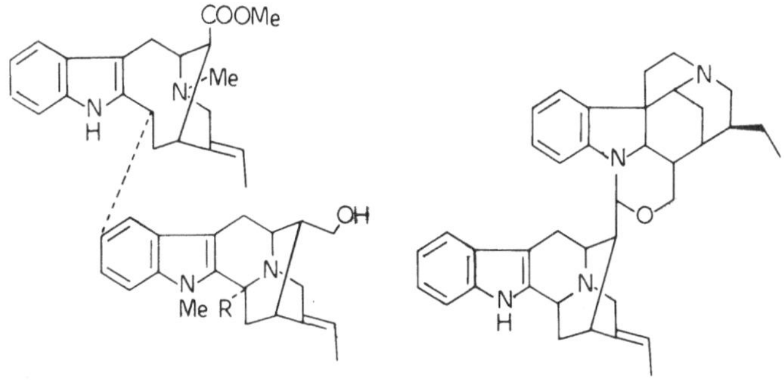

(**101**) R=H   Accedinisine
(**102**) R=OH Accedinine

(**103**) Geissolosimine

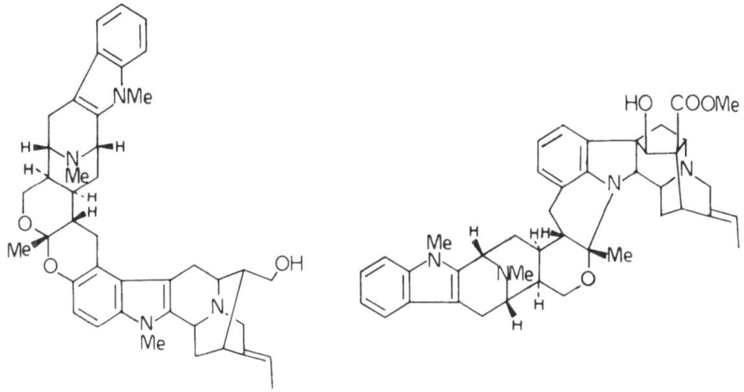

(**104**) Macralstonidine

(**105**) Alstonisidine

Fig. 13

3. Occurrence

The many sarpagine-ajmaline type indole alkaloids are widely distributed in nature. At present 103 plant species in 25 genera have been reported to contain alkaloids of this group. According to the classification presented by HESSE (*4*) (Table 1) these belong mainly to the family Apocynaceae; exceptions are *Gardneria nutans* (Loganiaceae) and five species of the genus *Strychnos* (Strychnaceae).

Table 1. *Botanical Classification of Plants Containing Sarpagine-Ajmaline Type Alkaloids*

A. Family:	Apocynaceae		
I. Subfamily:	Plumerioideae		
1. Tribe:	Carisseae		
	Hunteria	*Picralima*	
	Melodinus	*Pleiocarpa*	
4. Tribe:	Tabernaemontaneae		
	Gabunia	*Peschiera*	
	Hazunta	*Tabernaemontana*	
	Pandaca	*Voacanga*	
5. Tribe:	Alstonieae		
	Alstonia	*Geissospermum*	*Vinca*
	Aspidosperma	*Gonioma*	
	Catharanthus	*Rhazya*	
	Diplorrhyncus	*Tonduzia*	
6. Tribe:	Rauvolfieae		
	Cabucala	*Rauwolfia*	
	Ochrosia	*Vallesia*	
B. Family:	Loganiaceae.	genus: *Gardneria*	
C. Family:	Strychnaceae	genus: *Strychnos*	

The distribution of the alkaloids among different species is presented in Table 2.

References, pp. 334—346

Table 2. *Occurrence of Sarpagine-Ajmaline Type Alkaloids in Different Plants*

Source	Compound	No.	Ref.
Apocynaceae			
Alstonia			
boonei De Wild.	Akuammidine	(15)	(19)
constricta Muell.	17-O-Trimethoxybenzoylquebrachidine	(51)	(20)
	Vincamajine	(52)	(20)
	17-O-Trimethoxycinnamoylvincamajine	(55)	(20)
lanceolifera Moore	10-Methoxy-N$_a$-methylakuammidine	(17)	(21)
	17-O-Trimethoxycinnamoylvincamajine	(55)	(21)
	10-Hydroxy-17-O-trimethoxybenzoylvincamajine	(56)	(21)
	10-Hydroxy-17-O-trimethoxycinnamoylvincamajine	(57)	(21)
	Methoxyvincamajine	(58)	(21)
	10-Methoxy-17-O-trimethoxybenzoylvincamajine	(59)	(21)
	10-Methoxy-17-O-trimethoxycinnamoylvincamajine	(60)	(21)
macrophylla Wall.	Affinisine	(26)	(22)
	Macrosalhine	(46)	(46)
	Benzoylvincamajine	(54)	(24)
	Macralstonidine	(104)	(17)
muelleriana Domin	Alstonisidine	(105)	(25)
odontophora Boiteau	Quebrachidine	(50)	(26)
	Vincamajine	(52)	(26)
scholaris R. Br.	Akuammidine	(15)	(27)
somersetensis Bailey	Macralstonidine	(104)	(28)
spectabilis R. Br.	N$_a$-Methylsarpagine	(37)	(29)
	Quebrachidine	(50)	(29)
	Vincamajine	(52)	(29)
	Macralstonidine	(104)	(29)
Aspidosperma			
dasycarpon DC.	Polyneuridine aldehyde	(3)	(30)
peroba Allem *ex* Sald. =	Polyneuridine	(4)	(32)
polyneuron Muell. Arg.	Normacusine B	(22)	(32)
	Macusine B	(27)	(31)

Table 2 (continued)

Source	Compound	No.	Ref.
quebracho-blanco Schlecht.	Akuammidine	(15)	(33)
	Acetylakuammidine	(16)	(35)
	Quebrachidine	(50)	(34)
spegazzinii Molf. ex Meyer	Macusine C	(19)	(36)
	Spegatrine	(38)	(36)
	Ajmaline	(92)	(36)
Cabucala	Vellosimine	(20)	(37)
erythrocarpa var.	Quebrachidine	(50)	(37)
erythrocarpa (Vatke) MGF			
striolata Pichon	Quebrachidine	(50)	(38)
Catharanthus	Pericyclivine	(18)	(39)
lanceus Pichon	Akuammidine	(15)	(40)
longifolius Pichon	Pericyclivine	(18)	(40)
	Normacusine B	(22)	(40)
roseus G. Don	Pericyclivine	(18)	(41)
	Lochnerine	(36)	(42)
Diplorrhyncus	Normacusine B	(22)	(43)
condylocarpon Muell. Arg.			
Gabunia	Pericyclivine	(18)	(44)
odoratissima Stapf.			
Geissospermum	Vellosimine	(20)	(15)
vellosii Allem.	Normacusine B	(22)	(15)
	Geissolosimine	(103)	(15)
Gonioma	Akuammidine	(15)	(45)
kamassi E. Mey			
Hazunta	Akuammidine	(15)	(46)
modesta modesta Pichon	Polyneuridine	(4)	(46)
var. brevituba MGF.			
var. divaricata MGF.			
modesta var. modesta			
subvar. montana	Pericyclivine	(18)	(47)
Hunteria	Eburnaphylline	(32)	(48)
eburnea Pichon			

Genus	Species	Alkaloid		
Melodinus	*australis* (Muell.) Pierre	Akuammidine	(15)	(49)
	balansae var. *paucivenosus* (Moore) Boiteau	Ajmaline	(92)	(59)
	celastroides Baill.	Akuammidine	(15)	(50)
Ochrosia	*nakaiana* Koidz.	Akuammidine	(15)	(51)
Pandaca[a]	*ochrascens*	Akuammidine	(15)	(52)
Peschiera	*affinis* Muell. Arg.	Affinisine	(26)	(53)
		Akuammidine	(15)	(55)
	laeta Mart.	Normacusine B	(22)	(54)
Picralima	*klaineana* Pierre	Akuammidine	(15)	(56)
	nitida Stapf.	Akuammidine	(15)	(57)
Pleiocarpa	*mutica* Benth.	N_a-Methylsarpagine methosalt	(40)	(58)
	pycnantha Stapf.	Macusine B	(27)	(62)
	talbotii Wernham	Normacusine B	(22)	(61)
		Talpinine	(44)	(60)
	tubicina Stapf.	Macusine B	(27)	(62)
		N_a-Methylsarpagine methosalt	(40)	(62)
Rauwolfia	*balansae* spp. *balansae* Boiteau	Raucaffrinoline	(49)	(63)
		Vinorine	(61)	(63)
		Vomilenine	(62)	(63)
		Dihydrovomilenine	(66)	(63)
	spp. *schumanniana* var. *basicola* Boiteau	Raucaffrinoline	(49)	(63)
		Vinorine	(61)	(63)
		Vomilenine	(62)	(63)
		Dihydrovomilenine	(66)	(63)
	beddomei Hook	Sarpagine	(35)	(64)
	boliviana MGF.	Ajmaline	(92)	(65)
		Sarpagine	(35)	(69)
		Vomifoline	(45)	(70)
		Perakine	(48)	(67, 70)
		Raucaffrinoline	(49)	(68)
	caffra Sond.	Raucaffricine	(63)	(67)
		Ajmaline	(92)	(66)

Table 2 (continued)

Source	Compound	No.	Ref.
cambodiana Pierre	Ajmaline	(92)	(71)
canescens L.	Sarpagine	(35)	(72)
	Ajmaline	(92)	(72)
chinensis Hemsl.	Ajmaline	(92)	(73)
confertiflora Pichon	Rauflorine	(82)	(74)
cumminsii Stapf.	Pericyclivine	(18)	(77)
	Normacusine B	(22)	(78)
	O-Methylnormacusine B	(25)	(76)
	O-Methylnormacusine B N_b-oxide	(25) N_b-oxide	(78)
	Sarpagine	(35)	(77)
	Vomifoline	(45)	(78)
	Nortetraphyllicine	(67)	(76)
	Tetraphyllicine	(69)	(77)
	Norseredamine	(77)	(76)
	Dihydronorpurpeline	(78)	(75)
	Seredamine	(79)	(77)
	Trimethoxybenzoylseredamine	(80)	(76)
	Endolobine	(83)	(76)
	Vomalidine	(86)	(77)
	Normitoridine	(87)	(76)
	Norpurpeline	(88)	(76)
	Mitoridine	(89)	(75)
	Purpeline	(90)	(78)
	Ajmaline	(92)	(77)
	Ajmalan diacetate	(97)	(77)
decurva Hook.	Sarpagine	(35)	(79)
degeneri Sherff.	Tetraphyllicine	(69)	(80)
	Ajmaline	(92)	(80)
densiflora Benth.	Sarpagine	(35)	(81)
	Ajmaline	(92)	(82)
discolor[b]	Quebrachidine	(50)	(83)

Species	Alkaloid		
fruticosa Burck.	Ajmaline	(92)	(84)
heterophylla Roem. Schult. = *hirsuta* Jacq.	Sarpagine	(35)	(86)
indecora Woodson	Ajmaline	(92)	(85)
ligustrina R. & S.	Ajmaline	(92)	(87)
	Sarpagine	(35)	(87)
	Sarpagine	(92)	(88)
	Ajmaline	(92)	(88)
macrophylla Stapf.	Normacusine B	(22)	(90)
	Norajmaline	(91)	(89)
	Ajmaline	(92)	(89)
mannii Stapf.	Vincamajine	(52)	(91)
mauiensis Sherff.	Tetraphyllicine	(69)	(80)
	Mauiensine	(70)	(80)
	Ajmalidine	(84)	(92)
	Sandwicine	(93)	(80)
	Neosarpagine	(41)	(93)
micrantha Hook	Ajmaline	(92)	(94)
mombasiana Stapf.	Normacusine B	(22)	(95)
	Dihydroakuammidine	(31)	(96)
	Sarpagine	(35)	(96)
	Vomifoline	(45)	(95)
	Nortetraphyllicine	(67)	(96)
	Tetraphyllicine	(69)	(96)
	Endolobine	(83)	(95)
	Vomalidine	(86)	(96)
	Norpurpeline	(88)	(95)
	Purpeline	(90)	(95)
	Ajmaline	(92)	(96)
nitida Facq. Enum.	Vellosimine	(20)	(97)
	N_a-Methylvellosimine	(21)	(97)
	Normacusine B	(22)	(97)
	Sarpagine	(35)	(97)
	Lochnerine	(36)	(97)
	Vomifoline	(45)	(97)
	Raucaffrinoline	(49)	(97)
	Nortetraphyllicine	(67)	(97)

Table 2 (continued)

Source	Compound	No.	Ref.
	Tetraphyllicine	(69)	(97)
	17-O-Acetylnortetraphyllicine	(72)	(97)
	Ajmalidine	(85)	(97)
	Norajmaline	(91)	(97)
	Ajmaline	(92)	(97)
obscura Schum.	Tetraphyllicine	(69)	(98)
	Rauvomitine	(71)	(100)
	Vomalidine	(86)	(99)
	Norajmaline	(91)	(100)
	Ajmaline	(92)	(98)
	17-O-Trimethoxybenzoylajmaline	(98)	(100)
	12-Methoxyajmaline	(98)	(99)
oregiton MGF.	Normacusine B	(22)	(102)
	Vomifoline	(45)	(102)
	Nortetraphyllicine	(67)	(102)
	Tetraphyllicine	(69)	(103)
	Ajmaline	(92)	(101)
perakensis King. Gamble	Normacusine B	(22)	(105)
	Sarpagine	(35)	(104)
	Vomifoline	(45)	(105)
	Perakine	(48)	(104)
	Vinorine	(61)	(105)
	Ajmaline	(92)	(104)
reflexa Koerd. Valet	Reflexine	(81)	(106, 107)
	Rauflexine	(84)	(106, 107)
salicifolia Griseb.	Vellosimine	(20)	(108)
	Ajmalidine	(85)	(108)
sandwicensis DC.	Tetraphyllicine	(69)	(80)
	Sandwicine	(93)	(80)
schueli Speg.	Ajmaline	(92)	(109)
sellowii Muell.	Tetraphyllicine	(69)	(110)
	Ajmalidine	(85)	(110)
	Ajmaline	(92)	(110)

Species	Alkaloid		
serpentina Benth.	Rauwolfinine	(12)	(111)
	Sarpagine	(35)	(112)
	Tetraphyllicine	(69)	(113)
	Ajmaline	(92)	(5)
	Isoajmaline	(94)	(6)
sevenetii Boiteau	Raucaffrinoline	(49)	(63)
	Vinorine	(61)	(63)
	Vomilenine	(62)	(63)
	Dihydrovomilenine	(66)	(63)
spathulata Boiteau	Raucaffrinoline	(49)	(63)
	Vinorine	(61)	(63)
	Vomilenine	(62)	(63)
	Dihydrovomilenine	(66)	(63)
suaveolens Moore	Polyneuridine	(4)	(114)
	Normacusine B	(22)	(114)
	Lochnerine	(36)	(114)
	Tetraphyllicine	(69)	(114)
	Norajmaline	(91)	(114)
	Ajmaline	(92)	(114)
sumatrana Jack	Norseredamine	(76)	(115)
	Ajmaline	(92)	(83)
tetraphylla L.	Tetraphyllicine	(69)	(116)
	Ajmaline	(92)	(117)
verticillata Chevalier	Vellosimine	(20)	(119)
	Vomifoline	(45)	(119)
	Ajmaline	(92)	(118)
volkensii Stapf.	Polyneuridine	(4)	(120)
	Normacusine B	(22)	(120)
	O-Acetylnormacusine B	(23)	(121)
	Sarpagine	(35)	(122)
	Vomifoline	(45)	(120)
	O-Acetylpreperakine	(47)	(120)
	Perakine	(48)	(122)
	Raucaffrinoline	(49)	(122)
	Nortetraphyllicine	(67)	(120)
	Tetraphyllicine	(69)	(120)

Table 2 (continued)

Source	Compound	No.	Ref.
vomitoria Afz.	17-O-Acetyltetraphyllicine	(73)	(122)
	Ajmaline	(92)	(120)
	17-O-Acetylajmaline	(96)	(122)
	Vellosimine	(20)	(137)
	Normacusine B	(22)	(135)
	Sarpagine	(35)	(126)
	Vomifoline	(45)	(133)
	Perakine	(48)	(128)
	Raucaffrinoline	(49)	(140)
	Quebrachidine	(50)	(139)
	Vomilenine	(62)	(129)
	Nortetraphyllicine	(67)	(135)
	Norrauvomitine	(68)	(134)
	Tetraphyllicine	(69)	(132)
	Rauvomitine	(71)	(124, 125)
	Hydroxynortetraphyllicine	(74)	(135)
	Norseredamine	(77)	(135)
	Seredamine	(79)	(13)
	Vomalidine	(86)	(127)
	Norpurpeline	(88)	(135)
	Mitoridine	(89)	(13)
	Purpeline	(90)	(13)
	Ajmaline	(92)	(123)
	Sandwicine	(93)	(131)
	Isoajmaline	(94)	(136)
	Isosandwicine	(95)	(131)
	17-O-Acetylajmaline	(96)	(130, 132)
	Ajmalinol	(100)	(138)
	Ajmaline	(92)	(151)
yunnanensis Tsiang			
Rhazya stricta Decaisne	Akuammidine	(15)	(141)
Tabernaemontana accedens Muell. Arg.	Accedinisine	(101)	(16)
brachyantha Stapf.	Normacusine B	(22)	(142)

fuchsiaefolia DC.	Voachalotine	(5)	(144)
	Affinisine	(26)	(143)
holstii Schum.	Pericyclivine	(18)	(145)
johnstonii Pichon	Pericyclivine	(18)	(146)
olivacea Muell. Arg.	Akuammidine	(15)	(147)
Tonduzia longifolia (DC.) MGF.	Vincamajine	(52)	(149)
	Ajmaline	(92)	(148)
Vallesia dichotoma Ruiz et Pav.	Akuammidine	(15)	(150)
glabra Link	Akuammidine	(15)	(139)
Vinca difformis Pourret	Akuammidine	(15)	(152)
	Vellosimine	(20)	(153)
	Sarpagine	(35)	(154)
	Vincamajine	(52)	(155)
	Vincamedine	(53)	(156)
erecta Regel et Schmalh.	Akuammidine	(15)	(157)
	Normacusine B	(22)	(158)
	O-Benzoylnormacusine B	(24)	(162)
	Ervincidine	(29)	(161)
	Vincarine	(50')	(159)
	Methoxyvinorine	(64)	(160)
herbacea Waldst. *et* Kit.	Vincamajine	(52)	(163)
libanotica Zucc.	Quebrachidine	(50)	(164)
	Vincamajine	(52)	(164)
major L.	Lochvinerine	(9)	(170)
	10-Methoxyvellosimine	(33)	(170)
	Majvinine	(34)	(171)
	Sarpagine	(35)	(169)
	Vincamajine	(52)	(166)
	Vincamedine	(53)	(165)
	Majorinine	(65)	(172)
	Vincamajoreine	(75)	(168)
minor L.	Majoridine	(76)	(167)
	Vinorine	(61)	(173)

Table 2 (continued)

Source	Compound	No.	Ref.
Voacanga			
chalotiana Pierre ex Stapf.	Voachalotine	(5)	(174)
	Dehydrovoachalotine	(7)	(176)
	17-O-Acetyl-19,20-dihydrovoachalotine	(8)	(178)
	Voamonine	(42)	(177)
	Voacoline	(43)	(175)
Loganiaceae			
Gardneria			
mutans Sieb. et Zucc.	Gardnerine	(10)	(179, 180, 181)
	Hydroxygardnerine	(11)	(182)
	Gardnutine	(13)	(179, 180, 181)
	Hydroxygardnutine	(14)	(179, 180, 181)
Strychnaceae			
Strychnos			
amazonica Krukoff	Macusine B	(27)	(183)
decussata Gilg.	Macusine B	(27)	(184)
	O-Methylmacusine B	(28)	(184)
rubiginosa DC.	Normacusine B	(22)	(185)
toxifera Schomb.	Macusine A	(6)	(186)
	Macusine B	(27)	(186)
	Macusine C	(19)	(187)
	Lochneram	(39)	(188)
usambarensis Gilg.	O-Methylmacusine B	(28)	(189)
	Dihydro-O-methylmacusine B	(30)	(189)

a The genus *Pandaca* has been placed in *Tabernaemontana*. Thus, *P. ochrascens* is considered synonymous with *T. ochracea* Benth. *ex* Muell.
b *Rauwolfia discolor* is unknown to Index Kewensis.

4. Biogenesis

4.1. Barger-Hahn-Robinson Theory

The first biogenetic proposal for the sarpagine-ajmaline type indole alkaloids was put forth by ROBERT ROBINSON in 1956 (*190*) as an extension of the BARGER-HAHN model (*191, 192*) (Scheme 1). According to this theory, condensation of tryptamine with phenylalanine derived phenylpy-ruvic acid gives the 1-benzyl-tetrahydrocarboline (106), which captures a molecule of CH_2O and cyclizes to the carbinolamine (107). WOODWARD-cleavage then gives the corynantheal-derivative (108), and nucleophilic attack of the indole nucleus on the aldehyde carbon gives the intermediate (109). A series of oxidation-reduction steps leads to the iminium-ketone [here depicted as the carbinolamine (110)], which cyclizes to ajmalone (*193*) [=ajmalidine (85), a natural compound]. Reduction finally furnishes ajmaline (92).

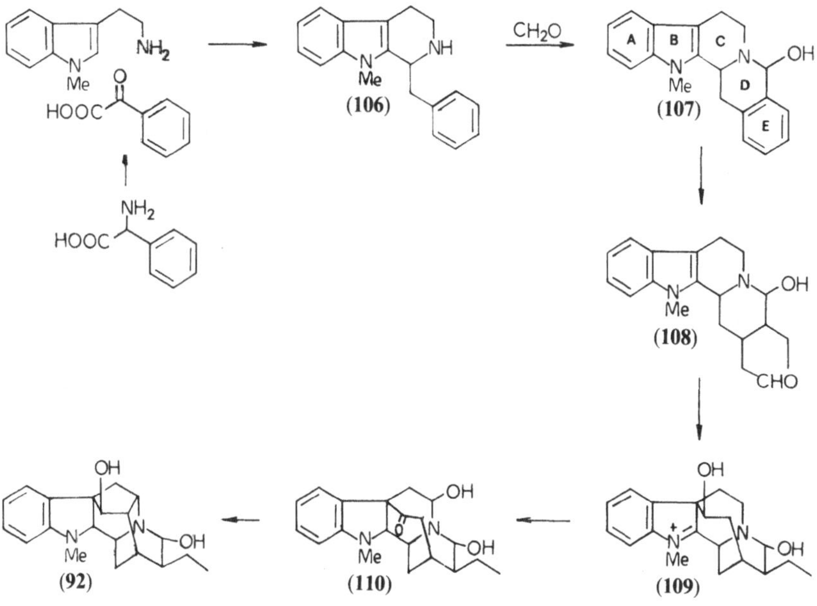

Scheme 1

4.2. Wenkert-Bringi Hypothesis

The BARGER-HAHN model of indole alkaloid biogenesis was found to be unsatisfactory for explaining three interdependent difficulties: a) the state of oxidation of ring E, b) the absolute configuration of C_{15} and c) the origin of the carbomethoxy group attached to C_{16}. The ROBINSON extension was found to be equally inadequate, especially in light of the stereochemical results obtained.

DAVIS had demonstrated that shikimic acid (111) and prephenic acid (126) are the natural progenitors of aromatic amino acids (194). WENKERT and BRINGI applied this finding to indole alkaloid biosynthesis (Scheme 2) (195, 196).

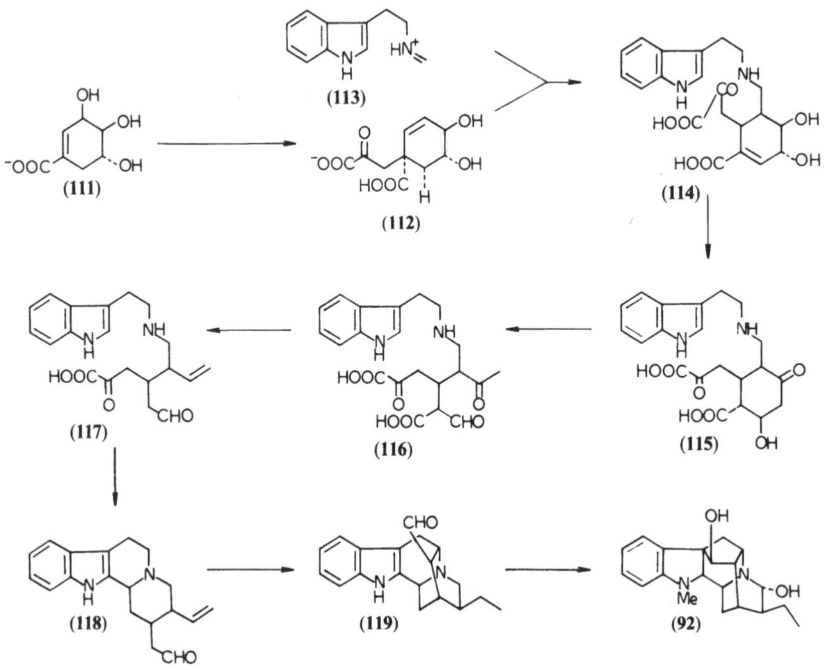

Scheme 2

According to their postulate, shikimic acid (111) reacts with pyruvate to form dehydroquinic acid (112), condensation of which with the iminium ion (113) formed from tryptamine and CH_2O gives the allylic alcohol (114). Dehydration and rehydration then furnish aldol (115), which undergoes a retro-aldol type cleavage to aldehydoketone (116). Reduction and decarboxylation lead to the olefinic compound (117), which cyclizes to corynantheine (118). Regiospecific oxidation to an iminium ion and cyclization lead, after reduction, to demethyl ajmalal-B (119). The biogenesis of ajmaline (92) is completed by ring-closure between the nucleophilic C_7 of the indole ring system and the C_{16} aldehyde carbon, N_a-methylation and finally C_{21} oxidation.

4.3. Thomas-Wenkert Hypothesis

In 1961 THOMAS (197) advanced a more direct route to the indole alkaloids based on the idea that the nine or ten carbon unit is derived from a cyclopentanoid monoterpene unit related to the alkaloids actinidine (120) and skytanthine (121). The carboxyl group at C_{16} in the corresponding indole alkaloids was also noted to have a precedent among cyclopentanoid monoterpenes, e. g. in genepin (122) and asperuloside (123).

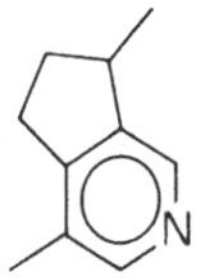

(120) Actinidine 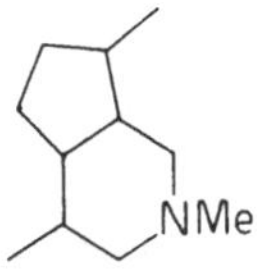   (121) Skytanthine

Fig. 14

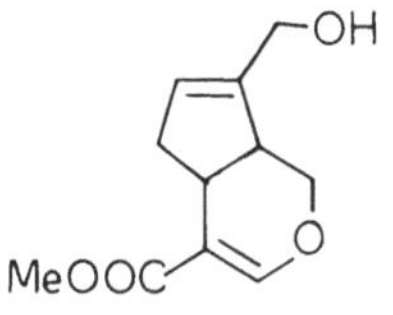

(122) Genepin 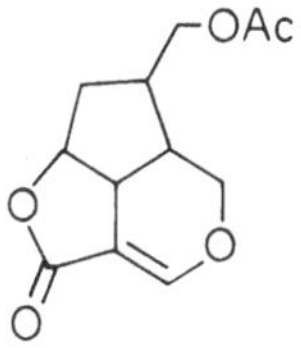   (123) Asperuloside

Fig. 15

Accordingly, THOMAS stated: "a) the ten-carbon unit (125) may be derived from two units of mevalonate (124) *via* a cyclopentanoid monoterpene.... The carbomethoxy group would be derived from either carbon 2 or 3 of mevalonate. Neither ring E substituent would be specifically derived from one-carbon units, as required by the earlier hypotheses; b) the alkaloids containing the acyclic ten-carbon unit (125) would represent the skeletal precursors of the homocyclic ring E group ... c) the ten-carbon series would be expected to represent the parent structures, giving rise to the nine-carbon series by decarboxylation." (197) The route is presented in Scheme 3.

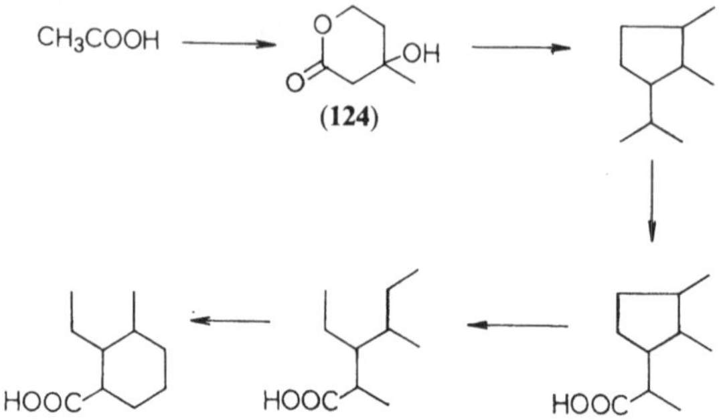

Scheme 3

Wenkert (*198*) expanded his prephenic acid hypothesis to lead to the "*seco*-prephenate formaldehyde" (SPF) unit (**127**), which was very suitable for the non-tryptophan precursor of the indole alkaloids. Rearrangement, hydration, retro-aldol reaction and finally formylation of prephenic acid (**126**) would lead to the ten carbon SPF-unit (**127**) (Scheme 4).

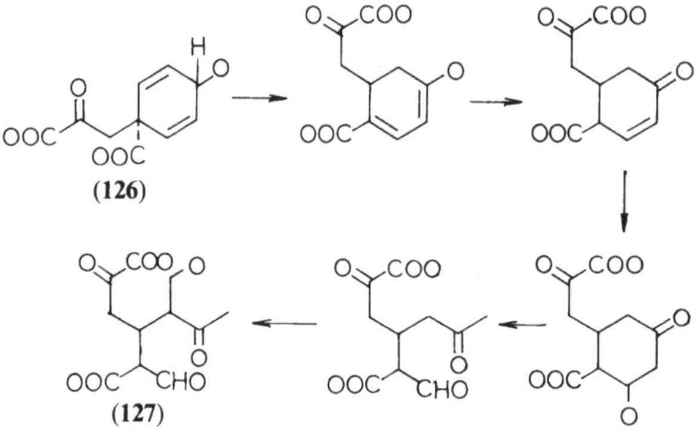

Scheme 4

Thus, the prephenic acid pathway leads essentially to the same ten carbon precursor as the monoterpenoid hypothesis, only the history of the fragment is different.

4.4. Leete's Postulate

Through tracer studies Leete was able to show that tryptophan is a true precursor of ajmaline (*199, 200*) and also apparently confirmed that C_{21} is derived from a one-carbon fragment (*201*). Further tracer studies employing labelled sodium acetate (*202*) and malonate (*203*) led to the proposal

(Scheme 5) that the intermediate (128), derived from acetylcoenzyme A, malonylcoenzyme A and formaldehyde is incorporated in ajmaline (92) with the loss of the carboxyl group at C_{16}. The intermediate is exactly the same as the one proposed by WENKERT, i. e. the SPF unit (127), thus offering an alternative to the prephenic acid pathway of WENKERT and the monoterpenoid theory of THOMAS.

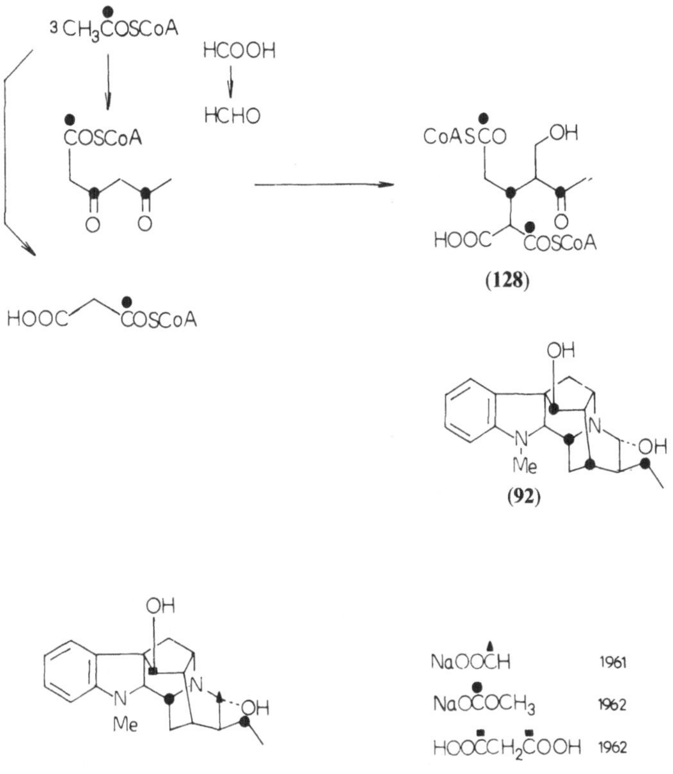

Scheme 5

4.5. Battersby's Observations

BATTERSBY's feeding experiments with *R. serpentina* were not in exact agreement with the results obtained by LEETE. The ajmaline isolated from plants fed with [^{14}C]-formate was degraded and decarbonoajmaline (129) still contained not less than 96% of the original activity (*204*). In 1965 time was ripe for the abandonment of the acetate-malonate pathway (*205, 206*) and further work (*207, 208*) led to the establishment of loganin (**130**) as the natural precursor of the indole alkaloids. Specifically, ajmaline obtained from *R. serpentina* fed with [1-³H]-loganin was labelled with tritium at C_{21}.

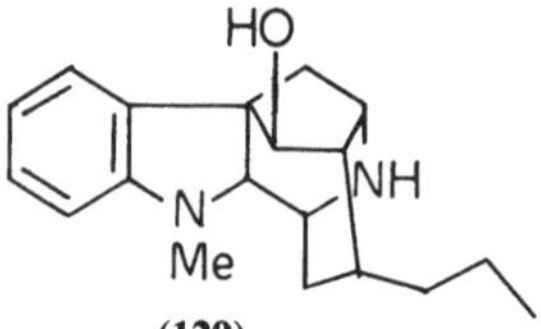

(**129**)

Fig. 16

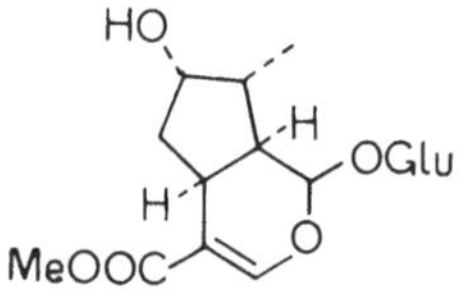

(**130**) Loganin

Fig. 17

4.6. Mevalonoid Origin of the Non-Tryptophan Unit

The ten-carbon moiety of the indole alkaloids was finally proved to originate from terpenoid precursors and proposed to derive from two molecules of mevalonate (**124**) by way of geraniol (**131**) → deoxyloganin (**132**) → loganin (**130**) → secologanin (**133**) ⇌ sweroside (**134**) and vincoside (**135**) (Scheme 6) (*209*).

Vincoside (**135**) was originally assumed to have the 3α (*S*) configuration (*210, 211*) but subsequently this was revised to 3β (*R*) (*212 – 214*). A special mechanism for the inversion of C_3 during the formation of the 3α (*S*) series of alkaloids was clearly required, since the hydrogen at C_3 of vincoside is retained (*213*). BROWN (*215*) proposed such a mechanism (Scheme 7): cleavage of the C_3-N_b bond in the immonium intermediate (**137**) would be facilitated by the electron donating N_a. Ring closure of the imine (**138**) would then occur from the side of the double bond to give the more stable structure (**139**), which would subsequently be reduced to (**140**). For a more thorough discussion of the controversy over the C-3 stereochemistry of vincoside (**135**) and isovincoside = strictosidine (**136**), the interested reader is referred to a review by CORDELL (*216*).

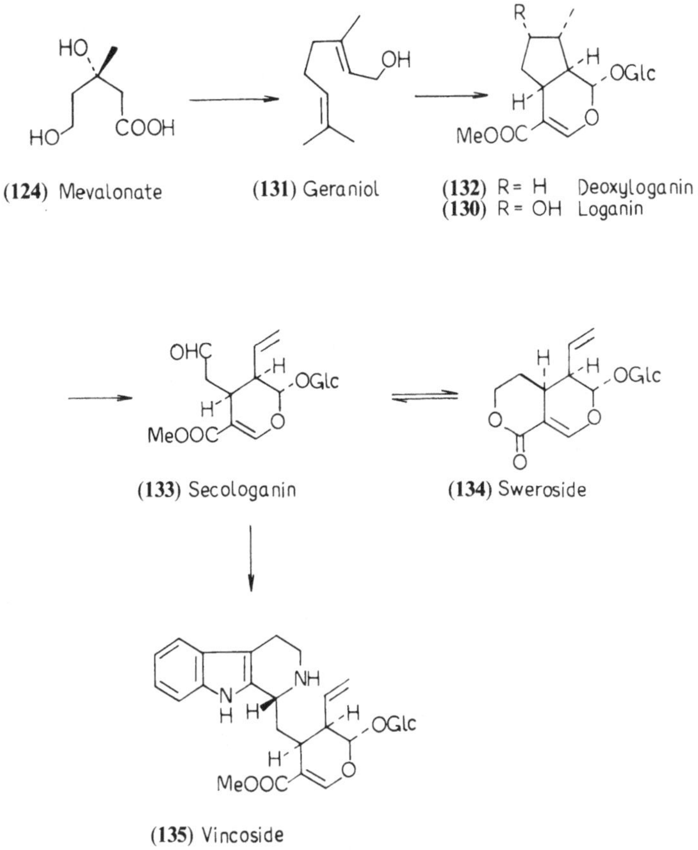

(124) Mevalonate   (131) Geraniol   (132) R = H   Deoxyloganin
                                      (130) R = OH Loganin

(133) Secologanin   (134) Sweroside

(135) Vincoside

Scheme 6

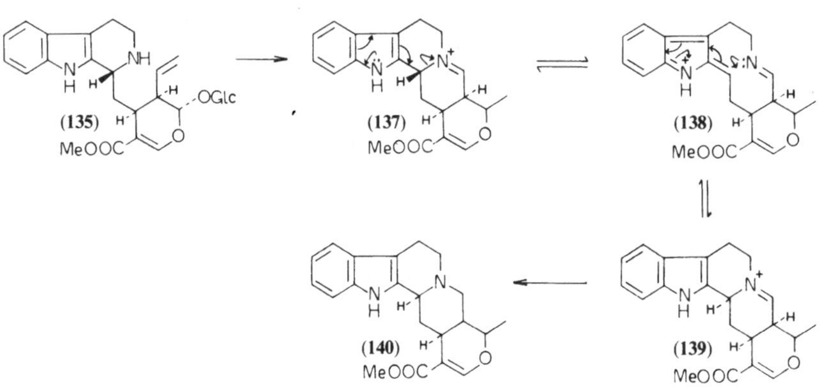

(135)   (137)   (138)
MeOOC   MeOOC   MeOOC

(140)   (139)
MeOOC   MeOOC

Scheme 7

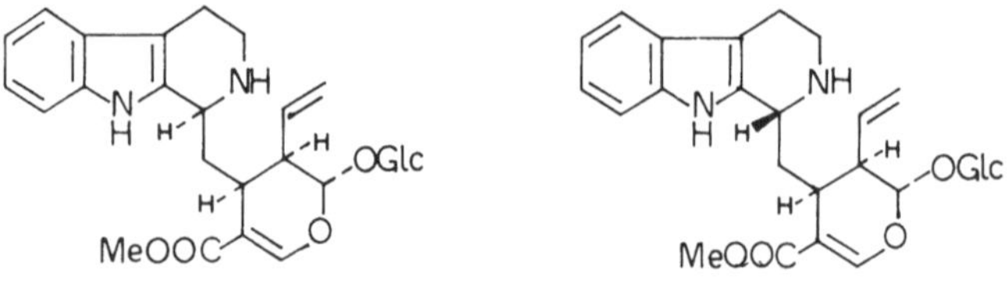

1969–71	*Vincoside*	Isovincoside
1971–77	Strictosidine = Isovincoside	*Vincoside*
1977–	*Strictosidine*	Vincoside

Fig. 18

The mevalonoid origin of the nine or ten carbon unit was firmly established by several groups almost simultaneously (*217–219*). The intermediate closest to the indole alkaloids, vincoside (**135**), soon had a serious rival: SMITH (*220*) was able to isolate strictosidine (**136**), the C_3 epimer of vincoside (**135**) and suggested that strictosidine is in fact the crucial intermediate in the biogenesis of these alkaloids.

Later SMITH observed that strictosidine and isovincoside are one and the same compound and further that the stereochemistry previously assigned to vincoside in fact corresponds to that of strictosidine (*213*). However this finding, and the demonstration that vallesiachotamine (**141**) (*213, 220*) and isovallesiachotamine (**142**) (*213*) are produced on mild hydrolysis of (**136**), were ignored until STÖCKIGT and ZENK (*221*) showed that in *Catharanthus roseus* only strictosidine (**136**) is used in the biosynthesis of the Corynanthé type alkaloids. Similar results were obtained almost simultaneously by SCOTT's group (*222*) using *C. roseus* cell-free culture.

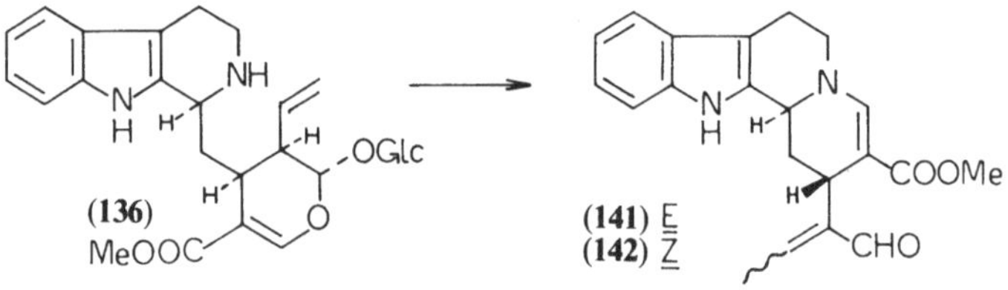

Equation 3

ZENK (*223*) also proved that strictosidine is the common precursor for monoterpenoid indole alkaloids of both the 3α and 3β series, and that in the formation of alkaloids with the 3β configuration the hydrogen at C_3 of strictosidine (**136**) is lost, rendering it unnecessary to assume special mechanisms (*211, 214*) in the inversion step.

4.7. Van Tamelen's Theory

During his studies on the oxidative decarboxylation of α-amino acids (*224*), VAN TAMELEN suggested that in the naturally occurring 5α-carboxystrictosidine (**143**) (*225*) preservation of the tryptophan carboxyl would be advantageous in that the α-amino acid unit could be used for specifically generating, through an oxidative decarboxylation process, the Δ4(5)-iminium ion (**145**) needed for the ensuing annulation step (Scheme 8) (*226, 227*). Based on this assumption, he generated an elegant synthesis leading to ajmaline (*vide infra*) (*227, 228*).

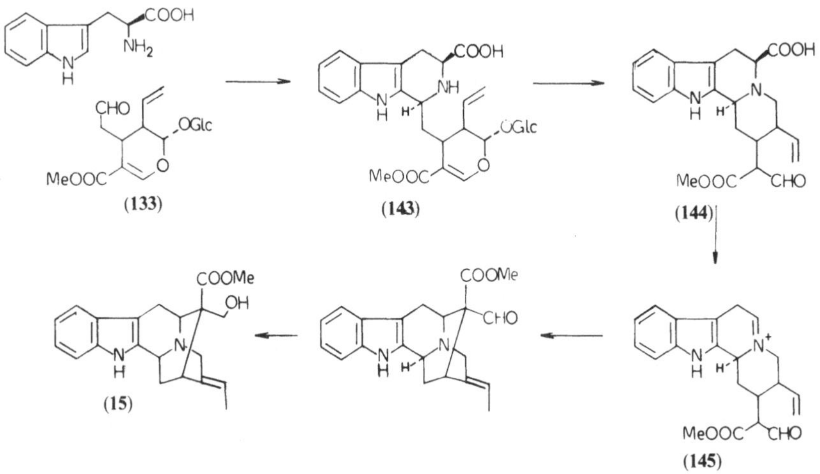

Scheme 8

4.8. Stöckigt's Work

VAN TAMELEN's proposal (*227*) that 5α-carboxystrictosidine (**143**) (*225*) and 5α-carboxyvincoside (**146**) play a significant role in the biosynthesis of the sarpagine-ajmaline type indole alkaloids was studied by STÖCKIGT using *Rauwolfia vomitoria*, *Vallesia glabra* and *Voacanga africana* (*139*). The incorporation of doubly labelled (**143**) and (**146**) was always less than 0.01%, a result taken to indicate that the carboxy-compounds are not involved in the biosynthesis of the sarpagine-ajmaline type alkaloids. Thus, the involvement of strictosidine (**136**) *via* the intermediates of the ajmalicine pathway (*229*) involving double bond isomerisation at N_b to (**145**) was suggested as an alternative possibility. However, as has been pointed out several times, lack of incorporation cannot be taken as an indication of non-involvement of the intermediate in question.

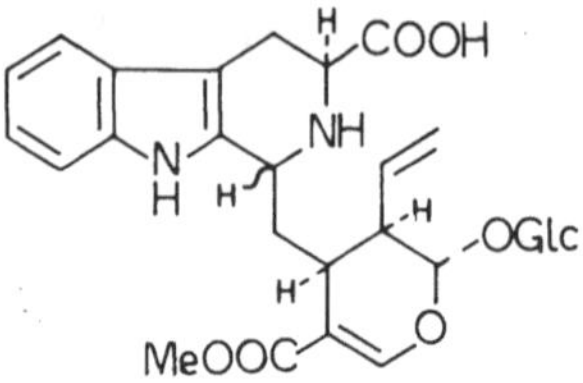

(143) 3α H
(146) 3β H

Fig. 19

4.9. Court's Observations

In the course of screening the alkaloids from several *Rauwolfia* species COURT formulated a view of the origin of several sarpagine-ajmaline type alkaloids. At first he suggested that 5α-carboxycorynantheine (**144**) was the common precursor of this group of alkaloids (**134**), but later reverted to the non-carboxylated stem molecule (**147**) (Scheme 9) (*96*). The general features of COURT's hypothesis are as follows: i) demethylcorynantheine (**147**) is the potential precursor of all these alkaloids, ii) the sarpagine type alkaloids are formed first and a further ring closure leads to the ajmaline group (*216*). Furthermore, working on *R. vomitoria*, COURT noticed that the N_a-methylated alkaloids tetraphyllicine (**69**) and seredamine (**79**) together with the N-methylated 12-hydroxy alkaloid, mitoridine (**89**), occur in the roots (*134*). This suggested to him that these alkaloids are formed by N_a-methylation and not from N_a-methylated precursors (*136*).

COURT has based his postulates mainly on isolated material. However, certain key intermediates involved in the metabolic pathway can escape even the most experienced eye merely because they are present in only low steady state concentration (*230*), a classical situation encountered in the elucidation of metabolic pathways.

Interlude: Enzymic Transformations

The enzymic transformations pertaining to alkaloid biogenesis have been reviewed (*282*), although only superficially at the molecular level. It is not possible in this review to delve into the detailed mechanisms of even the most important enzymic transformations related to indole alkaloid biogenesis, but a few pertinent reaction types will be illustrated to provide some practical tools for the discussion that follows. The "standard" biochemical operations such as decarboxylation, $C=C$ and $C=O$ reductions and the corresponding oxidations, allylic and benzylic oxygenations and the several

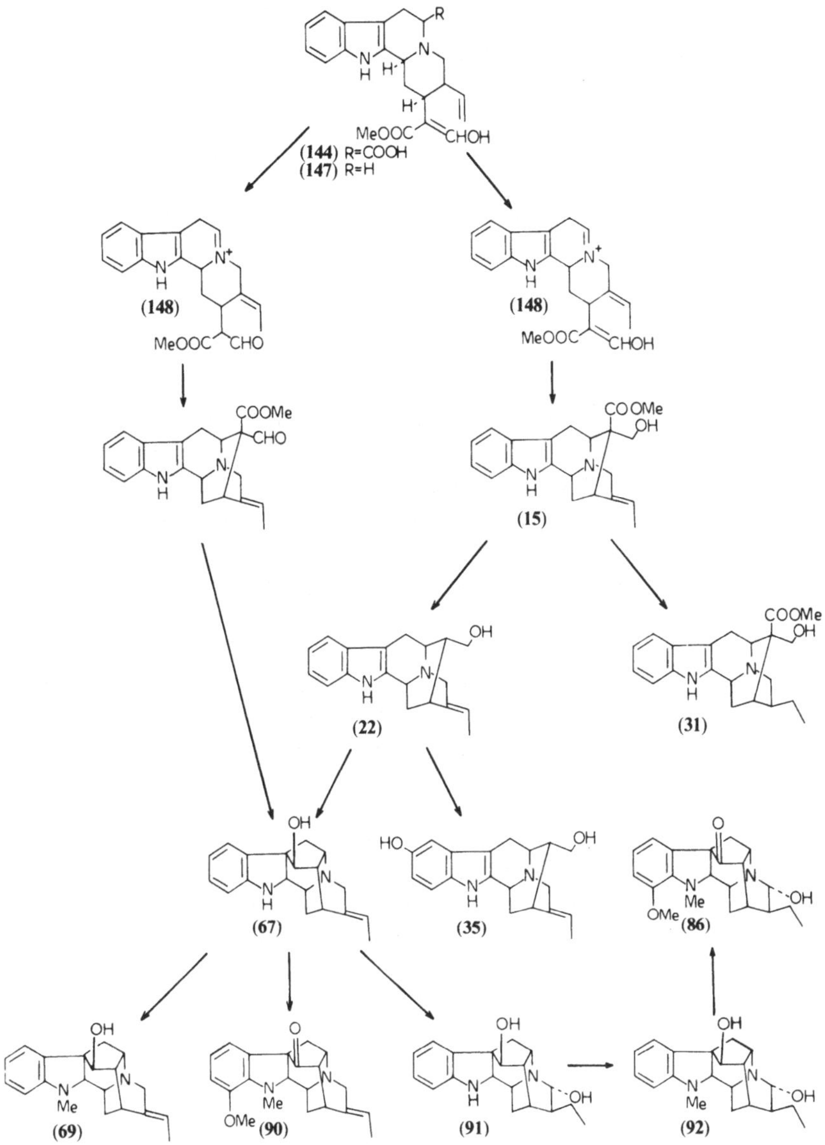

Scheme 9

acylations will be covered only briefly and the interested reader is referred to recent textbooks of bioorganic chemistry.

The most beautiful feature of enzymic transformations is their over-whelming selectivity. To produce a certain conversion the desired substrate is selected and bound stereospecifically on the active site of the enzyme.

Several models of this enzyme-substrate complex have been proposed and recently an account based on quantum mechanical calculations has appeared (283). After the substrate is bound to the active site, the transformation in question is effected both regio- and stereoselectively in most cases. The enzyme-substrate interaction can include hydrogen bonding, hydrophobic interactions, electrostatic interactions and dipole-dipole interactions. Certain enzymes, such as hydrolytic enzymes and aminotransferases, form covalent bonds with the substrate.

Another characteristic of enzyme mediated reactions is the controllability of the production of certain intermediates or the final product. It has been shown (231) that during alkaloid biosynthesis in *Catharanthus roseus*, vindoline exerts an inhibitory action on S-adenosylmethionine: loganic acid methyltransferase, an enzyme involved at an early stage of alkaloid production and responsible for the esterification of loganic acid to loganin. Vindoline (150) showed no appreciable inhibitory effect on the membrane-bound cytochrome P-450 dependent monooxygenase in the same plant. Catharanthine, on the other hand, was found to be a noncompetitive inhibitor with respect to both substrates, geraniol and NADPH.

The experimental work on the transformations of alkaloids has been done mainly on plant material or tissue cultures. Barton et al. (232) have observed that *Rauwolfia serpentina* Benth. plants are capable of methylating norajmaline and hydroxylating 21-deoxyajmaline to ajmaline.

Methylation

Stuart and Kutney (239) have studied alkaloid production in cell free extracts from mature *C. roseus* plants and shown that S-adenosylmethionine (SAM) (149) is utilised in the biosynthesis of vindoline (150). Incorporation of the ^{14}C-methyl group of SAM was fairly low and by way of explanation it was noted (231) that vindoline itself (as well as catharanthine) has been shown to significantly inhibit an 18-fold purified

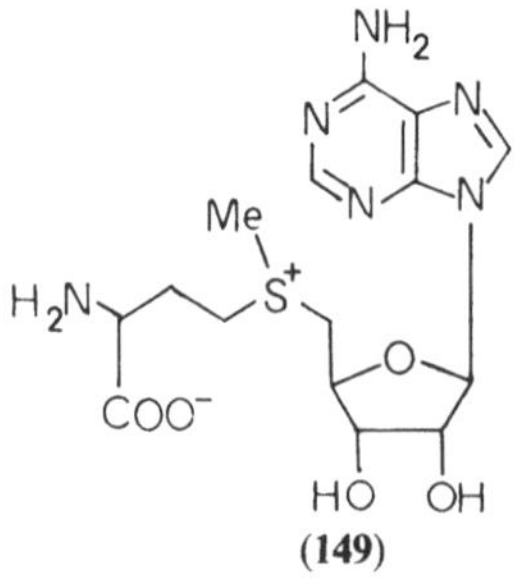

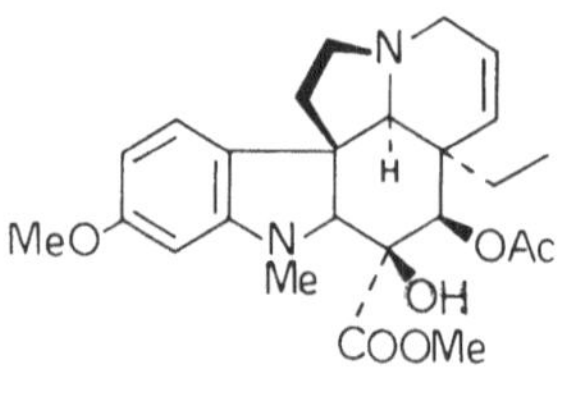

Fig. 20

loganic acid methyl transferase (*234*). The inhibitory action thus constitutes a plausible mode of end product inhibition, a phenomenon often encountered in metabolism. Thus, *S*-adenosylmethionine (**149**) is an effective methyl transfer cofactor, although other one carbon transfer agents such as tetrahydrofolate derivatives may in certain cases be implied.

Oxidative Transformations

Cytochrome P-450 dependent oxygenases are capable of catalysing the oxidation of tertiary amines (*235, 236*) by dioxygen and NADPH. This observation was employed in the liver microsome mediated rearrangement of dregamine (**151**) to 20-*epi*-ervatamine (**152**) by POTIER *et al.* (*237*) (Scheme 10). KUTNEY (*238*) had also suggested the intermediacy of the N-oxide (**153**) in a biological equivalent of the modified Polonovski reaction (*239 – 242*) during the biosynthesis of the anti-cancer natural products leurosine, catharine and vinblastine (*243, 244*).

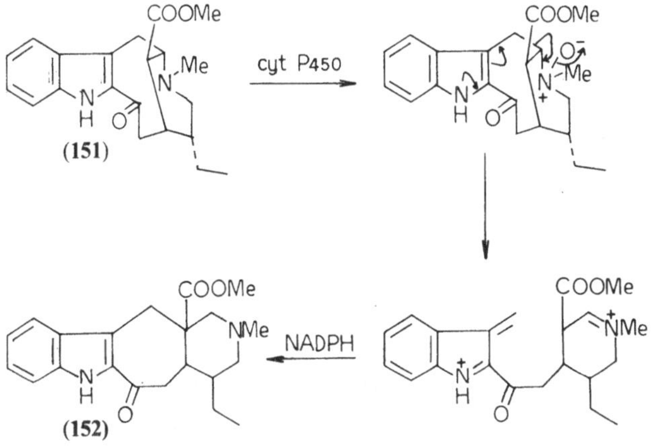

Scheme 10

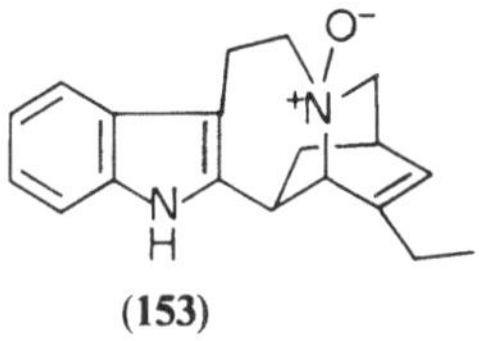

(153)

Fig. 21

A clear application of this biological equivalent of the modified Polonovski reaction (*239 – 242*) to the biosynthesis of the sarpagine-ajmaline type alkaloids would be formation of the quinuclidine moiety by

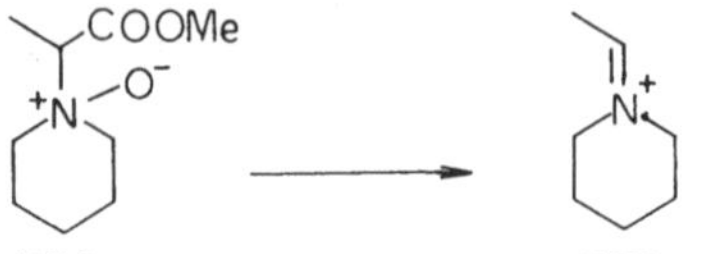

(154) (155)

Equation 4

ring closure through $C_{16} - C_5$ bond making. LOUNASMAA and KOSKINEN have shown (*245*) in model experiments that the modified Polonovski reaction easily generates a decarboxylated iminium ion (**155**) with an exocyclic double bond from the corresponding α-piperidino acid N-oxide (**154**) and further that ring closure of a suitably substituted piperidine derivative to the quinuclidine system by this path is a facile process. These observations are supportive of VAN TAMELEN's hypothesis *(vide supra)* (Scheme 8) that the carboxyl group of tryptophan is retained intact until the crucial ring formation step. By analogy, a similar N-oxidation — dehydration — hydration process can be rationalized for the oxygenation of C_{21} in the already formed sarpagine or ajmaline skeleton (Scheme 11).

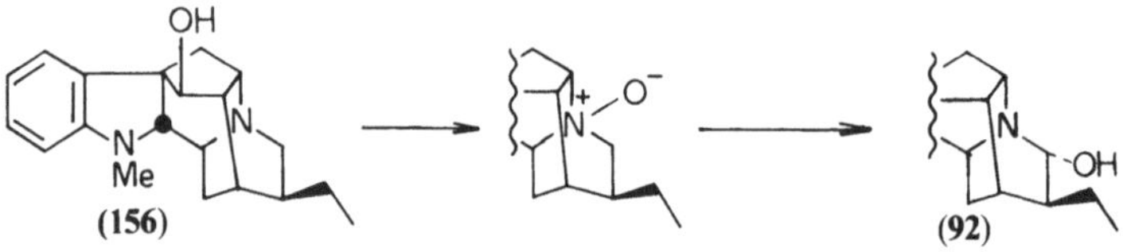

(156) (92)

Scheme 11

In fact, BARTON (*232*) has shown that deoxyajmaline (**156**) is a true precursor for ajmaline (**92**) by feeding experiments with *R. verticillata*.

Reduction

The standard $C = C$ and carbonyl reductions being well covered in textbooks of bioorganic chemistry only the reduction of imine double bonds will be mentioned here. The reduction proceeds by means of a nicotinamide cofactor dependent hydrogenase, and hydrogen bonding of the C_{17} hydroxyl group places the dihydropyridine moiety in an appropriate position to deliver the hydride (pro-*S* is implied) from the *si* face (i. e. from above) to the double bond to generate the $2R$ configuration.

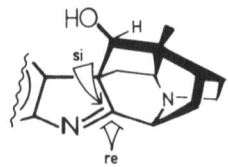

Fig. 22

In the case of the epimeric 2S series, the carbomethoxy group intact at C_{16} possibly causes electronic or purely sterical constraints to hinder the favourable attachment of the substrate and cofactor. Therefore the donation of hydride will be from the less hindered *re* face of the molecule (i. e. from below) to generate the 2S configuration.

4.10. The Concise Biogenetic Representation

4.10.1. Sarpagine Group

From what is known about the biogenesis of the Corynanthé type alkaloids, one can be sure that the sarpagine-ajmaline type alkaloids are derived from tryptophan and a ten carbon monoterpene unit arising from mevalonate. Whether the intermediate is corynantheine (**147**) or 5α-carboxycorynantheine (**144**) is still an open question, but whichever it is, the resulting $\Delta 4^{(5)}$-iminium ion (**148**) leads, after attack of the nucleophile at C_{16} on the electrophilic C_5, to the sarpagine ring system. At this stage two isomeric forms of the product can be formed: polyneuridine aldehyde (**3**) and the corresponding 16S isomer (**157**), provisionally named akuammidine aldehyde (Scheme 12).

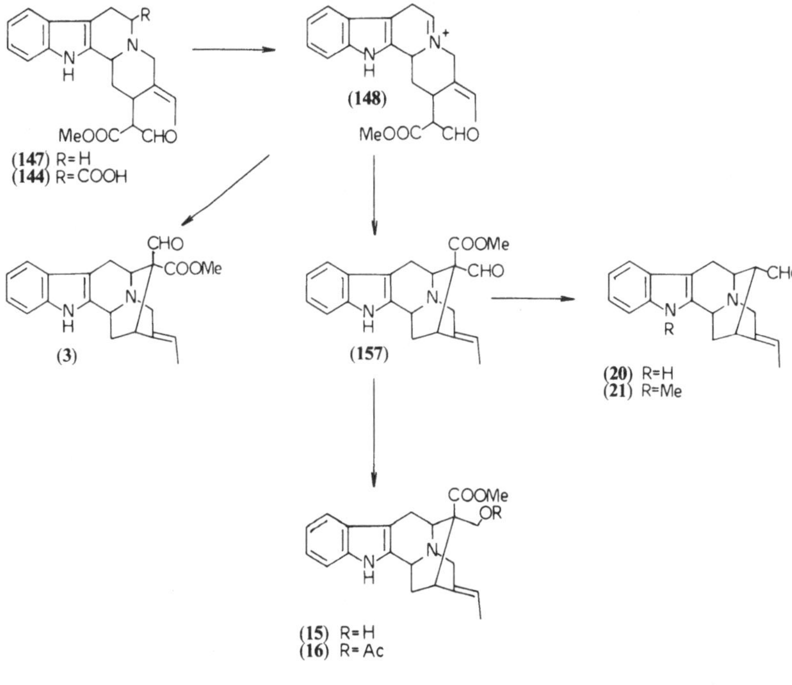

Scheme 12

Vellosimine (**20**) is derived by decarboxylation from the last mentioned 16*S* isomer (**157**). Trivial N_a-methylation of vellosimine gives N_a-methylvellosimine (**21**). Further reduction of the aldehyde function leads to normacusine B (**22**) which is the pivotal intermediate for the eight alkaloids shown in Scheme 13.

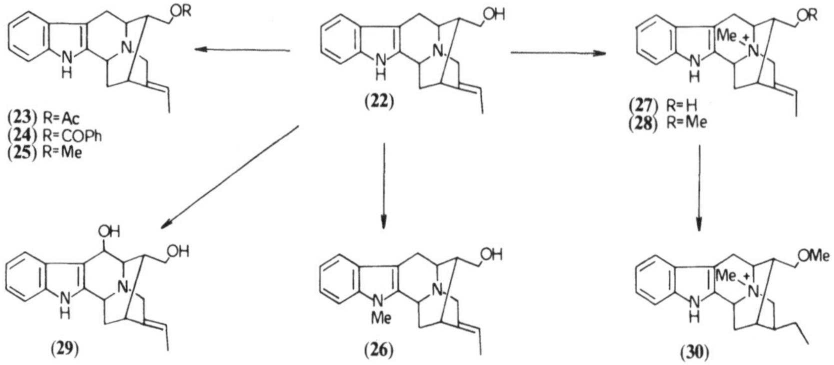
OR
N
H
(23) R=Ac
(24) R=COPh
(25) R=Me
OH
(22)
Me+
(27) R=H
(28) R=Me
(29)
Me
(26)
OMe
(30)

Scheme 13

On the other hand, Ar-hydroxylation of vellosimine (**20**) affords the hydroxyvellosimine (**158**) which on standard biochemical transformations gives the nine naturally occurring sarpagine type alkaloids shown in Scheme 14.

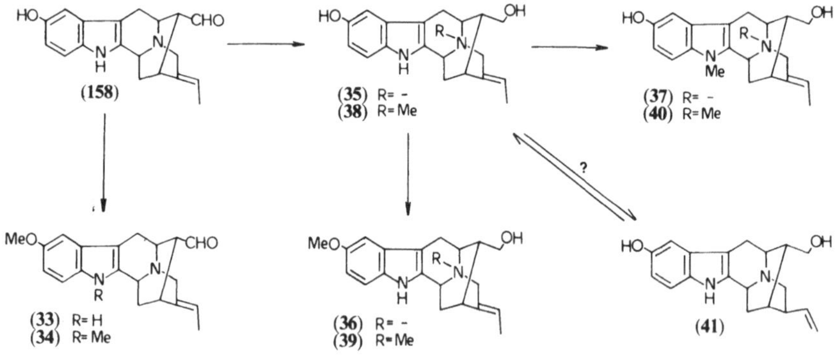
HO
CHO
(158)
(35) R= -
(38) R=Me
(37) R= -
(40) R=Me
?
MeO
(33) R=H
(34) R=Me
(36) R= -
(39) R=Me
(41)

Scheme 14

The third possible transformation of vellosimine (**20**), viz. reduction of the C_{17} aldehyde function, leads to the naturally occurring akuammidine (**15**) which again can serve as a precursor for the six alkaloids shown in Scheme 15.

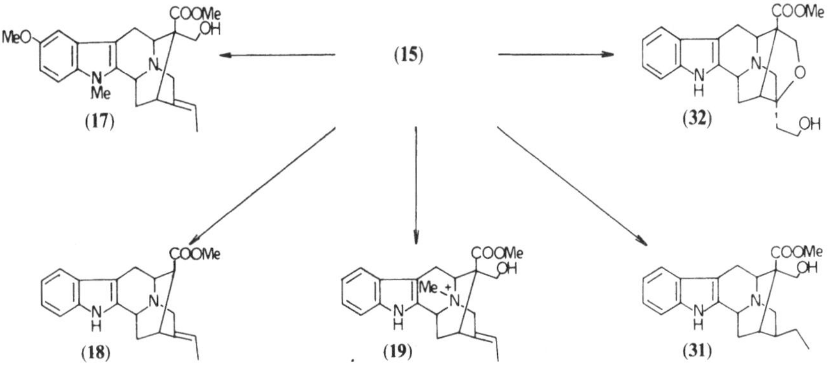

Scheme 15

The 16R isomer, polyneuridine aldehyde (3), in addition to being a plausible precursor of the ajmaline type alkaloids *(vide infra)*, can be reduced to polyneuridine (4). N_a-methylation of (4) gives voachalotine (5), a plausible precursor of three structurally similar natural products (Scheme 16).

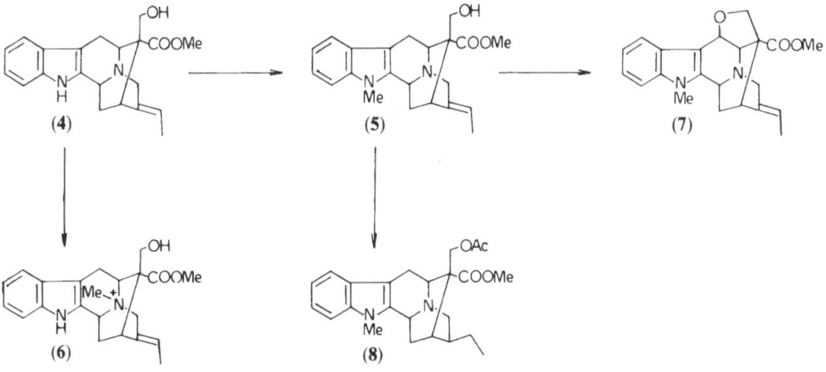

Scheme 16

Alternatively, decarboxylation and reduction of polyneuridine aldehyde (3) lead to the hydroxy compound (159) which can serve as the precursor for the five Ar-hydroxylated alkaloids shown in Scheme 17. Interestingly, the biogenesis of the rare rauwolfinine (12) can be rationalised as occurring *via* the hydroxycompound (159) and radical mediated ring formation, a process not unlike the one proposed for the biogenesis of preakuammicine by WENKERT (*246*).

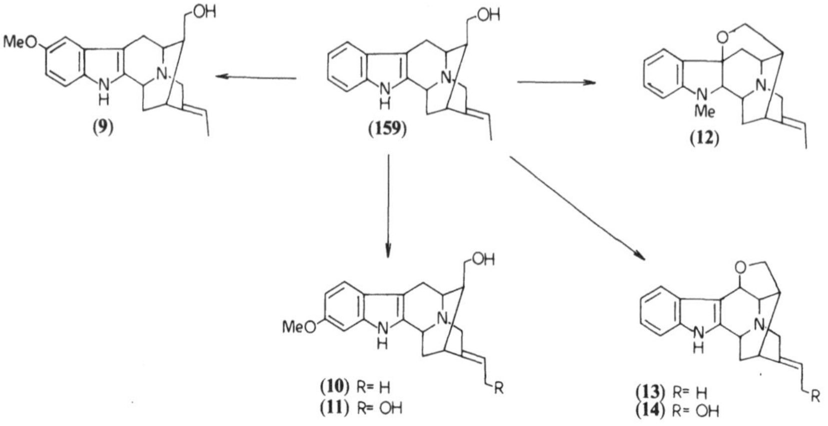

Scheme 17

The biogenesis of voamonine and voacoline can also be explained as starting from strictosidine (136) or its 5-carboxyderivative (143) *via* cathenamine (166) and dihydrocathenamine (140) type intermediates. $C_5 - C_{16}$ ring closure followed by reduction, oxygenation and formation of the cyclic acetal moiety gives voamonine (42) and its N_a-methyl derivative voacoline (43) (Scheme 18).

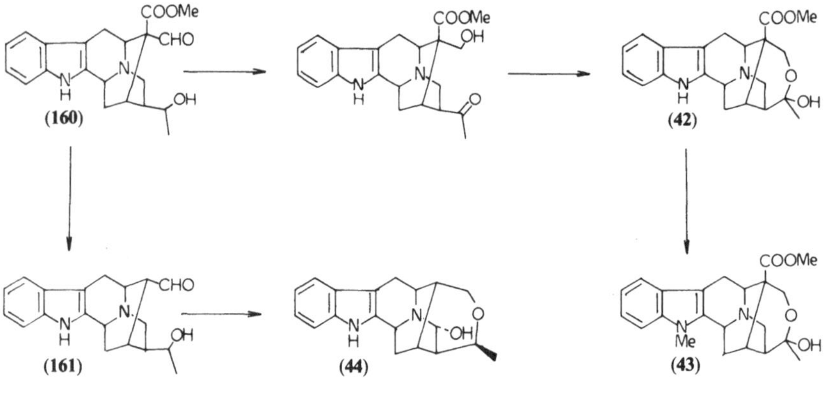

Scheme 18

Vomifoline, O-acetylpreperakine and macrosalhine (Scheme 19) are plausibly derived from vellosimine (20) *via* 21-hydroxylation and a rearrangement process similar to the one presented for the vomilenine → perakine transformation in Scheme 23 *(vide infra)*. A series of standard operations leads to the dialdehyde (162) which on reduction and acetal formation gives vomifoline (45). Reduction of the C_{17} aldehyde function furnishes the alcohol (163), which on acetalisation and N-methylation gives macrosalhine (46). On the other hand, acylation of (163) generates O-acetylpreperakine (47).

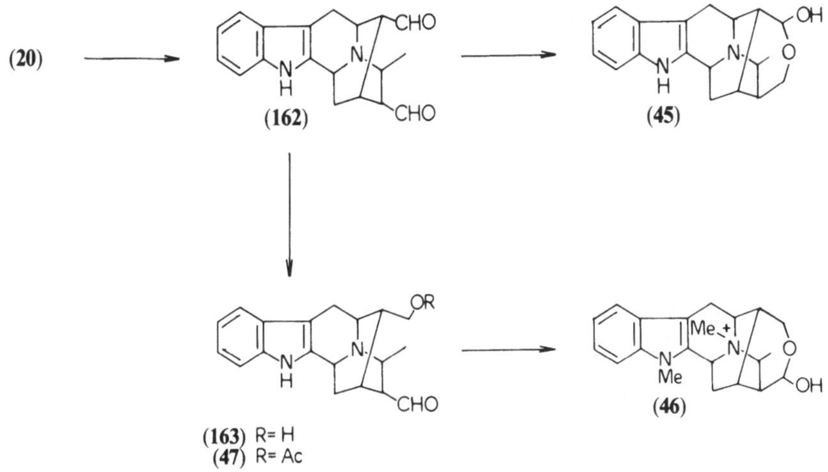

Scheme 19

The generation of talpinine, a unique alkaloid in the sarpagine group with a 21-hydroxyl function, can be rationalised as occurring along the lines suggested for voamonine and voacoline in Scheme 18. Thus, hydrolysis and decarboxylation of (160) produce the aldehydoalcohol (161), which after acetalisation, reduction and C_{21}-oxygenation gives birth to talpinine (44).

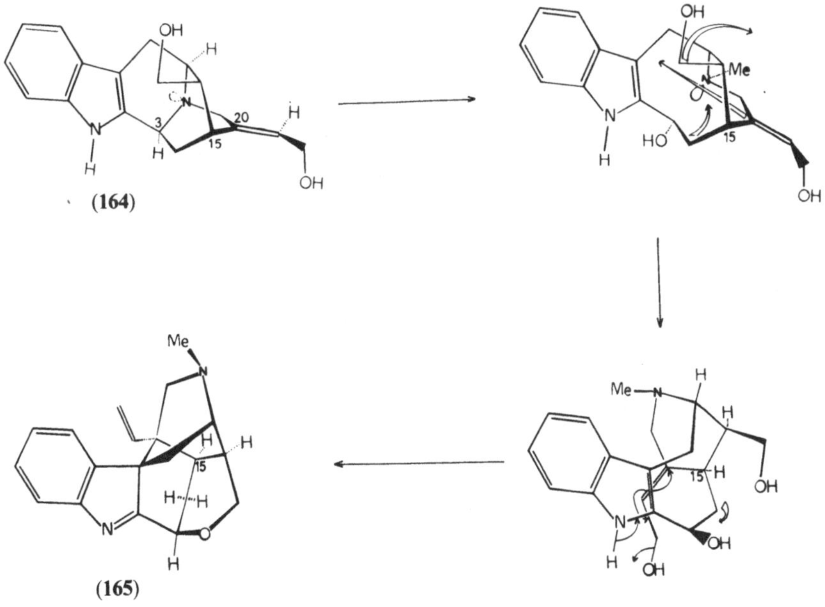

Scheme 20

At this point, it should be remembered that the sarpagine skeleton is not necessarily the end product of the metabolic pathway. Oxidation of C_3 can lead to the widely distributed 2-acylindole alkaloids. Moreover, koumine (165) might be generated from an 18-hydroxygardnerine type sarpagine derivative (164) *via* an intuitively facile process (Scheme 20) (*10*).

4.10.2. Ajmaline Group

The biosynthesis leading to ajmaline type alkaloids follows the same lines as to sarpagine type alkaloids up to polyneuridine aldehyde (3), which is apt to cyclise to the hexacyclic carbomethoxyindolenine (167) (Scheme 21) (*284*). Cyclisation of (3) accompanied by decarboxylation (either before or after ring closure) leads to the desacetylvinorine (168).

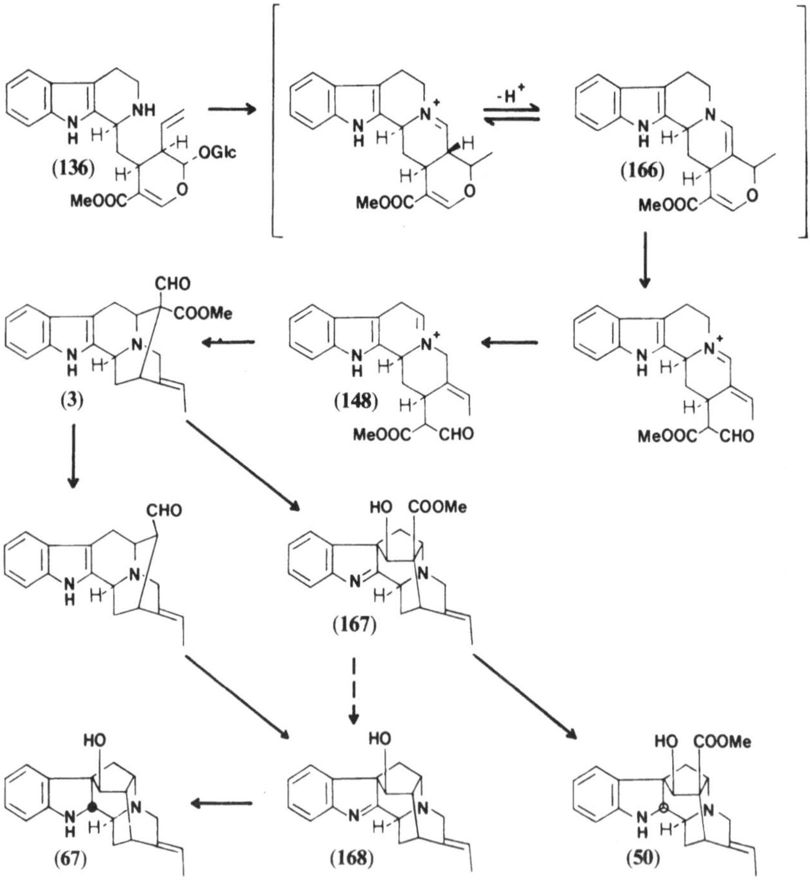

Scheme 21

Cyclisation of the five membered ring, as mentioned before, produces the 17R stereochemistry. Hydrogenation of the indolenine double bond of (167) from the least hindered side *(vide supra)* produces quebrachidine (50), a stem kernel for the 2S alkaloids, and the desacetylvinorine (168) serves as a precursor for the indolenine alkaloids of the vinorine subgroup. On the other hand, NAD(P)H assisted hydrogenation of (168) produces nortetraphyllicine (67), a stem kernel for the 2R alkaloids.

Vinorine (61) can easily be transformed to the five other alkaloids of the subgroup depicted in Scheme 22. Hydroxylation at C_{21} gives vomilenine (62) and its glucoside raucaffricine (63). Dihydrovomilenine (66) is derived from vomilenine (62) by hydrogenation. On the other hand, Ar-hydroxylation of (61) followed by methylation gives methoxyvinorine (64), and its hydroxylation at C_{21} leads to majorinine (65).

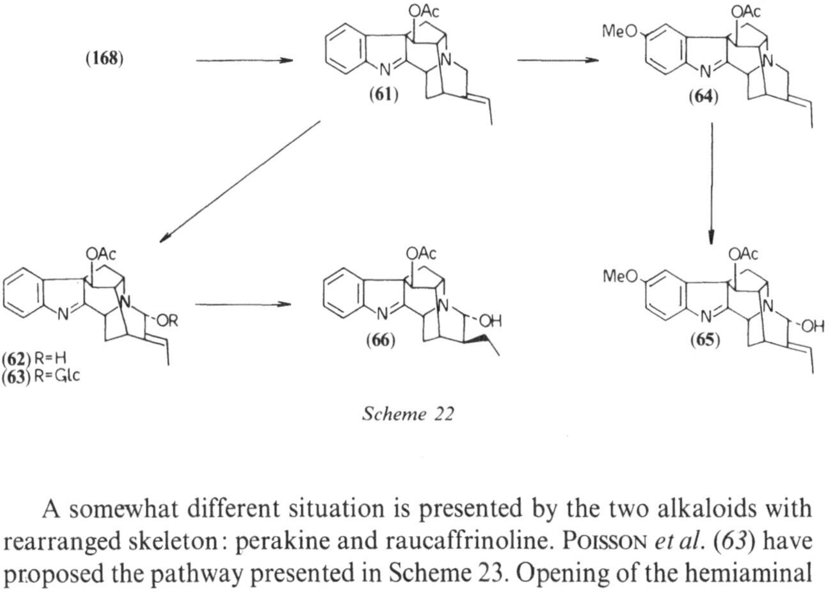

Scheme 22

A somewhat different situation is presented by the two alkaloids with rearranged skeleton: perakine and raucaffrinoline. POISSON *et al.* (*63*) have proposed the pathway presented in Scheme 23. Opening of the hemiaminal ring and recyclisation in a 1,4-manner gives perakine (48), which after reduction generates raucaffrinoline (49).

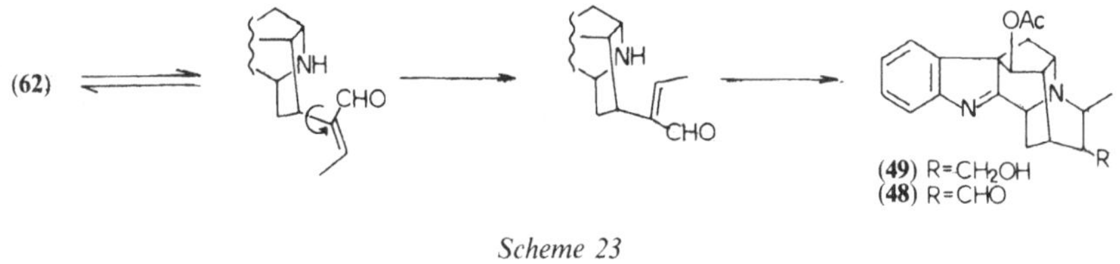

Scheme 23

Quebrachidine (**50**) serves as a precursor for the ten alkaloids presented in Scheme 24. Esterification with trimethylgallic acid gives 17-*O*-trimethoxybenzoylquebrachidine (**51**). Alternatively, N_a-methylation leads to vincamajine (**52**), a precursor for the three *O*-acylated alkaloids vincamedine (**53**) (R=Ac), *O*-benzoylvincamajine (**54**) (R=COPh) and 17-*O*-trimethoxycinnamoylvincamajine (**55**) [R= −CH=CH−C_6H_2 (OMe)$_3$]. Ar-hydroxylation of vincamajine produces (**169**), which on methylation gives methoxyvincamajine (**58**). 17-*O*-Acylation gives 10-hydroxy-17-*O*-trimethoxybenzoylvincamajine (**56**) and 10-hydroxy-17-*O*-trimethoxycinnamoylvincamajine (**57**). *O*-Methylation of each leads to the corresponding 10-methoxy alkaloids (**59**) and (**60**).

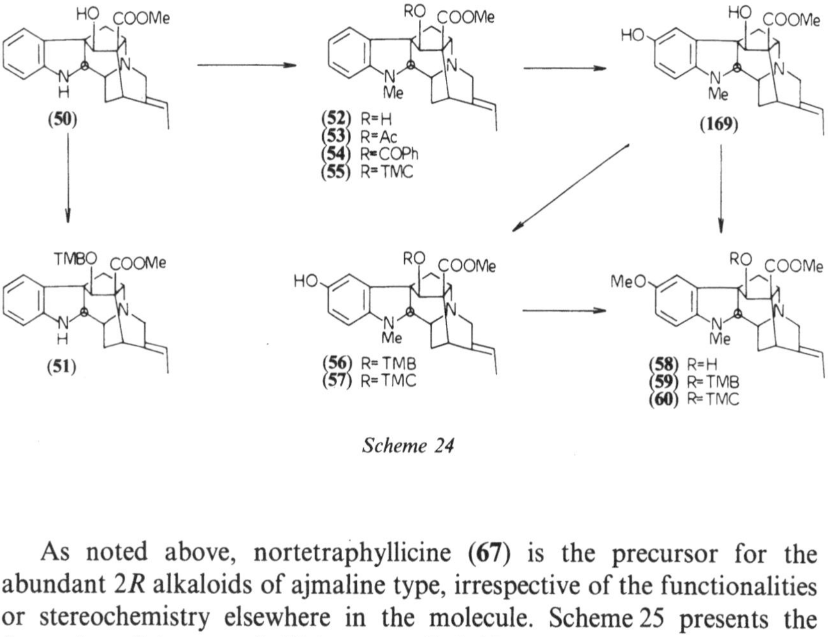

Scheme 24

As noted above, nortetraphyllicine (**67**) is the precursor for the abundant 2*R* alkaloids of ajmaline type, irrespective of the functionalities or stereochemistry elsewhere in the molecule. Scheme 25 presents the formation of the tetraphyllicine type alkaloids. Features common to these alkaloids are 2*R*, 17*R* stereochemistry and the *E*-19,20-double bond. Acylation of nortetraphyllicine (**67**) gives 17-*O*-acetylnortetraphyllicine (**72**) and norrauvomitine (**68**). N_a-Methylation generates tetraphyllicine (**69**). Rauvomitine (**71**) can be construed to arise either by N_a-methylation of norrauvomitine (**68**) or, as 17-*O*-acetyltetraphyllicine (**73**), by acylation of tetraphyllicine (**69**). As before, Ar-hydroxylation is a facile process, which generates hydroxynortetraphyllicine (**74**), and, after *N*-methylation and *O*-methylation, vincamajoreine (**75**). Majoridine (**76**) is the result of acetylation of (**75**).

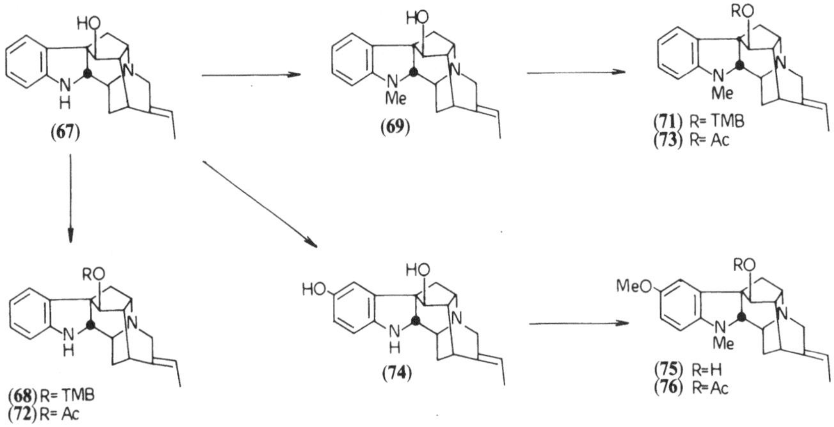

Scheme 25

In Scheme 25, the parent compound nortetraphyllicine (67) was initially methylated and acylated. An alternative Ar-hydroxylation followed by O-methylation generates norseredamine (77), an obvious precursor for seredamine (79) and trimethoxybenzoylseredamine (80) (Scheme 26). 12-Methoxyajmaline (99) is conveniently derived from seredamine (79).

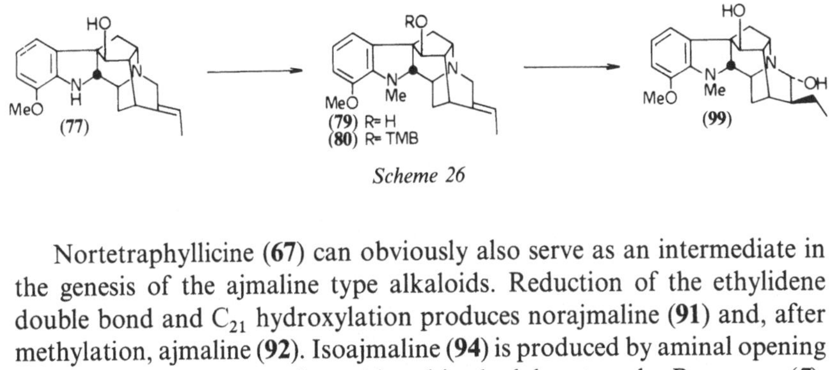

Scheme 26

Nortetraphyllicine (67) can obviously also serve as an intermediate in the genesis of the ajmaline type alkaloids. Reduction of the ethylidene double bond and C_{21} hydroxylation produces norajmaline (91) and, after methylation, ajmaline (92). Isoajmaline (94) is produced by aminal opening and reclosure, a process also achieved in the laboratory by ROBINSON (7). Simple acylation reactions provide 17-acetylajmaline (96), 17-O-tri-methoxybenzoylajmaline (98) and ajmalan diacetate (97) (Scheme 27).

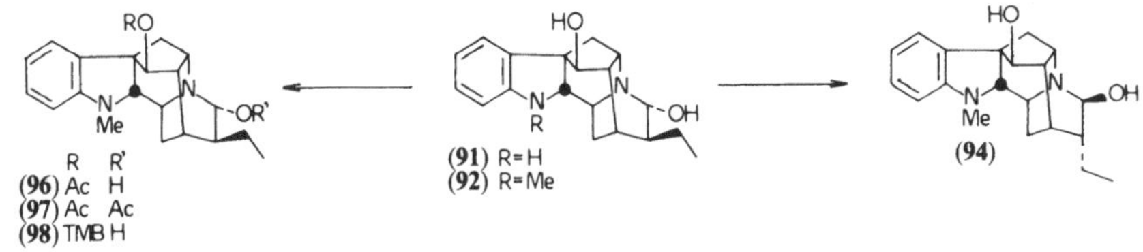

Scheme 27

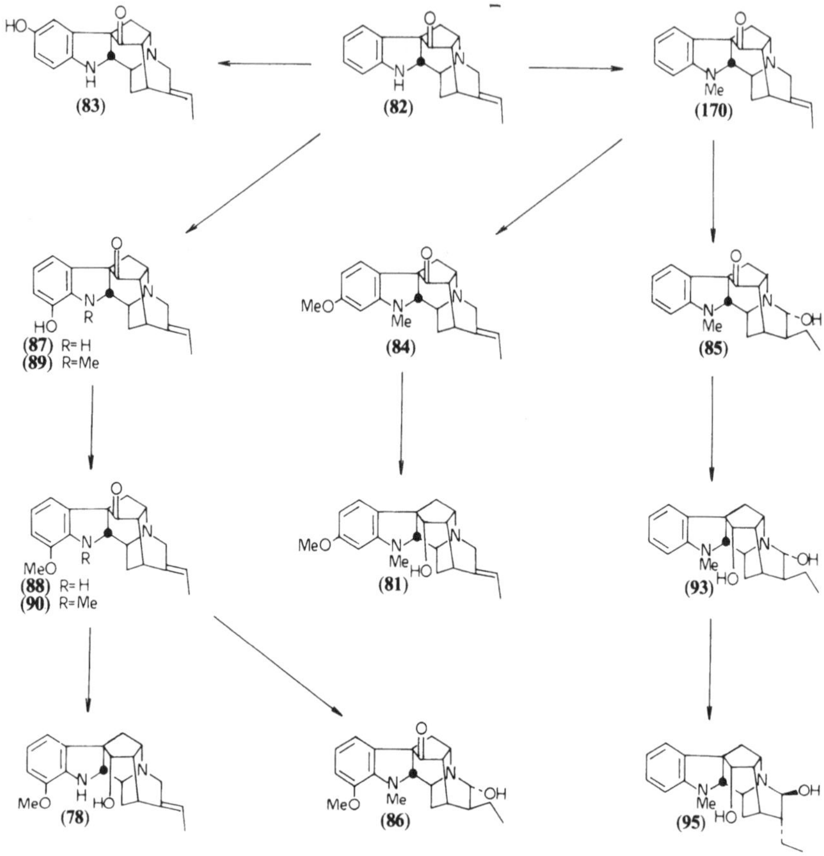

Scheme 28

The last of the possible one-step modifications of the nortetraphyllicine stem nucleus involves oxidation of nortetraphyllicine (67) at C_{17} to carbonyl oxidation level with production of rauflorine (82) (Scheme 28). Ar-Hydroxylation of rauflorine gives endolobine (83) and normitoridine (87) and therefrom the N_a-methyl derivative mitoridine (89). O-Methylation of (87) generates norpurpeline (88), the precursor of purpeline (90) and dihydronorpurpeline (78). On C_{21} hydroxylation and reduction norpurpeline (88) is converted to vomalidine (86). On the other hand, N_a-methylation of rauflorine gives birth to (170) which after Ar-hydroxylation and O-methylation provides rauflexine (84). Hydride reduction of rauflexine (84) generates a hydroxyl group at C_{17}. As in laboratory reductions (13), the donation of hydride occurs from the less hindered face of the carbonyl group to generate 17S stereochemistry in the product, reflexine

(81). The ketone (170) can also be oxygenated at C_{21} and reduced to ajmalidine (85), which on similar hydride reduction provides sandwicine (93), epimeric to ajmaline (92) at C_{17}. Ring opening and re-closure as in the case of ajmaline generates isosandwicine (95).

A last mention is left for the hydroxyajmaline derivative ajmalinol (100) (138) isolated from *Rauwolfia vomitoria* Afz. in 1979. The phenolic nature of the compound was deduced from chemical and UV, IR, MS and ^{1}H NMR spectroscopic data and the hydroxyl group was suggested from ^{13}C NMR data to be at C_{11} (138). As KOSKINEN and LOUNASMAA have pointed out (247), however, the ^{13}C NMR data given in the original paper cannot be ascribed to the proposed compound, or to any Ar-hydroxylated indoline compound, though it is obvious from other data, especially MS, that the indoline portion does contain an oxygen substituent (see section 7.2.2. for the mass spectra and 7.1.2. for the ^{13}C NMR spectra of the ajmaline type alkaloids). This contradiction has led us to consider the position of the oxygen substituent as unresolved.

4.10.3. Bisindole Alkaloids

The biogenesis of the bisindole alkaloids follows the generally accepted lines, i. e. the monomeric units are formed first and "dimerisation" is a late step. Thus, accedinisine (101) is the product of condensation of affinisine (26) and vobasinol (171) as shown below. Geissolosimine (103) is feasibly derived from vellosimine (20) and geissoschizoline (172). Condensation of macroline (173) with N_a-methylsarpagine (37) provides macralstonidine (104) (249).

Equation 5

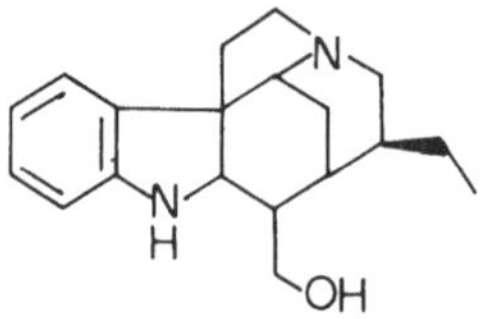

(172) 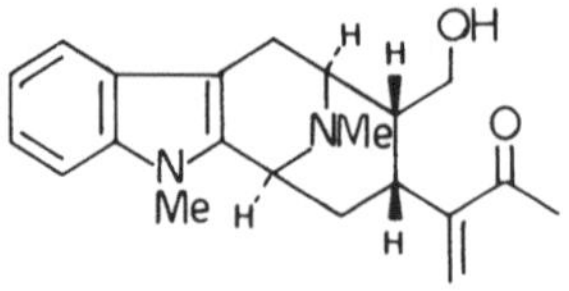   (173)

Fig. 23

Finally, macroline (173) and quebrachidine (50) are potential precursors of alstonisidine (105) (249).

5. Chemistry

Most of the chemical transformations of the sarpagine-ajmaline type alkaloids were discovered during elucidation of the structures of the earliest compounds. Thus, it is no surprise that the most important reactions are described in the two reviews by TAYLOR (2a, b) and in the series *The Alkaloids* (3).

Two interesting subjects will be discussed here, *viz.* the biomimetic type synthesis of macroline from a sarpagine type indole alkaloid and the photochemical oxidations carried out on ajmaline.

5.1. Conversion of Normacusine B to Macroline

ESMOND and LE QUESNE succeeded in converting normacusine B (22) into macroline (173) in a biomimetic laboratory synthesis (248) (Scheme 29). Macroline (173) itself, although shown (249) to be a direct precursor of the bisindole alkaloids villalstonine, alstonisidine, macralstonine and macralstonidine as well as of the monomeric base alstonerine, has not been encountered as a natural product. Thus (173), or an equivalent ketoammonium salt such as (174), was proposed as an intermediate occurring in low steady-state concentrations in the biosynthetic pathway. Strong support for this proposal was given by the synthesis of macroline (173) *via* a derivative of (174).

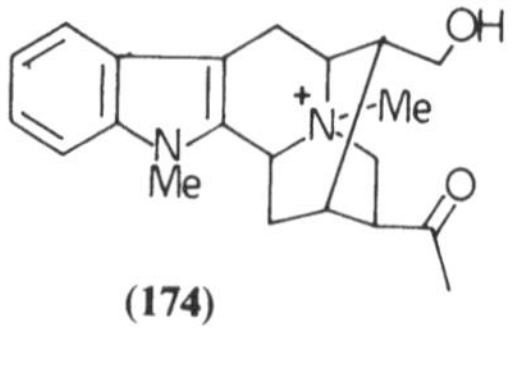

(174)

Fig. 24

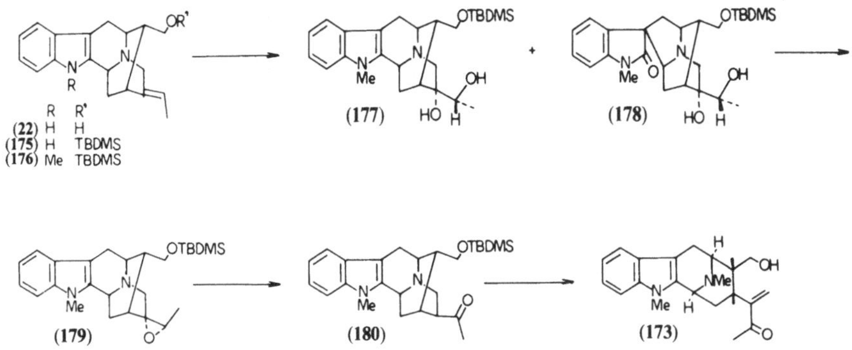

Scheme 29

The hydroxyl group of normacusine B (22) was protected as the TBDMS ether. N-Methylation of (175) gave (176) which was stereoselectively osmylated to give a mixture of the diol (177) and the spirooxindole (178). The diols (177) and (178) were converted to the corresponding epoxides. At this stage the desired epoxide (179) could be separated and rearranged to the ketone (180), an analog of (174). Further treatment of (180) with Me_2SO_4 followed by $Bu_4N^+F^-$ gave only macroline (173).

5.2. Photochemistry

KHUONG-HUU and GOUTAREL (250, 251) have studied sensitised photochemical reactions of ajmaline. In the absence of oxygen, ajmaline (92) merely demethylates to norajmaline (91) as shown in eq. 6.

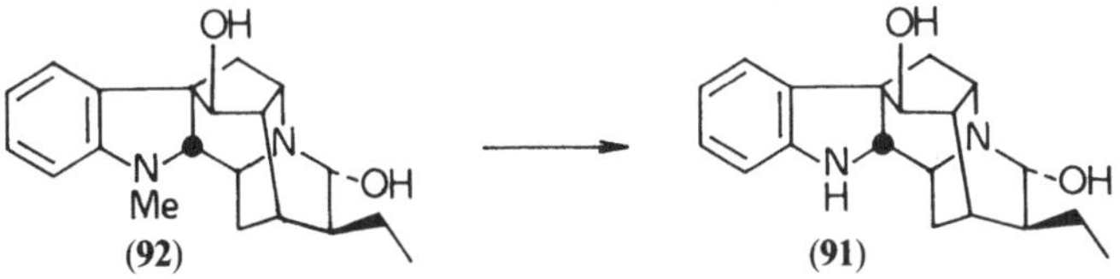

Equation 6

However, in the presence of KCN and eosine (251) the aminonitrile (181) is produced (eq. 7). The process is rationalised (251) as occurring via 2-oxidation. In a similar experiment with methylene blue as sensitizer, the 17-blocked acetylajmaline (182) provided the 2-hydroxy-derivative (184) in excellent yield (eq. 8).

Equation 7

(182) 2βH R=H
(183) 2αH R=Ac

(184) R=H
(185) R=Ac

Equation 8

Interestingly, when 2-epi-diacetylajmaline (183) was subjected to the same conditions, the same (2-epi) stereochemistry was observed in the product (185). The results indicate that the stereochemistry at C_2 does not influence the site of photo-oxygenation as it does in the case of chemical oxygenation with MnO_2 (251).

6. Syntheses

6.1. Total Syntheses of Ajmaline

Only three total syntheses of ajmaline have been published and are presented here in chronological order.

6.1.1. Masamune Synthesis

The first total synthesis of ajmaline was performed by Masamune et al. (252) in 1967 (Scheme 30). Condensation of the Mg chelate of ethyl hydrogen Δ^3-cyclopentenylmalonate (185) with N-methyl-3-indolylacetyl chloride (186) provided the ketoester (187) in 80% yield. This was converted to the corresponding methyloxime, which in turn was reduced with $LiAlH_4$ to give the epimeric α,γ-diamino alcohols (188a) and (188b) in 70% yield. Benzoylation to (189) and $OsO_4/NaIO_4$ cleavage then provided (190) in 40 and 50% yield of the isomers. Conversion to the cyanides (191) and ethylation with EtI furnished the monoethyl compounds (192) in 42–47% yield. Removal of the benzoyl protective group provided the hydroxy compounds (193) which were identical with degradation products of ajmaline. The alcohols (193) were oxidized with DMSO and Ac$_2$O to afford the epimeric aldehydes (194) which were readily equilibrated in the presence of alumina. Acid-catalysed cyclization of (194a) (16S) provided (195) in

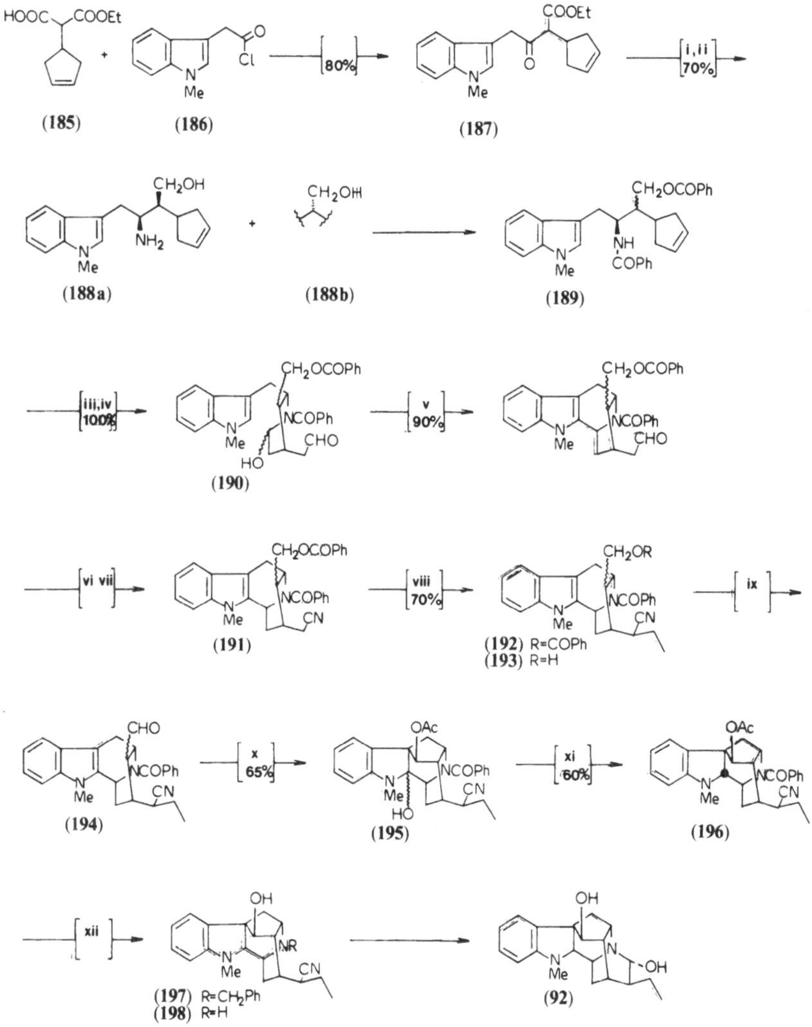

Scheme 30. Reagents: i MeONH$_2$; ii LiAlH$_4$; iii OsO$_4$; iv NaIO$_4$; v AcOH, 50°; vi HONH$_2$; vii PhCOCl; viii Ph$_3$C$^-$Na$^+$, EtI; ix Sc$_2$O, DMSO; x HCl, AcOH, Ac$_2$O; xi H$_2$-PtO$_2$; xii LiAl(OEt)$_3$H

65% yield, and **(195)** was hydrogenated to **(196)** in 60% yield. Reduction [LiAl(OEt)$_3$H] to the benzyl derivative **(197)** and hydrogenolysis then gave the secondary amine **(198)**. Since the latter had previously been converted to ajmaline with LiAlH$_4$, the total synthesis of ajmaline was considered complete. The total yield of the synthesis remains well below 5%. (Yields were not given for all steps in the original paper, and the use of relay substances further obscures the overall productivity of the route.)

6.1.2. Mashimo and Sato Synthesis

Just two years after MASAMUNE's ajmaline synthesis, MASHIMO and SATO succeeded in the synthesis of isoajmaline (*253, 254*), a naturally occurring compound stereoisomeric with ajmaline at C$_{20}$ and C$_{21}$. Their synthesis of isoajmaline and also of ajmaline (*255*) itself converged to the MASAMUNE synthesis in the final steps.

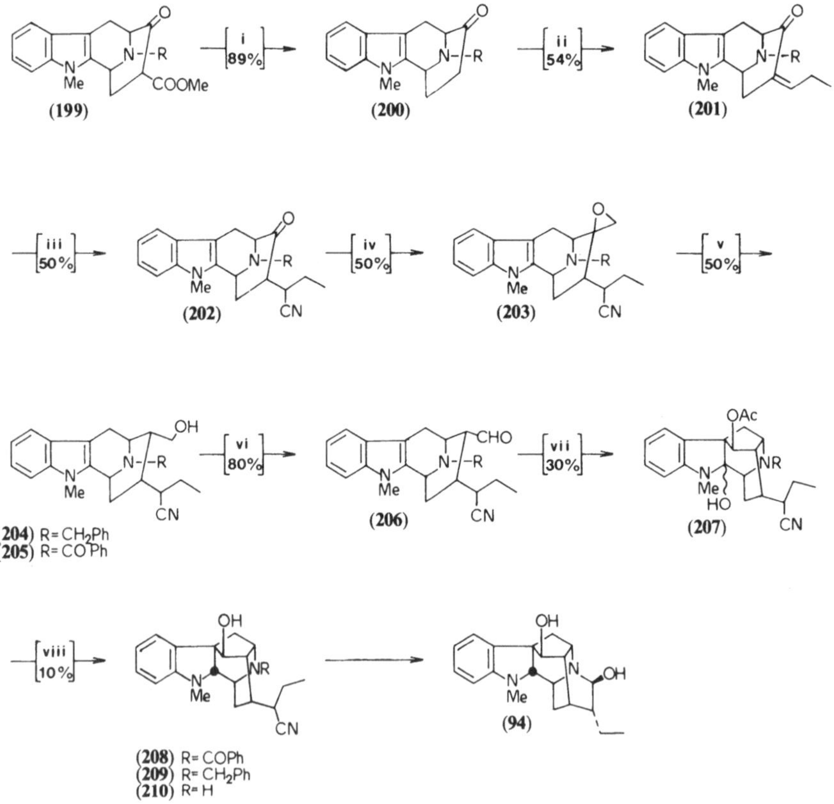

Scheme 31. Reagents: i HCl, AcOH; ii Triton-B, CH$_3$CH$_2$CHO; iii KCN; iv Me$_2$$\overset{+}{S}OCH_2^-$; v AlH$_3$; vi Ac$_2$O, DMSO; vii HCl, AcOH, Ac$_2$O; viii H$_2$-PtO$_2$

We shall first consider the synthesis of isoajmaline (*253, 254*) outlined in Scheme 31.

The synthesis started with the previously described β-ketoester (**199**), which was subjected to ketone fission (HCl-AcOH, Δ, 80%) to provide the ketone (**200**). Condensation of this with *n*-propanal by means of Triton-B afforded the propylidene derivative (**201**) in 54% yield. Hydrocyanation (KCN-NH$_4$Cl, 54%), conversion to the oxirane (**203**) according to the Corey method (MeSOCH$_2^-$, 50%) and reductive oxirane cleavage with AlH$_3$ provided the carbinol (**204**) (54% yield). Debenzylation (H$_2$-Pd/C, 58%), dibenzoylation and selective *O*-debenzoylation then afforded the carbinol (**205**), isomeric with the MASAMUNE intermediate (**193**). The remaining steps of the synthesis of isoajmaline closely parallel those of the earlier ajmaline synthesis. Oxidation of (**205**) with DMSO-Ac$_2$O gave (**206**) in 80% yield. Equilibration, acid catalyzed cyclization to (**207**) (30% yield) and hydrogenolysis (PtO$_2$-Pt, HCl, 6N, 10%) furnished (**208**), which was alkylated with the Meerwein reagent and the product reduced with NaBH$_4$ to give (**209**) (85% yield). Reductive debenzylation of (**209**) furnished (**210**), which had previously been converted to isoajmaline (**94**) by ROBINSON. The overall yield of the route was only 0.017% according to the original report and the full paper.

MASHIMO and SATO (*255*) also achieved the synthesis of ajmaline. Preparation of the Masamune intermediate (**191**) was performed in the following manner (Scheme 32): Alkylation of the pyrrolidine-enamine (**211**) of the ketone (**200**) with chloroacetonitrile afforded the nitrile (**212**) in 50% yield. Corey epoxide formation led to (**213**) in 70% yield and reductive oxirane cleavage gave the carbinol (**214**) in 80% yield. Reductive debenzylation (Pd/C, 80%) and dibenzoylation then furnished (**191**).

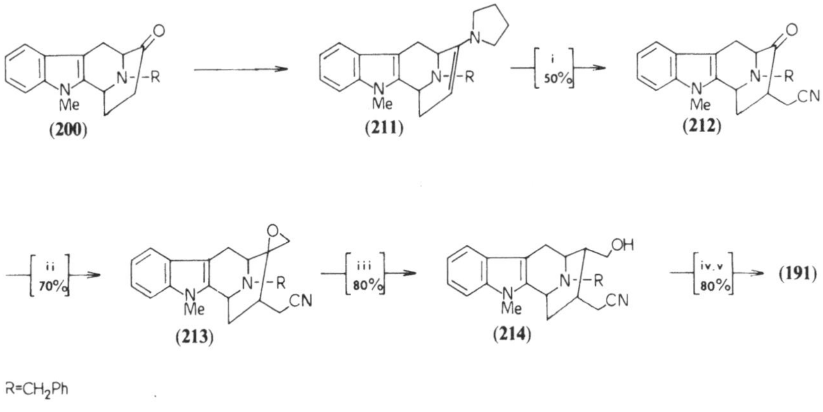

R=CH$_2$Ph

Scheme 32. Reagents: i ClCH$_2$CN; ii Me$_2$$\overset{+}{S}OCH_2^-$; iii HAlCl$_2$ or AlH$_3$; iv H$_2$-Pd/C; v PhCOCl

6.1.3. The Biogenetic-Type Synthesis of van Tamelen and Olivier

In 1970 VAN TAMELEN (228) succeeded in performing a synthesis closely paralleling the biogenetic scheme postulated for the biogenesis of sarpagine type indole alkaloids. Their synthesis of ajmaline proceeded as outlined in Scheme 33 (227).

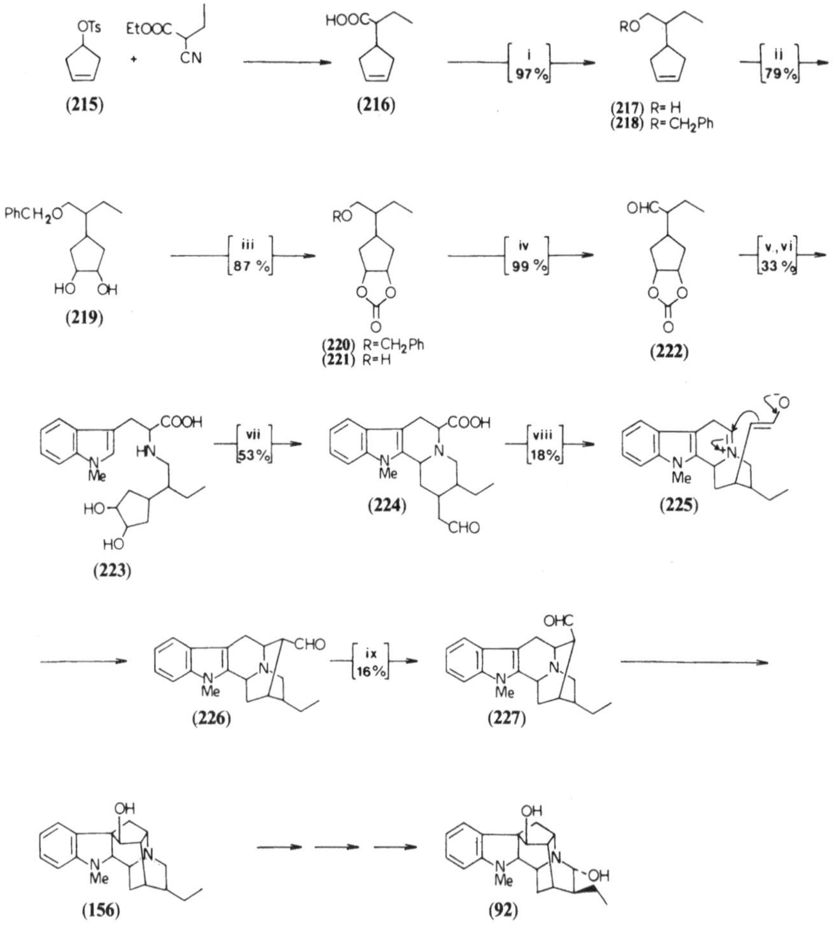

Scheme 33. Reagents: i LiAlH₄; ii OsO₄; iii (MeO)₂CO, NaOMe; iv CrO₃·py; v Trp, H₂-Pd/C; vi KOH, MeOH; vii NaOAc, HIO₄; viii DCC, TsOH; ix NaOac, AcOH

The Δ³-cyclopentenyl tosylate (215) was converted to dl-α-(Δ³-cyclopentenyl)-butyric acid (216) which was reduced with LiAlH₄ to the alcohol (217) in 94−97% yield. Benzylation (PhCH₂Cl, 100°, KOH, 83%) and osmium tetroxide oxidation (79%) then furnished the diol (219) which was protected as the carbonate (**220**) (87% yield on transesterification with Me$_2$CO$_3$ in the presence of NaOMe). Catalytic debenzylation (H$_2$-Pd/C in EtOH, 95%) and Collins oxidation of the alcohol furnished aldehyde (**222**) in 50% overall yield from (**220**). Reductive alkylation of *N*-methytryptophan with the aldehyde (**222**) gave the *N*-substituted amino acid (**223**) in 51% yield. Metaperiodate glycol cleavage gave the corresponding dialdehyde, which spontaneously cyclized to the tetracycle (**224**) in 53% yield from (**222**). The aldehydoacid was now submitted to decarbonylation induced by DCC, furnishing an iminium salt. This readily cyclized to *dl*-deoxyajmalal-B (**226**) in 18% yield in a reaction closely mimicking the biosynthetic pathway. The synthetic aldehyde was resolved using D-camphorsulfonic acid and the resolved base as well as its sulfonate salt proved to be identical in all respects with authentic specimens derived from natural ajmaline. The completion of the synthesis depended, once again, on relay operations. Deoxyajmalal-B (**226**) can be equilibrated with deoxy ajmalal-A (**227**) (15% A to 85% B). The latter had earlier been converted to deoxyajmaline (**156**) by TAYLOR *et al.* (*280*) and functionalization of (**156**) at C-21 had been achieved by the phenylchloroformate ring opening — oxidative ring closing method of HOBSON and MCCLUSKEY (*281*).

6.2. Partial Synthesis of Talpinine

GARNICK and LE QUESNE (*249*) have transformed alstonisine (**228**), a natural base readily obtainable from *Alstonia muelleriana,* to talpinine (**44**) by means of LiAlH$_4$ reduction followed by acid catalyzed isomerization (eq. 9). The reaction, although proceeding in poor yield, was considered biomimetic.

Equation 9

6.3. Synthesis of the Ring Skeleton

During the total synthesis of dregamine (**151**) and 20-epidregamine, KUTNEY (*256*) was able to devise a method for the synthesis of the ring skeleton of sarpagine. The synthesis followed the lines presented in Scheme 34. Hydride reduction of L-(−)-tryptophan followed by tosylation gave the ditosylate (**229**). Displacement with cyanide and deprotection gave the β-aminonitrile (**230**). Formylation and polyphosphate ester catalysed cyclisation led to the unstable 3(*S*)-β-carboline (**231**). Acid catalysed annelation of (**231**) with 3-methylenepentan-2-one gave (**232**) (21%), its 3-isomer (**233**) (15%) and the C_{20} epimeric pair of compounds (39%). Compound (**233**) could be epimerised to (**232**). Finally, treatment of (**232**) with lithium diethylamide in THF at 0° furnished the more stable 16(*R*) nitrile (**234**) together with the 16(*S*) epimer in a ca 2:1 ratio.

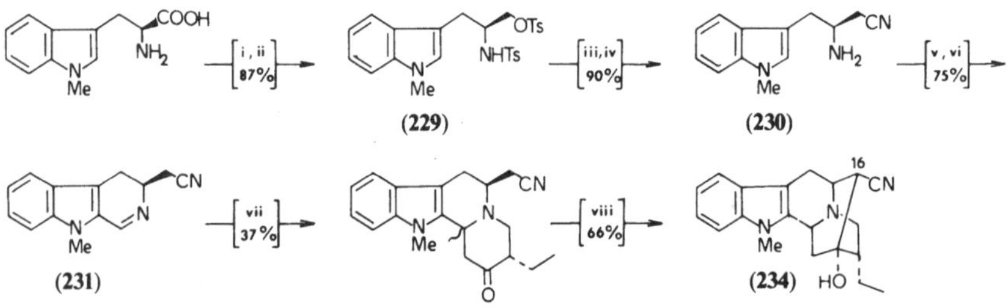

Scheme 34. Reagents: i LiAlH₄; ii TsCl; iii KCN, MeOH; iv Na, NH₃; v HCOOMe, NaOMe; vi PPE; vii (structure), MeOH, 70°; viii LiNEt₂, THF

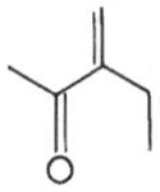

6.4. Bisindole Alkaloids

Alstonisidine (*257, 258*) and accedinisine (*16*) have been synthesised in a biomimetic fashion from the corresponding monomeric units. These syntheses have recently been reviewed (*259*).

7. Spectroscopy

Only the two most important spectroscopic methods for sarpagine-ajmaline type indole alkaloids, ^{13}C NMR and mass spectrometry, will be discussed in this review.

7.1. Carbon-13 NMR Spectroscopy

The reported ^{13}C NMR spectra for sarpagine and ajmaline type alkaloids are presented together with some general observations about the effects of substituents and stereochemistry on the spectral parameters.

The indoline nucleus of the ajmalinoid natural products is a structural moiety common to several *Aspidosperma* alkaloids whose ^{13}C NMR spectra have been interpreted fully (*260*). Examples of the N_a-alkylindole unit containing a single methoxy substituent at C_{10}, C_{11} and C_{12} have been reported. Each presents a characteristic pattern of aromatic methine shifts which is only minimally dependent on the nature of the C_2, C_7 and N_a-alkyl substituents. Methine carbons situated *ortho* to either the methoxy or the indoline nitrogen are subject to strong shielding (ca. 12 – 16 ppm) and consequently resonate above 112 ppm. Thus, determination of the number of aromatic methines that fulfils this criterion unambiguously identifies the location of the methoxy group.

7.1.1. Sarpagine Group

^{13}C NMR spectra of dehydrovoachalotine (**7**) (*178*), 17-*O*-acetyl-19,20-dihydrovoachalotine (**8**) (*178*), gardnerine (**10**) (*182*), hydroxygardnerine (**11**) (*182*), hydroxygardnutine (**14**) (*182*), akuammidine (**15**) (*261*), pericyclivine (**18**) (*41*), *O*-acetylnormacusine B (**23**) (*185*), 17-*O*-acetylvoachalotine (**235**) (*178*), diacetylvoachalotinol (**236**) (*178*), 16-epigardnerine (**237** (*182*), and 17-deoxy-16-epigardnerine (**238**) (*182*) have been reported and are summarized in Table 3.

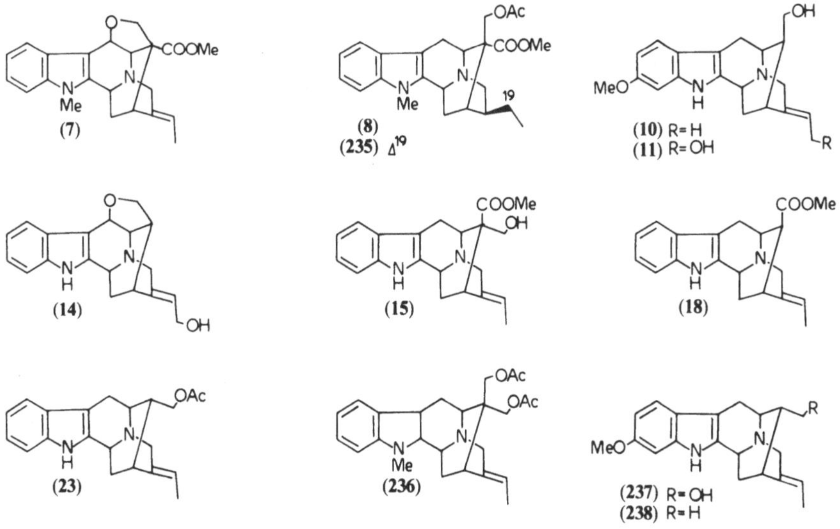

Fig. 25

Table 3. ^{13}C NMR Spectra of Sarpagine Type Compounds

C	**(7)**[a] (178)	**(8)**[a] (178)	**(10)**[b] (182)	**(11)**[b] (182)	**(14)**[c] (182)	**(15)**[a] (261)	**(18)**[a] (41)	**(23)**[a] (185)	**(235)**[a] (178)	**(236)**[a] (178)	**(237)**[b] (182)	**(238)**[a] (182)
2	143.6	137.6	138.3[d]	138.2[d]	141.7[d]	143.9	139.4	134.9	137.5	136.3	138.3[d]	137.6
3	47.0	47.4	50.5[e]	50.3[e]	59.1	60.5	43.8	49.9	47.8	48.4	51.2	50.1[d]
5	61.5	53.8	53.0[e]	53.0[e]	59.1	62.8	55.9	54.5	53.8	57.3	51.2	59.7[d]
6	72.6	23.4	23.4	23.2	70.6	29.1	26.8[d]	26.9	23.0	22.9	27.9	26.1
7	103.5	104.9	106.0	105.8	102.7	110.2	105.5	103.9	104.7	105.0	104.1	104.6
8	126.5	126.7	f	f	120.5	132.2	127.0	127.4	126.3	126.4	123.0	122.3
9	119.1	118.3	118.8	118.8	118.3	122.7	119.2[e]	117.9	118.3	118.6	118.7	118.5
10	120.1	119.3	108.7	108.7	108.4	123.6	117.4[e]	121.1	119.2	119.5	108.7	108.7
11	121.7	121.3	156.3	156.4	155.3	125.5	121.3	119.1	121.3	121.6	156.4	156.0
12	109.2	109.0	95.8	96.0	95.4	116.3	110.9	110.8	108.9	109.2	96.0	95.4
13	137.7	138.8	137.9[d]	137.6[d]	136.6[d]	142.2	137.5	136.3	138.3	138.0	137.8[d]	137.6
14	29.0	27.1	27.9	27.9	27.8	34.2	24.2	33.2	28.6	28.5	34.4	34.1
15	31.0	32.7	27.3	27.9	27.4	34.3	27.2[d]	27.6	31.3	31.3	28.4	32.5
16	53.9	48.5	43.6	43.6	47.9	55.0	50.3	40.8	52.0	41.8	45.3	36.3
17	68.3	67.2	60.5	60.5	63.9	73.4	172.9	66.1	65.4	62.2	64.7	18.7
18	12.7	12.5	13.0	58.0	56.7	16.3	12.9	12.7	12.8	12.7	13.0	12.6
19	116.5	32.7	112.6	120.0	119.5	120.6	114.4	116.9	116.7	117.3	115.7	116.1
20	135.9	40.5	142.3	144.0	141.9	144.0	136.6	138.1	136.1	136.3	139.2	136.4
21	55.4	56.2	56.9	56.6	55.1	56.5	53.0	55.9	55.8	55.9	56.6	56.3
22	175.9	176.3				178.4			174.8	68.4		
–COOCH₃	52.1	51.9				55.1	50.8		52.0			
–N–CH₃	29.0	29.2							29.2	29.2		
Ar–OCH₃			55.6	55.6	55.5						55.6	55.8
–COCH₃		169.9						170.8	169.9	170.3[d]		
–COCH₃										171.1[d]		
–COCH₃		20.7						20.9	20.7	20.8		
–COCH₃										20.8		

[a] In $CDCl_3$. [b] In pyridine-d_5. [c] In DMSO-d_6. [d,e] Signals may be interchanged within each vertical column. [f] Hidden in the solvent absorption.

Assignment (*182*) of C_6 and C_{14} of the gardnerine-type alkaloids (**10, 11, 14, 237, 238**) was based on the following considerations: i) the methylene frequency observed at 23.4 ppm in the spectrum of (**10**) was shifted to 70.6 ppm in the spectrum of hydroxygardnutine (**11**), which has an oxygen function at C_6. Thus the signal at 23.4 ppm must correspond to C_6; ii) the methylene resonance at 27.9 ppm in both (**10**) and (**11**) is shifted to lower field in the spectra of the C_{16} epimers (**237**) and (**238**). The change can be regarded as the result of diminution of the γ-effect exerted by the substituent on C_{16}.

7.1.2. Ajmaline Group

^{13}C NMR spectra have been reported for raucaffrinoline (**49**) (*63*), vincamajine (**52**) (*107*), vomilenine (**62**) (*63*), dihydrovomilenine (**66**) (*63*), vincamajoreine (**75**) (*107*), majoridine (**76**) (*107*), rauflexine (**84**) (*107*), ajmaline (**92**) (*262*), sandwicine (**93**) (*262*), isoajmaline (**94**) (*262*) and isosandwicine (**95**) (*262*) and are summarized in Table 4.

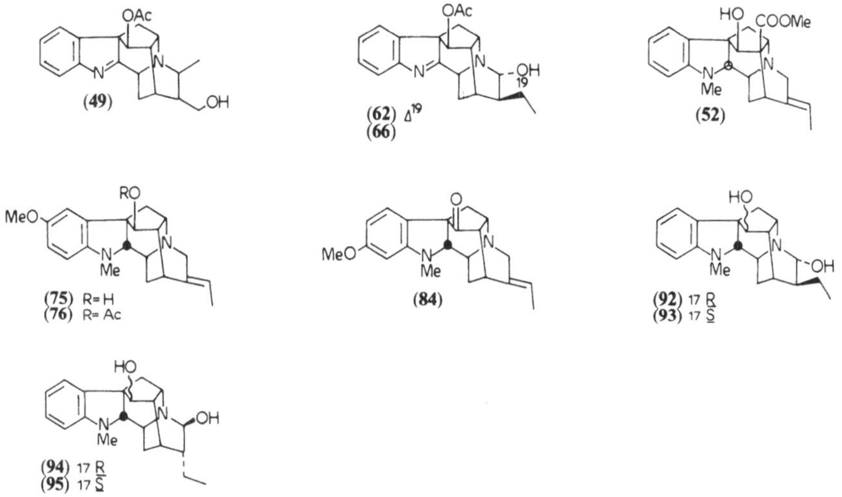

Fig. 26

CHATTERJEE *et al.* (*107*) used ^{13}C NMR data to study the aromatic substitution pattern assigned to the indolic bases (**75**), (**76**) and (**92**) from *Rauwolfia reflexa* and found the previously assigned pattern to require revision. In (**75**) and (**76**) the methylene carbons C_6, C_{14} and C_{21} are unaffected by the structural difference at C_{17}. The alteration of ring D of ajmaline (**92**) with respect to (**75**) perturbs the C_{14} resonance but not that of C_6.

Table 4. ^{13}C NMR Spectra of Ajmaline Type Compounds

C	(49)[a] (63)	(52)[b] (107)	(62)[a] (63)	(66)[a] (63)	(75)[b] (107)	(76)[a] (107)	(84)[a] (107)	(92)[a] (262)	(93)[a] (262)	(94)[a] (262)	(95)[a] (262)
2	183.6	74.4	183.6	183.0	79.6	79.6	78.4	79.3	75.7	78.7	76.3
3	57.2	52.7	54.3[c]	54.8	49.0[c]	49.3[c]	50.1[c]	43.0	44.8	47.5	48.1
5	51.2[c]	61.1	50.9[c]	49.8[c]	55.8[c]	55.9[c]	53.1[c]	52.8	53.0	48.0	48.6
6	37.5	35.0	36.4	37.4	34.9	36.1	35.3	34.8	34.9	34.1	35.1
7	65.0	56.5	65.1	65.1	54.9	53.6	57.8	56.1	54.0	55.6	54.5
8	136.5	129.7	136.1	136.1	134.4	133.3	121.6	133.3	131.8	133.7	131.8
9	123.8	124.2	123.9	123.7	110.2	110.0	122.5	122.8	120.2	123.1	119.6
10	125.4	118.5	125.8	125.4	153.0	153.0	103.8	119.0	118.7	118.5	118.8
11	128.6	127.6	128.9	128.7	111.4	111.1	160.1	127.1	127.1	126.5	127.3
12	120.9	108.4	121.1	121.1	109.2	109.6	97.5	109.5	109.7	108.6	109.5
13	156.5	153.8	156.3	156.6	147.7	147.8	155.1	153.6	153.3	153.4	153.9
14	21.7	21.4	26.3	27.8	29.2	29.4	31.5	31.4	31.0	22.3	22.3
15	26.6	29.6	28.2	27.5	27.9	27.8	28.5	28.3	27.2	29.1	28.2
16	49.7[c]	59.6	49.0[c]	47.2[c]	51.9	50.1	50.3[c]	45.2	34.4	53.0	43.4
17	78.2	73.9	77.5	78.7	76.0	79.1	214.0	77.3	70.8	75.9	72.3
18	18.3	12.3	13.0	11.9	12.5	12.8	12.9	12.2	12.1	12.3	12.4
19	53.2	116.1	119.4	26.0	114.2	114.3	115.7	25.4	25.0	25.6	25.8
20	45.7	135.6	131.0	42.0	138.6	139.3	137.3	48.0	48.7	54.2	45.6
21	61.7	54.7	82.5	87.5	54.6	55.2	55.7	88.1	88.1	87.5	88.1
22		172.8									
–COOCH$_3$		51.1									
–N–CH$_3$		33.8			34.8	35.1	34.2	34.0	34.4	34.0	34.5
Ar–OCH$_3$					55.6	55.5	55.3				
–CO̲CH$_3$	169.9		169.7	169.8		169.9					
–COC̲H$_3$	d		21.1	21.1		21.1					

[a] In CDCl$_3$. [b] In 5:1 CDCl$_3$:CD$_3$OD. [c] Signals may be interchanged within each vertical column. [d] Signal not reported.

DANIELI et al. (262) have analysed the ^{13}C NMR spectra of the ajmaline stereoisomers and observed that C_{21} is the most deshielded methine. The C_{17} methine can be differentiated from C_2 by the acetylation shift of the former. Of the remaining methines, C_{15} is shifted upfield and C_3 and C_5 are distinguished from C_{16} and C_{20} by the large β-effect (11 – 13 ppm) after N_b-methiodide formation. C_3 is slightly upfield of C_5. The criterion for the differentiation of C_{16} from C_{20} is based on the observation that C_{16}, in contrast to C_{20}, exhibits sharp one-bond components in SFORD spectra due to few long-range interactions and no second order couplings. As regards the methylene signals, C_{19} is recognised by its invariance on methiodide formation. The configurational changes at C_{17}, C_{20} and C_{21} are accompanied by significant and diagnostic shift changes. In particular, the inversion of C_{17} causes a moderate shift of C_{17} itself, a slight movement of C_7 and a marked downfield shift of C_{16}. On the other hand, concomitant inversion at C_{20} and C_{21} merely deshields C_3 and C_{16} and shields C_5 and C_{14} owing to the loss and presence of γ-steric interactions, and leaves C_{19} invariant as a consequence of the replacement of the γ-effect exerted by C_{16} by one of comparable magnitude exerted by C_{14} (262).

The work of DANIELI et al. also led to the reassignment of parts of the ^{13}C NMR spectra of ajmaline (92) and isoajmaline (94), which were previously misassigned.

7.2. Mass Spectrometry

Mass spectra of the sarpagine-ajmaline type alkaloids have been well reviewed (2, 263), and only the most important points will be considered here.

7.2.1. Sarpagine Group

The mass spectrometric behaviour of the sarpagine type alkaloids does not differ significantly from that of the Corynanthé type alkaloids (263). A characteristic loss of hydrogen to give the M-1 peak is facilitated by a series of rearrangements (Scheme 35): the $C_5 - C_{16}$ bond is broken, the C_6 hydrogen is rearranged and the C_3 hydrogen is lost (264, 265).

Another set of fragmentation reactions (Scheme 36) leads to the intermediate mass peaks at m/e 185 and m/e 184. The $C_3 - C_{14}$ bond of the radical ion species (239) is cleaved homolytically and the resulting pyridinium ion (240) can fragment to the m/e 184 and m/e 185 species (264).

In the case of the C_{16} stereoisomers akuammidine (15) and poly-neuridine (3) (Scheme 37), the cleavage of the side chain at C_{15}, facilitated by the radical stabilising action of the carbomethoxy group at C_{16}, gives rise to the indoloquinolizine unit at m/e 249 (153).

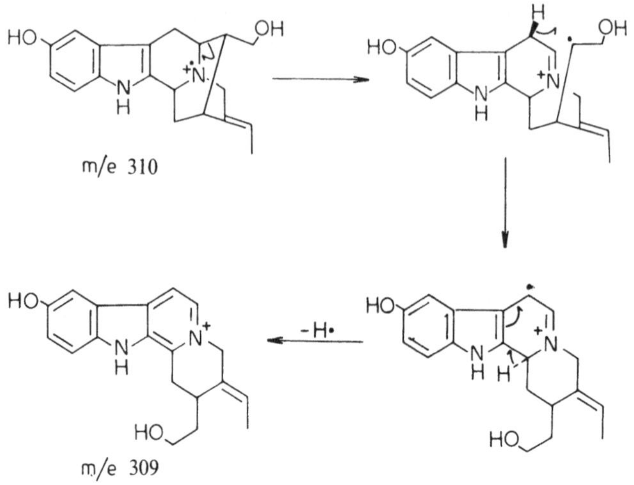

Scheme 35

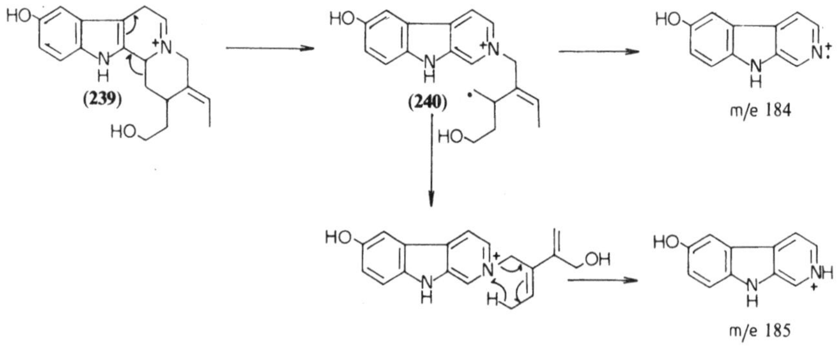

Scheme 36

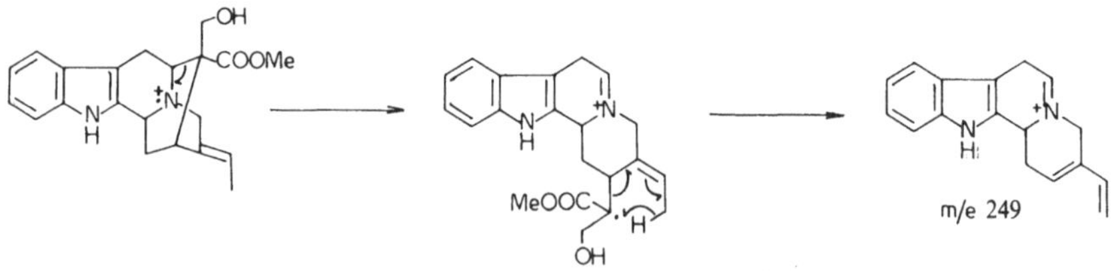

Scheme 37

Thermal loss of water in the mass spectrometer is very pronounced for polyneuridine (**3**) (eq. 10) and absent for akuammidine (**15**). This difference can be rationalised by the involvement of the N_a-hydrogen during the elimination process. Noteworthy, too, is the fact that this peak is highly characteristic of the stereochemistry at C_{16} (*153*).

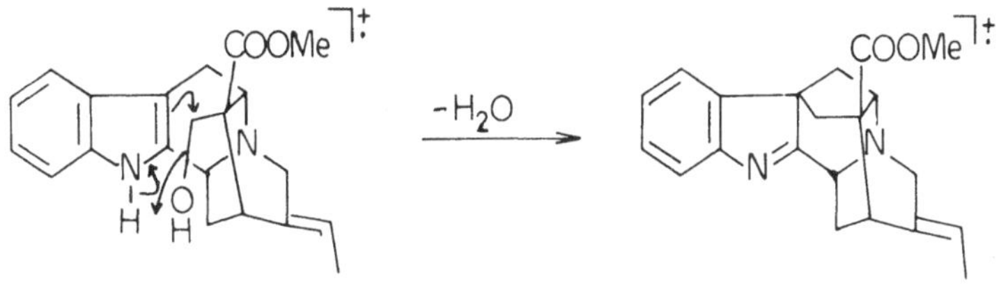

$$-H_2O$$

m/e 352   m/e 334

Equation 10

7.2.2. Ajmaline Group

The mass spectral behaviour of ajmaline type alkaloids has been studied by BIEMANN (*266*). He proposed that the mass spectra of these alkaloids be divided into three different groups: i) the ajmaline group, ii) the ajmalidine group (17-keto) and iii) the quebrachidine group (2 α H) (*263, 266*).

i) The Ajmaline Group

The mass spectral behaviour of these alkaloids is exemplified by the fragmentation of ajmaline itself (Scheme 38). The losses of 15 and 18 mass units correspond to loss of a methyl group and water, respectively. The loss of 29 mass units is attributed for the most part to the loss of CHO, caused by the opening of the carbinolamine grouping in the gas phase. Most characteristic of the stereochemistry at C_2 is the EI-induced cleavage of the five-membered ring with migration of the β-hydrogen from C_2 to C_{17} and formation of a carbon skeleton of the sarpagine type.

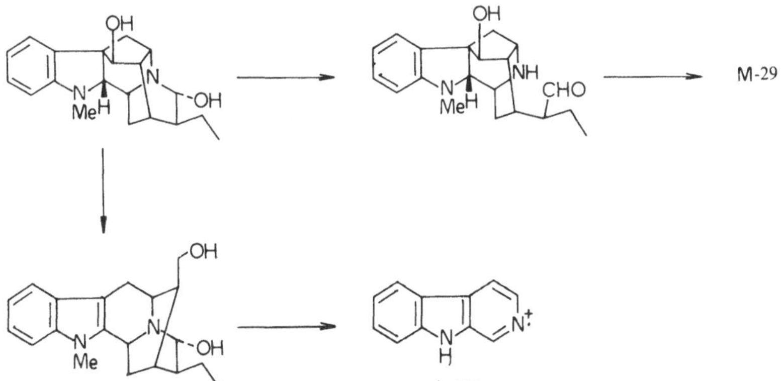

M-29

m/e 182

Scheme 38

Changes such as epimerisation at C_{20}, removal of the 21-hydroxyl group, epimerisation of the 17-hydroxyl group or unsaturation of the ethyl side chain have practically no effect on the basic fragmentation pattern outlined in Scheme 39.

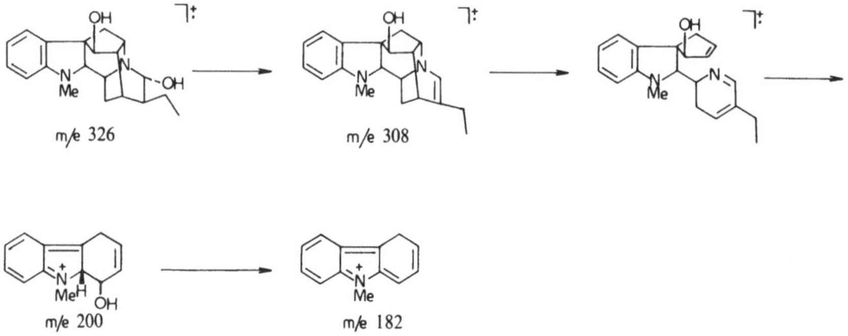

Scheme 39

ii) The Ajmalidine Group

Although the changes just mentioned affect the mass spectrum only in a minor way, oxidation of the C_{17} hydroxyl to a carbonyl group drastically changes fragmentation of the ring system (*267*). The carbonyl group at C_{17} can be eliminated as CO. Cleavage of the $C_2 - C_3$ bond leads directly to m/e 144 ($C_{10}H_{10}N$). The genesis of the peak at m/e 198 ($C_{13}H_{12}NO$) in the mass spectrum (*267*) of ajmalidine is analogous to the one at m/e 200 in ajmaline. The fragmentation of ajmalidine (**85**) is shown in Scheme 40 (*266*).

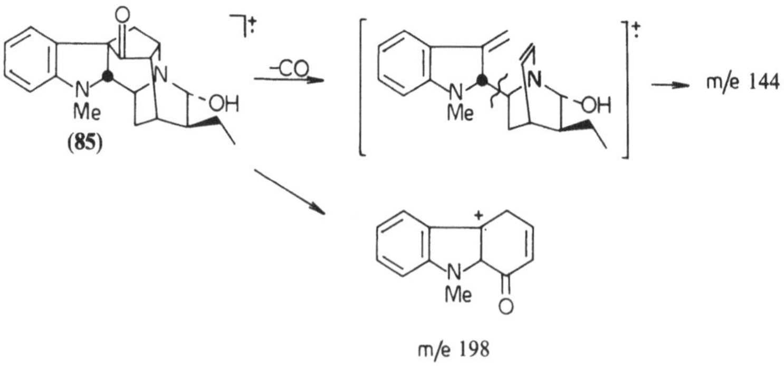

Scheme 40

iii) The Quebrachidine Group

The most drastic change in the fragmentation is produced by epi-merisation at C_2 (*268*). The peaks at m/e 182 and 183 are practically absent in the MS of 21-deoxy-2-epi-ajmalan, while a new one appears at m/e 166, usually the most intensive one in the spectrum (Scheme 41) (*266*).

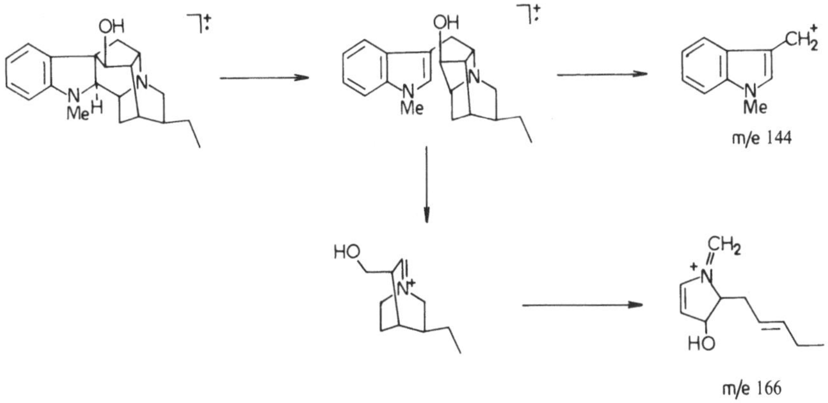

Scheme 41

Thus, the spatial arrangement of the atoms and bonds around C_2 has a pronounced effect on the mass spectra of the different C_2 epimers. In the β-epimers (e. g. ajmaline), the hydrogen at C_2 is sufficiently close to C_{17} to lead to its migration and formation of the aromatic indole system. In contrast, in the 2-epi series (2α), this hydrogen is far away from C_{17} and the $C_2 - C_3$ to $C_3 - C_{17}$ rearrangement is the only available process for aromatisation.

8. Pharmacology

Of the sarpagine-ajmaline type alkaloids, only ajmaline (**92**) has established an important place for itself in therapy. Ajmaline was introduced in the treatment of arrhythmia in 1959 by KLEINSORGE (*269*) and its mode of action has been elucidated. It acts principally to reduce the maximal upstroke velocity of the cardiac action potential without affecting the resting potential (*270*). The reduction of the upstroke rate is dependent on the frequency of stimulation, suggesting that ajmaline inactivates the sodium carrier system (*270*). In a study of the metabolic effects of ajmaline CASTELLUCCI (*271*) observed that ajmaline stimulates glycolysis thus increasing lactate and pyruvate but decreasing FDP and ATP concentrations. The finding that the consumption was not affected was interpreted to indicate that the action was occurring at mitochondrial level. However, the metabolic rate *in vivo* was not affected and thus the increase in glycolysis was considered solely as a side effect, unrelated to the main pharmacological effect on the membranes, i. e. local anaesthetic action inhibiting glucose uptake (*271*). The catecholamine content of the hearts of guinea pigs was likewise not decreased by the use of ajmaline (**92**) or its *N*-methyl or *N*-propyl derivatives (*272*).

Of the chemical modifications of ajmaline, *N*-propylajmaline (*273*) and 4-(3′-diethylamino-2′-hydroxypropyl)-ajmaline hydrogen tartrate (*274*) were shown to be more potent than ajmaline itself.

Among other pharmacological effects, normacusine B (**22**) has sedative and ganglion blocking activity (*275*), gardnutine (**13**), hydroxygardnutine (**14**) and gardnerine (**10**) have ganglion blocking effects (*276 – 278*), lochnerine (**36**) has hypoglycemic activity comparable to tolbutamide (*279*) and pericyclivine (**18**) reportedly has weak cytotoxic activity against rat leukemia P-388 (*146*).

References

1.   *Rauwolfia* alkaloids have been subjected to review:
    a) CHATTERJEE, A.: *Rauwolfia* Alkaloids. Fortschr. Chem. org. Naturstoffe **10**, 390 (1953).
    b) CHATTERJEE, A., S. C. PAKRASHI, and G. WERNER: Recent Developments in the Chemistry and Pharmacology of *Rauwolfia* Alkaloids. Fortschr. Chem. org. Naturstoffe **13**, 346 (1956).
    c) PAKRASHI, S. C., and B. ACHARI: *Rauwolfia* Alkaloids in Retrospect. J. Sci. Ind. Res. **27**, 58 (1968).

2. a) TAYLOR, W. I.: The Ajmaline-Sarpagine Alkaloids. In: The Alkaloids (R. H. F. MANSKE, ed.), vol. VIII, 785. New York: Academic Press. 1965.
    b) — The Ajmaline-Sarpagine Alkaloids. In: The Alkaloids (R. H. F. MANSKE, ed.), vol. XI, 41. New York: Academic Press. 1968.

3.   cf. also The Alkaloids, (Specialist Periodical Reports) [ed. J. E. SAXTON (vols. 1 – 5), ed. M. F. GRUNDON (vols. 6 – 11)]. London: The Chemical Society.

4.   HESSE, M.: Alkaloid Chemistry. New York: Wiley-Interscience. 1981.

5.   SIDDIQUI, S., and R. H. SIDDIQUI: Chemical Examination of the Roots of *Rauwolfia serpentina*, Benth. J. Ind. Chem. Soc. **8**, 667 (1931). Chem. Abstr. **26**, 1288 (1932).

6.   VAN ITALLIE, L., and A. J. STEENHAUSER: *Rauwolfia serpentina* Benth. Pharm. Weekblad **69**, 334 (1932). Ref. Chem. Abstr. **26**, 3257 (1932).

7.   ANET, F. A. L., D. CHAKRAVARTI, R. ROBINSON, and E. SCHLITTLER: Ajmaline, Part I. J. Chem. Soc. **1954**, 1242.

8.   WOODWARD, R. B.: Neuere Entwicklungen in der Chemie der Naturstoffe. Angew. Chem. **68**, 13 (1956).

9.   Chemical Abstracts **1981**. Chemical Substance Index **94**.

10.   LOUNASMAA, M., and A. KOSKINEN: A Plausible Biogenetic Proposal for Koumine. Planta Med. **44**, 120 (1982).

11.   KUTNEY, J. P.: Studies in Indole Alkaloid Synthesis. A General Synthetic Route to 2-Acylindole Alkaloids and Related Compounds. Heterocycles **8**, 813 (1977).

12. KINGSTON, D. G. I., and O. EKUNDAYO: 2-Acylindole Alkaloids. J. Nat. Prod. **44**, 509 (1981).
13. POISSON, J., P. R. ULSHAFER, L. E. PASZEK, and W. I. TAYLOR: Mitoridine, sérédamine et purpéline, nouveaux alcaloïdes du *Rauwolfia vomitoria* Afz. Bull. Soc. Chim. Fr. **1964**, 2683.
14. YULDASHEV, P. K., and S. Y. YUNUSOV: Structure of Vincarine. Khim. Prir. Soedin **1**, 110 (1965). Chem. Abstr. **63**, 8428a (1965).
15. RAPOPORT, H., and R. E. MOORE: Alkaloids of *Geissospermum vellosii*. Isolation and Structure Determinations of Vellosimine, Vellosiminol and Geissolosimine. J. Org. Chem. **27**, 2981 (1962).
16. ACHENBACH, H., and E. SCHALLER: Über einige Bisindolalkaloide aus *Tabernaemontana accedens*. Ber. **109**, 3527 (1976).
17. WALDNER, E. E., M. HESSE, W. I. TAYLOR, and H. SCHMID: Über die Konstitution des Macralstonidins. Helv. Chim. Acta **50**, 1926 (1967).
18. COOK, J. M., and P. W. LE QUESNE: The Structure of Alstonisidine, a Novel Dimeric Indole Alkaloid of *Alstonia muelleriana* Domin. J. Org. Chem. **36**, 582 (1971).
19. CROQUELOIS, G., N. KUNESCH, M. DEBRAY, and J. POISSON: Alcaloïdes de l'*Alstonia boonei* De Wild. Plant. Méd. Phytotér. **6**, 122 (1972).
20. CROW, W. D., N. C. HANCOX, S. R. JOHNS, and J. A. LAMBERTON: New Alkaloids of *Alstonia constricta*. Aust. J. Chem. **23**, 2489 (1970).
21. LEWIN, G., N. KUNESCH, A. CAVÉ, T. SÉVENET, and J. POISSON: Alcaloïdes d'*Alstonia lanceolifera*. Phytochem. **14**, 2067 (1975).
22. BANERJI, A., M. CHAKRABARTY, and B. MUKHERJEE: Minor Indole Alkaloids of *Alstonia macrophylla*. Phytochem. **11**, 2605 (1972).
23. KHAN, Z. M., M. HESSE, and H. SCHMID: Die Struktur des quartären Alkaloides Macrosalhin. Helv. Chim. Acta **50**, 1002 (1967).
24. MUKHERJEE, B., A. B. RAY, A. CHATTERJEE, and B. C. DAS: O-Benzoylvincamajine: a New Alkaloid from the Leaves of *Alstonia macrophylla* Wall. Chem. Ind. (London) **1969**, 1387.
25. ELDERFIELD, R. C., and R. E. GILMAN: Alkaloids of *Alstonia muelleriana*. Phytochem. **11**, 339 (1972).
26. VERCAUTEREN, J., G. MASSIOT, T. SÉVENET, J. LÉVY, L. LE MEN-OLIVIER, and J. LE MEN: Alcaloïdes des feuilles et écorces de tronc d'*Alstonia odontophora*. Phytochem. **18**, 1729 (1979).
27. RASTOGI, R. C., R. S. KAPIL, and S. P. POPLI: Picralinal − A Key Alkaloid of Picralima Group from *Alstonia scholaris* R. Br. Experientia **26**, 1056 (1970).
28. SHARP, T. M.: The Alkaloids of *Alstonia* Barks, Part II. *A. macrophylla*, Wall., *A. somersetensis*, F. M. Bailey, *A. verticillosa*, F. Muell., *A. villosa*, Blum. J. Chem. Soc. **1934**, 1227.
29. HART, N. K., S. R. JOHNS, and J. A. LAMBERTON: Tertiary Alkaloids of *Alstonia spectabilis* and *Alstonia glabriflora* (Apocynaceae). Aust. J. Chem. **25**, 2739 (1972).
30. JOULE, J. A., M. OHASHI, B. GILBERT, and C. DJERASSI: Alkaloid Studies − LIII. The Structures of Nine New Alkaloids from *Aspidosperma dasycarpon* DC. Tetrahedron **21**, 1717 (1965).
31. FISH, F., M. QAISUDDIN, and J. B. STENLAKE: Isolation of Macusine B from *Aspidosperma peroba* F. Allem. ex Sald. Chem. Ind. (London) **1964**, 319.
32. ANTONACCIO, L. D., N. A. PEREIRA, B. GILBERT, H. VORBRUEGGEN, H. BUDZIKIEWICZ, J. M. WILSON, L. J. DURHAM, and C. DJERASSI: Alkaloid Studies. XXXIII. Mass Spectrometry in Structural and Stereochemical Problems. VI. Polyneuridine, A New Alkaloid from *Aspidosperma polyneuron* and Some Observations on Mass Spectra of Indole Alkaloids. J. Am. Chem. Soc. **84**, 2161 (1962).
33. MARKEY, S., K. BIEMANN, and B. WITKOP: Isolation of Rhazidine and Akuammidine

from *Aspidosperma quebracho blanco*. The Structure of Rhazidine. Tetrahedron Lett. **1967**, 157.

34. TUNMANN, P., and J. RACHOR: Über neue Alkaloide aus der Rinde von *Aspidosperma quebracho blanco* Schlecht. Naturwiss. **47**, 471 (1960).

35. LYON, R. L., H. H. S. FONG, and N. R. FARNSWORTH: Biological and Phytochemical Evaluation of Plants XII: Isolation of Acetylakuammidine from *Aspidosperma quebracho blanco* Leaves. J. Pharm. Sci. **62**, 833 (1973).

36. ORAZI, O. O., R. A. CORRAL, and M. E. STOICHEVICH: Studies on Plants XI. Alkaloids of *Aspidosperma spegazzinii*. Can. J. Chem. **44**, 1523 (1966).

37. DOUZOUA, L., M. MANSOUR, M.-M. DEBRAY, L. LE MEN-OLIVIER, and J. LE MEN: Alcaloïdes du *Cabucala erythrocarpa* var. *erythrocarpa*. Phytochem. **13**, 1994 (1974).

38. BOMBARDELLI, E., A. BONATI, B. DANIELI, B. GABETTA, and G. MUSTICH: Alkaloids of *Cabucala striolata*. Fitoterapia **45**, 183 (1974).

39. FARNSWORTH, N. R., W. D. LOUB, R. N. BLOMSTER, and M. GORMAN: Pericyclivine, a New *Catharanthus* Alkaloid. J. Pharm. Sci. **53**, 1558 (1964).

40. RASOANAIVO, P., N. LANGLOIS, and P. POTIER: Alcaloïdes du *Catharanthus longifolius*. Phytochem. **11**, 2616 (1972).

41. MUKHOPADHYAY, S., and G. A. CORDELL: *Catharanthus* Alkaloids XXXVI. Isolation of Vincaleukoblastine (VLB) and Periformyline from *Catharanthus trichophyllus* and Pericyclivine from *Catharanthus roseus*. J. Nat. Prod. **44**, 335 (1981).

42. JANOT, M.-M., and J. LE MEN: Sur les alcaloïdes cristallisés du *Lochnera (Vinca) rosea*. (L.) Reichb. ou *Catharanthus roseus* G. Don. C. R. Acad. Sc. Paris, Sér. C **243**, 1789 (1956).

43. STAUFFACHER, D.: Alkaloide aus *Diplorrhyncus condylocarpon* (Muell. Arg.) Pichon ssp. *mossambicensis* (Benth.) Duvign. Helv. Chim. Acta **44**, 2006 (1961).

44. CAVA, M. P., S. K. TALAPATRA, J. A. WEISBACH, B. DOUGLAS, R. F. RAFFAUF, and J. L. BEAL: Gabunine: A Natural Dimeric Indole Derived from Perivine. Tetrahedron Lett. **1965**, 931.

45. KASCHNITZ, R., and G. SPITELLER: Anwendungen der Massenspektrometrie zur Strukturaufklärung von Alkaloiden, 7. Mitt.: Neue Alkaloide aus *Gonioma kamassi* E. Mey. Monatsh. Chem. **96**, 909 (1965).

46. BUI, A.-M., P. POTIER, M. URREA, A. CLASTRES, D. LAURENT, and M.-M. DEBRAY: Étude chimiotaxonomique de deux espèces nouvelles de *Hazunta* (Apocynaceae). Phytochem. **18**, 1329 (1979).

47. BUI, A.-M., B. C. DAS, and P. POTIER: Étude chimiotaxonomique de *Hazunta modesta*. Phytochem. **19**, 1473 (1980).

48. MORFAUX, A.-M., L. OLIVIER, and J. LE MEN: Alcaloïdes indoliques: structure de l'éburnaphylline, alcaloïde principal des feuilles de l'*Hunteria eburnea* Pichon (Apocynacées). Bull. Soc. Chim. Fr. **1971**, 3967.

49. LINDE, H. H. A.: Die Alkaloide aus *Melodinus australis* (F. Mueller) Pierre (Apocynaceae). Helv. Chim. Acta **48**, 1822 (1965).

50. RABARON, A., M. H. MEHRI, T. SÉVENET, and M. M. PLAT: Alcaloïdes de *Melodinus celastroides*. Phytochem. **17**, 1452 (1978).

51. SAKAI, S., N. AIMI, K. TAKAHASHI, M. KITAGAWA, K. YAMAGUCHI, and J. HAGINAWA: Studies of Plants Containing Indole Alkaloids. (4). Alkaloids in *Ochrosia nakaiana* Koidz. Yakugaku Zasshi **94**, 1274 (1974).

52. PANAS, J. M., B. RICHARD, C. SIGAUT, M.-M. DEBRAY, L. LE MEN-OLIVIER, and J. LE MEN: Alcaloïdes du *Pandaca ochrascens*. Phytochem. **13**, 1969 (1974).

53. WEISSBACH, J. A., R. F. RAFFAUF, O. RIBEIRO, E. MACKO, and B. DOUGLAS: Problems in Chemotaxonomy. I. Alkaloids of *Peschiera affinis*. J. Pharm. Sci. **52**, 350 (1963).

54. JAHODÁR, L., Z. VOTICKÝ, and M. P. CAVA: Geissoschizol in *Peschiera laeta*. Phytochem. **13**, 2880 (1974).

55. VOTICKÝ, Z., L. JAHODÁR, and M. P. CAVA: Alkaloids from *Peschiera laeta* Mart. Coll. Czech. Chem. Commun. **42**, 1403 (1977).

56. HENRY, T. A.: The Alkaloids of *Picralima klaineana* Pierre, Part II. J. Chem. Soc. **1932**, 2759.

57. LÉVY, J., J. LE MEN, and M.-M. JANOT: Structure de l'akuammidine, alcaloïde du *Picralima nitida* Stapf. C. R. Acad. Sc. Paris, Sér. C **253**, 131 (1961).

58. KHAN, Z. M., M. HESSE, and H. SCHMID: Quartäre Alkaloide aus *Pleiocarpa mutica* Benth. Helv. Chim. Acta **48**, 1957 (1965).

59. MEHRI, M. H., A. RABARON, T. SÉVENET, and M. M. PLAT: Plantes de Nouvelle Calédonie. Partie 26. Alcaloïdes du *Melodinus balansae* var. *paucivenosus*. Phytochem. **17**, 1451 (1978).

60. NARANJO, J., M. PINAR, M. HESSE, and H. SCHMID: Über die Indolalkaloide von *Pleiocarpa talbotii* Wernham. Helv. Chim. Acta **55**, 752 (1972).

61. PINAR, M., M. HESSE, and H. SCHMID: Die Isolierung weiterer Alkaloide aus *Pleiocarpa talbotii* Wernham. Helv. Chim. Acta **56**, 2719 (1973).

62. KHAN, Z. M., M. HESSE, and H. SCHMID: Notiz über die Isolierung quartärer Alkaloide aus *Pleiocarpa tubicina* Stapf. Helv. Chim. Acta **50**, 625 (1967).

63. LIBOT, F., N. KUNESCH, and J. POISSON: Structure complète de la raucaffrinoline et filiation avec la vomilénine. Phytochem. **19**, 989 (1980).

64. BOSE, S., S. K. TALAPATRA, and A. CHATTERJEE: The Alkaloids of *Rauwolfia beddomei*. J. Ind. Chem. Soc. **33**, 379 (1956); Chem. Abstr. **51**, 671 (1957).

65. IACOBUCCI, G., and V. DEULOFEU: Alkaloids of *Rauwolfia schueli* and *Rauwolfia boliviana*. Anales asoc. quím. arg. **46**, 143 (1958); Chem. Abstr. **53**, 3595 (1959).

66. SCHULER, B. O. G., and F. L. WARREN: Rauwolfia Alkaloids, Part I. Reserpine and Ajmaline from *Rauwolfia natalensis* Sond *(R. caffra)*. J. Chem. Soc. **1956**, 215.

67. KHAN, N. H., M. A. KHAN, and S. SIDDIQUI: Studies in the Alkaloids of *Rauwolfia caffra* Sonder. Part I. Isolation of Ajmalicine, Ajmaline, Raucaffrine and Three New Alkaloids, Raucaffricine, Raucaffriline and Raucaffridine. Pakistan J. Sci. Ind. Res. **8**, 23 (1964).

68. KHAN, M. A., and S. SIDDIQUI: Isolation and Structure of Raucaffrinoline − A New Alkaloid from *Rauwolfia caffra* Sonder. Experientia **28**, 127 (1972).

69. HABIB, M. S., and W. E. COURT: Minor Alkaloids of *Rauwolfia caffra*. Phytochem. **12**, 1821 (1973).

70. − − Leaf Alkaloids of *Rauwolfia caffra*. Phytochem. **13**, 661 (1974).

71. KIDD, D. A. A.: Alkaloids of *Rauwolfia* species. Part IV. *Rauwolfia cambodiana* Pierre. J. Chem. Soc. **1958**, 2432.

72. KECK, J.: Über die Isolierung weiterer Inhaltsstoffe aus den Wurzeln von *Rauwolfia canescens* L. Naturwiss. **42**, 391 (1955).

73. YAMAGUCHI, K., and H. SHOJI: Determination of Alkaloids from *Rauwolfia serpentina*. VI. Alkaloids of *Rauwolfia chinensis*. Eisei Shikenjo Hôkoku **76**, 99 (1958); Chem. Abstr. **53**, 17419 (1959).

74. DANIELI, B., E. BOMBARDELLI, A. BONATI, and B. GABETTA: New Alkaloids from *Rauwolfia confertiflora*. Chim. Ind. (Milan) **53**, 1042 (1971).

75. IWU, M. M., and W. F. COURT: N_a-Demethyl-purpeline and N_a-demethyl-dihydropurpeline, new alkaloids from *Rauwolfia cumminsii* Stapf. Experientia **33**, 1268 (1977).

76. − − Leaf Alkaloids of *Rauwolfia cumminsii* Stapf. Planta Med. **33**, 360 (1978).

77. − − The Alkaloids of *Rauwolfia cumminsii*. Planta Med. **34**, 390 (1978).

78. − − Alkaloids of *Rauwolfia cumminsii* Stem. Phytochem. **17**, 1651 (1978).

79. ATAL, C. K.: A Phytochemical Investigation of *Rauwolfia decurva*. J. Am. Pharm. Assoc. **48**, 37 (1959).

80. GORMAN, M., N. NEUSS, C. DJERASSI, J. P. KUTNEY, and P. J. SCHEUER: Alkaloid Studies − XIX. Alkaloids of Some Hawaiian *Rauwolfia* Species: The Structure of Sandwicine and Its Interconversion with Ajmaline and Ajmalidine. Tetrahedron **1**, 328 (1957).

81. BHATTACHARJI, S., M. M. DHAR, and M. L. DHAR: Chemical Examination of *Rauwolfia densiflora* Benth. J. Sci. Ind. Res. **21B,** 454 (1962).

82. CHATTERJEE, A., and S. TALAPATRA: Alkaloids of the Roots of *Rauwolfia densiflora* Benth. and Hook, *Rauwolfia perakensis* King and Gamble, *Rauwolfia canescens* Linn. and *Rauwolfia serpentina* Benth. Naturwiss. **42,** 182 (1955).

83. COMBES, G., L. FONZES, and F. WINTERNITZ: Sur les alcaloïdes des *Rauwolfia* de Madagascar. Phytochem. **5,** 1065 (1966).

84. CHAUDHURY, N. A., and A. CHATTERJEE: Studies on the Alkaloids of *Rauwolfia sumatrana* (Miq.) Jack & *Rauwolfia fruticosa* Burck: Part I. J. Sci. Ind. Res. **18B,** 130 (1959).

85. DJERASSI, C., M. GORMAN, A. L. NUSSBAUM, and J. REYNOSO: Alkaloid Studies. IV. The Isolation of Reserpine, Serpentine and Ajmaline from *Rauwolfia heterophylla* Roem. and Schult. J. Am. Chem. Soc. **76,** 4463 (1954).

86. VERGARA, B. U.: Extraction of Reserpine and Other Alkaloids from Colombian *Rauwolfia hirsuta.* J. Am. Chem. Soc. **77,** 1864 (1955).

87. ISHIDATE, M., M. OKADA, and K. SAITO: Isolation of Alkaloids from *Rauwolfia.* Pharm. Bull. (Japan) **3,** 319 (1955); Chem. Abstr. **50,** 13369 (1956).

88. MÜLLER, J. M.: Über die Alkaloide von *Rauwolfia ligustrina* R. & S. Raugustin, ein neues reserpinähnliches Alkaloid. Experientia **13,** 479 (1957).

89. TIMMINS, P., and W. E. COURT: Alkaloids of *Rauwolfia macrophylla.* Phytochem. **13,** 281 (1974).

90. AMER, M. M. A., and W. E. COURT: Root Wood Alkaloids of *Rauwolfia macrophylla.* Planta Med. **43,** 94 (1981).

91. PATEL, M. B., J. POISSON, J. L. POUSSET, and J. M. ROWSON: Vincamajine, the Major Alkaloid of Leaves of *Rauwolfia mannii* Stapf. J. Pharm. Pharmacol. **17,** 323 (1965).

92. SCHEUER, P. J., M. Y. CHANG, and H. FUKAMI: Hawaiian Plant Studies. X. The Structure of Mauiensine. J. Org. Chem. **28,** 2641 (1963).

93. PILLAY, P. P., D. S. RAO, and S. B. RAO: Chemical Investigation of *Rauwolfia micrantha* Hook. f. J. Sci. Ind. Res. **19B,** 135 (1960).

94. COURT, W. E., W. C. EVANS, and G. E. TREASE: The Distribution of Alkaloids in *Rauwolfia caffra* Sond. and Related Species. J. Pharm. Pharmacol. **10,** 380 (1958).

95. IWU, M. M., and W. E. COURT: The Alkaloids of *Rauwolfia mombasiana* Leaves. Planta Med. **33,** 232 (1978).

96. — — Alkaloids of *Rauwolfia mombasiana* Stem Bark. Planta Med. **36,** 208 (1979).

97. AMER, M. A., and W. E. COURT: Alkaloids of *Rauwolfia nitida* Root Bark. Phytochem. **20,** 2569 (1981).

98. ROLAND, M.: Les alcaloïdes du *Rauwolfia obscura* K. Schum. J. Pharm. Belg. **14,** 347 (1959).

99. TIMMINS, P., and W. E. COURT: Root Bark Alkaloids of *Rauwolfia obscura.* Phytochem. **13,** 1997 (1974).

100. — — Further Alkaloids from the Roots of *Rauwolfia obscura.* Planta Med. **29,** 283 (1976).

101. COURT, W. E.: The Alkaloids of *Rauwolfia volkensii* Stapf. Can. J. Pharm. Sci. **3,** 70 (1968).

102. AKINLOYE, B. A., and W. E. COURT: Leaf Alkaloids of *Rauwolfia oreogiton.* Phytochem. **19,** 2741 (1980).

103. — — The Alkaloids of *Rauwolfia oreogiton.* Planta Med. **41,** 69 (1981).

104. KIANG, A. K., and A. S. C. WAN: Alkaloids in the Roots of *Rauwolfia perakensis.* J. Chem. Soc. **1960,** 1394.

105. KIANG, A. K., H. LEE, J. GOH, and A. S. C. WAN: Alkaloids in the Leaves and Stems of *Rauwolfia perakensis.* Lloydia **27,** 220 (1964).

106. CHATTERJEE, A., A. K. GHOSH, and M. CHAKRABARTY: Reflexine, A New Indole Alkaloid of *Rauwolfia reflexa.* Experientia **32,** 1236 (1976).

107. CHATTERJEE, A., M. CHAKRABARTY, A. K. GHOSH, E. W. HAGAMAN, and E. WENKERT: Indole Alkaloids of *Rauwolfia reflexa*. The Structures of Rauflexine and Reflexine. Tetrahedron Lett. **1978**, 3879.

108. SIERRA, P., and L. NOVOTNÝ: Alkaloids from *Rauwolfia salicifolia*. Planta Med. **42**, 108 (1981).

109. IACOBUCCI, G., and V. DEULOFEU: Alkaloids from *Rauwolfia schueli*. J. Org. Chem. **22**, 94 (1957).

110. PAKRASHI, S. C., C. DJERASSI, R. WASICKY, and N. NEUSS: Alkaloid Studies. IX. *Rauwolfia* Alkaloids. IV. Isolation of Reserpine and Other Alkaloids from *Rauwolfia sellowii* Muell. Argov. J. Am. Chem. Soc. **77**, 6687 (1955).

111. CHATTERJEE, A., and S. BOSE: A New Alkaloid from the Root of *Rauwolfia serpentina*. Science and Culture **17**, 139 (1951); Chem. Abstr. **46**, 9264 (1952).

112. STOLL, A., and A. HOFMANN: Sarpagin, ein neues Alkaloid aus *Rauwolfia serpentina* Benth. Helv. Chim. Acta **36**, 1143 (1953).

113. BOSE, S.: Serpinine, a minor Alkaloid of *Rauwolfia serpentina* Benth. Naturwiss. **42**, 71 (1955).

114. MAJUMDAR, S. P., J. POISSON, and P. POTIER: Alcaloïdes de *Rauwolfia suaveolens*. Phytochem. **12**, 1167 (1973).

115. HANAOKA, M., M. HESSE, and H. SCHMID: N_a-Demethylseredamin, ein neues Alkaloid aus *Rauwolfia sumatrana;* absolute Konfiguration von Seradamin. Helv. Chim. Acta **53**, 1723 (1970).

116. DJERASSI, C., and J. FISHMAN: Tetraphyllin and Tetraphyllicine, Two New Alkaloids from *Rauwolfia tetraphylla* L. Chem. Ind. (London) **1955**, 627.

117. DJERASSI, C., J. FISHMAN, M. GORMAN, J. P. KUTNEY, and S. C. PAKRASHI: Alkaloid Studies. XVI. Alkaloids of *Rauwolfia tetraphylla* L. The Structures of Tetraphylline and Tetraphyllicine. J. Am. Chem. Soc. **79**, 1217 (1957).

118. ARTHUR, H. R., and S. N. LOO: An Examination of *Rauwolfia verticillata* of Hong Kong – II. Phytochem. **5**, 977 (1966).

119. ARTHUR, H. R., S. R. JOHNS, J. A. LAMBERTON, and S. N. LOO: Alkaloids of *Rauwolfia verticillata* (Lour.) Bail. of Hong Kong. Identification of Vellosimine and Peraksine, and Demonstration from N. M. R. Data that Peraksine is a Mixture of Two Epimers. Aust. J. Chem. **21**, 1399 (1968).

120. AKINLOYE, B. A., and W. E. COURT: Stem Bark Alkaloids of *Rauwolfia volkensii*. Planta Med. **37**, 361 (1979).

121. — — Leaf Alkaloids of *Rauwolfia volkensii*. Phytochem. **19**, 307 (1980).

122. — — The Alkaloids of *Rauwolfia volkensii*. J. Ethnopharmacol. **4**, 99 (1981).

123. SCHLITTLER, E., H. SCHWARZ, and F. BADER: Isolierung von Alstonin aus afrikanischen *Rauwolfia*-Arten. Helv. Chim. Acta **35**, 271 (1952).

124. POISSON, J., R. GOUTAREL, and M.-M. JANOT: Présence dans les racines du *Rauwolfia vomitoria* Afz. de l'ester triméthoxybenzoïque d'un alcaloïde du type de l'ajmaline. C. R. Acad. Sc. Paris **241**, 1840 (1955).

125. HAACK, E., A. POPELAK, and H. SPINGLER: Rauvomitin, ein neues Alkaloid aus *Rauwolfia vomitoria* Afz. Naturwiss. **42**, 627 (1955).

126. POISSON, J., and R. GOUTAREL: Alcaloïdes du *Rauwolfia vomitoria* Afz.: Présence de l'isoréserpiline et de la sarpagine. Bull. Soc. Chim. Fr. **1956**, 1703.

127. HOFMANN, A., and A. J. FREY: Isolierung weiterer Alkaloide aus *Rauwolfia vomitoria* Afz. Vomalidin, ein neues Alkaloid der Ajmalin-Gruppe. Helv. Chim. Acta **40**, 1866 (1957).

128. ULSHAFER, P. R., M. F. BARTLETT, L. DORFMAN, M. A. GILLEN, E. SCHLITTLER, and E. WENKERT: Isolation and Structure of Perakine. Tetrahedron Lett. **1961**, 363.

129. TAYLOR, W. I., A. J. FREY, and A. HOFMANN: Vomilenin und seine Umwandlung in Perakin. Helv. Chim. Acta **45**, 611 (1962).

130. Muquet, M., J.-L. Pousset, and J. Poisson: A propos de l'acétyl-17 ajmaline naturelle. C. R. Acad. Sc. Paris, Sér. C **266**, 1542 (1968).
131. Ronchetti, F., G. Russo, E. Bombardelli, and A. Bonati: A New Alkaloid from *Rauwolfia vomitoria*. Phytochem. **10**, 1385 (1971).
132. Bombardelli, E., A. Bonati, and G. Russo: Un nuovo alcaloide della *Rauwolfia vomitoria*. Afz. Fitoterapia **38**, 126 (1967).
133. Pousset, J.-L., M. Debray, and J. Poisson: Alcaloïdes des feuilles de *Rauwolfia vomitoria:* Identité de la vomifoline avec la péraksine. Phytochem. **16**, 153 (1977).
134. Iwu, M. M., and W. E. Court: Root Alkaloids of *Rauwolfia vomitoria* Afz. Planta Med. **32**, 88 (1977).
135. Sabri, N. N., and W. E. Court: Stem Alkaloids of *Rauwolfia vomitoria*. Phytochem. **17**, 2023 (1978).
136. Paris, R.: Sur une Apocynacée africaine: le *Rauwolfia vomitoria* Afz. Ann. Pharm. Fr. **1**, 138 (1943).
137. Chatterjee, A., and S. Bandyopadhyay: Vellosimine, an Alkaloid of *Rauwolfia vomitoria*. Indian J. Chem. **18B**, 87 (1979).
138. Siddiqui, S., and A. Malik: Isolation and Structure of Ajmalinol − a New Alkaloid from *Rauwolfia vomitoria* Afzuelia. J. Chem. Soc. Pak. **1**, 1 (1979).
139. Stöckigt, J.: Non-involvement of 5α-Carboxystrictosidine and -vincoside in the Biosynthesis of Sarpagine- and Ajmaline-Type Alkaloids. Tetrahedron Lett. **1979**, 2615.
140. Amer, M. M., and W. E. Court: Leaf Alkaloids of *Rauwolfia vomitoria*. Phytochem. **19**, 1833 (1980).
141. Chatterjee, A., C. R. Ghosal, N. Adityachaudhury, and S. Ghosal: Alkaloids of *Rhazya stricta* Decaisne. Chem. Ind. (London) **1961**, 1034.
142. Patel, M. B., L. Thompson, C. Miet, and J. Poisson: Alcaloïdes de *Tabernaemontana brachyantha*. Phytochem. **12**, 451 (1973).
143. Defay, N., M. Kaisin, J. Pecher, and R. H. Martin: Alcaloïdes indoliques III. Structure de la voachalotine. Bull. Soc. Chim. Belges **70**, 475 (1961).
144. Achenbach, H.: Voachalotin und Affinisin, Nebenalkaloide in *Tabernaemontana fuchsiaefolia*. Tetrahedron Lett. **1966**, 4405.
145. Kingston, D. G. I., B. T. Li, and F. Ionescu: Plant Anticancer Agents III: Isolation of Indole and Bisindole Alkaloids from *Tabernaemontana holstii* Roots. J. Pharm. Sci. **66**, 1135 (1977).
146. Kingston, D. G. I., B. B. Gerhart, F. Ionescu, M. M. Mangino, and S. M. Sami: Plant Anticancer Agents V: New Bisindole Alkaloids from *Tabernaemontana johnstonii* Stem Bark. J. Pharm. Sci. **67**, 249 (1978).
147. Achenbach, H., and B. Raffelsberger: Alkaloide in *Tabernaemontana*-Arten, XII. Untersuchung der Alkaloide von *Tabernaemontana olivacea* — Condylocarpin-N-oxid, ein neues Alkaloid aus *T. olivacea*. Z. Naturforsch. **35b**, 885 (1980).
148. St André, A. F., B. Korzun, and F. Weinfeldt: *Rauwolfia* Alkaloids. XXIV. Note on the Alkaloids of *Tonduzia longifolia* (A. DC.) Mgf. J. Org. Chem. **21**, 480 (1956).
149. Goodwin, S., and E. C. Horning: Isolation of Vincamajine from *Tonduzia longifolia* (A. DC.) Markgraf. Chem. Ind. (London) **1956**, 846.
150. Walser, A., and C. Djerassi: Alkaloid-Studien LII. Die Alkaloide aus *Vallesia dichotoma* Ruiz et Pav. Helv. Chim. Acta **48**, 391 (1965).
151. Chen, W.-H., and Y.-C. Bai: Alkaloids of *Rauwolfia yunnanensis*. Yun-nan Chih Wu Yen Chiu **1**, 37 (1979); Chem. Abstr. **92**, 194470w (1980).
152. Gosset, J., J. Le Men, and M.-M. Janot: Isolement de l'akuammidine et d'un nouvel alcaloïde: la vincadifformine, des feuilles du *Vinca difformis* Pourr. Alcaloïdes des Pervenches. Ann. Pharm. Fr. **20**, 448 (1962).
153. Ohashi, M., H. Budzikiewicz, J. M. Wilson, C. Djerassi, J. Lévy, J. Gosset, J. Le Men, and M.-M. Janot: Mass Spectrometry in Structural and Stereochemical Problems —

XXXVI. Alkaloids of Periwinkles – 27. The Mass Spectra of Stereoisomers of the Sarpagine-Akuammidine Group. Tetrahedron **19**, 2241 (1963).

154. JANOT, M.-M., J. LE MEN, and C. FAN: Alcaloïdes du *Vinca difformis* Pourr. (Apocynacées). II. Présence de la sarpagine et d'un isomère de la vincamine. Ann. Pharm. Fr. **15**, 513 (1957).

155. STROUF, O., and K. KAVKOVA: Alkaloids of the Genera *Vinca* and *Catharanthus*. Chem. Listy **56**, 987 (1962); Chem. Abstr. **57**, 16673a (1962).

156. JANOT, M.-M., J. LE MEN, and Y. HAMMOUDA: Sur la vincamédine, alcaloïde cristallisé du *Vinca difformis* Pourr. (Apocynacées). C. R. Acad. Sc. Paris, Sér. C **243**, 85 (1956).

157. MALIKOV, V. M., P. K. YULDASHEV, and S. Y. YUNUSOV: Constitution of Ervine. Khim. Prir. Soedin. **2**, 338 (1966); Chem. Abstr. **66**, 65684h (1967).

158. SULTANOV, M. B., and T. SAIDKASYMOV: Pharmacology of Tombozine. Dokl. Akad. Nauk. Uz. SSR **22**, 41 (1965); Chem. Abstr. **63**, 13853f (1965).

159. YULDASHEV, P. K., and S. Y. YUNUSOV: Vincarine, a New Alkaloid from Roots of *Vinca erecta*. Dokl. Akad. Nauk. SSSR **154**, 1412 (1964); Chem. Abstr. **62**, 4409 (1964).

160. KHALMIRZAEV, M. M., V. M. MALIKOV, and S. Y. YUNUSOV: Isolation of the N-oxide of Vineridine and 16-Methoxy-vincadifformine from *Vinca erecta*. Khim. Prir. Soedin. **9**, 806 (1973); Chem. Abstr. **81**, 166399j (1974).

161. MALIKOV, V. M., M. R. SHARIPOV, and S. Y. YUNUSOV: Structure of Ervincidine. Khim. Prir. Soedin. **8**, 760 (1972); Chem. Abstr. **78**, 94810z (1973).

162. SHARIPOV, M. R., M. KHALMIRZAEV, V. M. MALIKOV, and S. Y. YUNUSOV: Isolation of O-Benzoyltombozine, N-Oxides of Kopsinine and Pseudokopsinine. Khim. Prir. Soedin. **10**, 413 (1974); Chem. Abstr. **81**, 152475u (1974).

163. ALIEV, A. M., and N. A. BABAEV: Isolation of Vincamajine from Herbaceous Periwinkle. Farmatsiya **25**, 30 (1976); Chem. Abstr. **85**, 106639k (1976).

164. AYNILIAN, G. H., N. R. FARNSWORTH, and J. TROJÁNEK: Alkaloids of *Vinca* Species. III. Isolation and Characterization of Indole Alkaloids from *Vinca libanotica*. Lloydia **37**, 299 (1974).

165. TROJÁNEK, J., and J. HODKOVÁ: Über Alkaloide V. Isolierung von Vincamedin aus dem großen Immergrün (*Vinca major* L.). Coll. Czech. Chem. Comm. **27**, 2981 (1962).

166. JANOT, M.-M., and J. LE MEN: Sur la vincamajine, quatrième alcaloïde cristallisé de la grande pervenche (*Vinca major* L. Apocynacées). C. R. Acad. Sci. Paris, Sér. C **241**, 767 (1955).

167. KAUL, J. L., and J. TROJÁNEK: On Alkaloids. XVI. Isolation and Characterization of Some New Alkaloids from *Vinca major*. Lloydia **29**, 26 (1966).

168. JANOT, M.-M., and J. LE MEN: Sur la vincamajoréine: troisième alcaloïde cristallisé de la Grande Pervenche *Vinca major* L. (Apocynacées). Ann. Pharm. Fr. **13**, 325 (1955).

169. FARNSWORTH, N. R., H. H. S. FONG, R. N. BLOMSTER, and F. J. DRAUS: Studies on *Vinca major*. II. Phytochemical Investigation. J. Pharm. Sci. **51**, 217 (1962).

170. BANERJI, A., and M. CHAKRABARTY: Lochvinerine: A New Indole Alkaloid of *Vinca major*. Phytochem. **13**, 2309 (1974).

171. — — Majvinine: A New Indole Alkaloid of *Vinca major*. Phytochem. **16**, 1124 (1977).

172. ILYASHENKO, L. I., V. M. MALIKOV, M. R. YAGUDAEV, and S. Y. YUNUSOV: Alkaloids from *Vinca major*. Khim. Prir. Soedin. **13**, 382 (1977); Chem. Abstr. **88**, 47454h (1978).

173. MEISEL, H., W. DÖPKE, and E. GRÜNDEMANN: Struktur und Stereochemie des Vinorins. Tetrahedron Lett. **1971**, 1291.

174. PECHER, J., N. DEFAY, M. GAUTHIER, J. PEETERS, R. H. MARTIN, and A. VANDERMEERS: Voachalotine, the Major Alkaloid of "Voacanga chalotiana" (Pierre *ex* Stapf). Chem. Ind. (London) **1960**, 1481.

175. LHOEST, G., R. DE NEYS, N. DEFAY, J. SEIBL, J. PECHER, and R. H. MARTIN: Alcaloïdes Indoliques VII, Structure de la Voacoline. Bull. Soc. Chim. Belges **74**, 534 (1965).

176. TIRIONS, G., M. KAISIN, J. C. BRAEKMAN, J. PECHER, and R. H. MARTIN: Indole

Alkaloids: XV: Dehydrovoachalotine, a minor base from *Voacanga chalotiana* Pierre *ex* Stapf. Chimia **22**, 87 (1968).

177. DENAYER-TOURNEY, M., J. PECHER, R. H. MARTIN, M. FRIEDMANN-SPITELLER, and G. SPITELLER: Alcaloïdes indoliques V. Structure de la voacarpine. Bull. Soc. Chim. Belges **74**, 170 (1965).

178. BOMBARDELLI, E., A. BONATI, B. GABETTA, E. MARTINELLI, G. MUSTICH, and B. DANIELI: 17-O-Acetyl-19,20-Dihydrovoachalotine, A New Alkaloid from *Voacanga chalotiana*. Phytochem. **15**, 2021 (1976).

179. HAGINIWA, J., S. SAKAI, A. KUBO, and T. HAMAMOTO: 10th Symposium on the Chemistry of Natural Products, Tokyo, 1966.

180. SAKAI, S., A. KUBO, and J. HAGINIWA: *Gardneria* Alkaloids. II. The Structures of Gardnerine, Gardnutine and Hydroxygardnutine. Tetrahedron Lett. **1969**, 1485.

181. SAKAI, S., A. KUBO, T. HAMAMOTO, M. WAKABAYASHI, K. TAKAHASHI, H. OHTANI, and J. HAGINIWA: The Structures of Gardnerine, Gardnutine and Hydroxygardnutine and the Absolute Configuration of Gardnerine. Chem. Pharm. Bull. **21**, 1783 (1973).

182. AIMI, N., K. YAMAGUCHI, S. SAKAI, J. HAGINIWA, and A. KUBO: *Gardneria* Alkaloids. XII. Carbon Magnetic Resonance Spectra of *Gardneria* Alkaloids. A Study on the Configuration of the Side Chain Double Bonds of Indole Alkaloids. Chem. Pharm. Bull. **26**, 3444 (1978).

183. GALEFFI, C., G. B. MARINI-BETTOLO, and E. M. DELLE MONACHE: *Strychnos amazonica* and *Strychnos brachiata* Alkaloids. XXVIII. *Strychnos* Alkaloids. Ann. Chim. (Rome) **63**, 849 (1973).

184. OLANIYI, A. A., W. N. A. ROLFSEN, and R. VERPOORTE: Quarternary Indole Alkaloids of *Strychnos decussata*. Planta Med. **43**, 353 (1981).

185. MARINI-BETTOLO, G. B., C. GALEFFI, M. NICOLETTI, and I. MESSANA: Alkaloids of *Strychnos rubiginosa*. Phytochem. **19**, 992 (1980).

186. BATTERSBY, A. R., R. BINKS, H. F. HODSON, and D. A. YEOWELL: Alkaloids of *Calabash-curare* and *Strychnos* Species. Part II. Isolation of New Alkaloids. J. Chem. Soc. **1960**, 1848.

187. BATTERSBY, A. R., and D. A. YEOWELL: Alkaloids of *Calabash-curare* and *Strychnos* Species. Part III. Structure and Absolute Stereochemistry of Macusine-A, Macusine-B and Macusine-C. J. Chem. Soc. **1964**, 4419.

188. ARNOLD, W., F. BERLAGE, K. BERNAUER, H. SCHMID, and P. KARRER: Über Lochneram, ein neues Calebassenalkaloid, und über C-Alkaloid M. Helv. Chim. Acta **41**, 1505 (1958).

189. ANGENOT, L.: De nouveaux alcaloïdes quarternaires du *Strychnos usambarensis*. Planta Med. **27**, 24 (1975).

190. ROBINSON, R.: Die Chemie des Ajmalins. Angew. Chem. **69**, 40 (1957).

191. BARGER, G., and C. SCHOLZ: Über Yohimbin. Ber. **81**, 1343 (1933).

192. HAHN, G., and H. WERNER: Synthese von Tetrahydroharman(4-Carbolin)-Systemen unter physiologischen Bedingungen. III. Mitteilung. Synthese von Yohimbin-Gerüsten. Ann. Chem. **520**, 123 (1935).

193. DJERASSI, C., M. GORMAN, S. C. PAKRASHI, and R. B. WOODWARD: The Structures of Tetraphyllicine, Ajmalidine and Rauvomitine. J. Am. Chem. Soc. **78**, 1259 (1956).

194. WEISS, U., C. GILVARG, E. S. MINGIOLI, and B. D. DAVIS: Aromatic Biosynthesis. XI. The Aromatization Step in the Synthesis of Phenylalanine. Science **119**, 774 (1954).

195. WENKERT, E., and N. V. BRINGI: A Stereochemical Interpretation of the Biosynthesis of Indole Alkaloids. J. Am. Chem. Soc. **81**, 1474 (1959).

196. WENKERT, E.: Alkaloid Biosynthesis. Experientia **15**, 165 (1959).

197. THOMAS, R.: A Possible Biosynthetic Relationship Between the Cyclopentanoid Monoterpenes and the Indole Alkaloids. Tetrahedron Lett. **1961**, 544.

198. WENKERT, E.: Biosynthesis of Indole Alkaloids, The *Aspidosperma* and *Iboga* Bases. J. Am. Chem. Soc. **84**, 98 (1962).

199. LEETE, E.: The Biogenesis of the *Rauwolfia* Alkaloids. I. The Incorporation of Tryptophan into Ajmaline. J. Am. Chem. Soc. **82**, 6338 (1960).

200. — Biogenesis of the *Rauwolfia* Alkaloids — II. The Incorporation of Tryptophan into Serpentine and Reserpine. Tetrahedron **14**, 35 (1961).

201. EDWARDS, P. N., and E. LEETE: Incorporation of Formate into Ajmaline. Chem. Ind. (London) **1961**, 1666.

202. LEETE, E., S. GHOSAL, and P. N. EDWARDS: Biosynthesis of the Non-Tryptophan Derived Portion of Ajmaline. J. Am. Chem. Soc. **84**, 1068 (1962).

203. LEETE, E., and S. GHOSAL: Further Studies on the Biosynthesis of the Non-Tryptophan Derived Portion of Ajmaline and Related Alkaloids. Tetrahedron Lett. **1962**, 1179.

204. BATTERSBY, A. R., R. BINKS, W. LAWRIE, G. V. PARRY, and B. R. WEBSTER: Biosynthesis of the Indole and *Ipecacuanha* Alkaloids. Proc. Roy. Soc. **1963**, 369.

205. — — — — — Alkaloid Biosynthesis. Part IX. The *Ipecacuanha* Alkaloids. J. Chem. Soc. **1965**, 7459.

206. LEETE, E., A. AHMAD, and I. KOMPIŠ: Biosynthesis of the *Vinca* Alkaloids. I. Feeding Experiments with Tryptophan-2-C^{14} and Acetate-1-C^{14}. J. Am. Chem. Soc. **87**, 4168 (1965).

207. BATTERSBY, A. R., R. T. BROWN, R. S. KAPIL, J. A. MARTIN, and A. O. PLUNKETT: Role of Loganin in the Biosynthesis of Indole Alkaloids. Chem. Comm. **1966**, 890.

208. BATTERSBY, A. R., R. S. KAPIL, J. A. MARTIN, and L. MO: Loganin as Precursor of the Indole Alkaloids. Chem. Comm. **1968**, 133.

209. KOMPIŠ, I., M. HESSE, and H. SCHMID: An Approach to the Biogenetic Classification of Indole Alkaloids. Lloydia **34**, 269 (1971).

210. BATTERSBY, A. R., A. R. BURNETT, and P. G. PARSONS: Alkaloid Biosynthesis. Part XV. Partial Synthesis and Isolation of Vincoside and Isovincoside: Biosynthesis of the Three Major Classes of Indole Alkaloids from Vincoside. J. Chem. Soc. (C) **1969**, 1193.

211. SCOTT, A. I.: Biosynthesis of the Indole Alkaloids. Accts. Chem. Res. **3**, 151 (1970).

212. BATTERSBY, A. R., and K. H. GIBSON: Further Studies on Rearrangements during Biosynthesis of Indole Alkaloids. Chem. Comm. **1971**, 902.

213. DE SILVA, K. T. D., G. N. SMITH, and K. E. H. WARREN: Stereochemistry of Strictosidine. Chem. Comm. **1971**, 905.

214. BLACKSTOCK, W. P., R. T. BROWN, and G. K. LEE: Configuration at C-3 in Vincoside. Chem. Comm. **1971**, 910.

215. BROWN, R. T., C. L. CHAPPLE, and R. PLATT: Biomimetic Inversion of C-3 in Monoterpenoid Indole Alkaloids. Chem. Comm. **1974**, 929.

216. CORDELL, G. A.: The Biosynthesis of Indole Alkaloids. Lloydia **37**, 219 (1974).

217. BATTERSBY, A. R., R. T. BROWN, R. S. KAPIL, A. O. PLUNKETT, and J. B. TAYLOR: Biosynthesis of Indole Alkaloids. Chem. Comm. **1966**, 46.

218. LOEW, P., H. GOEGGEL, and D. ARIGONI: A Monoterpene Precursor in the Biosynthesis of Indole Alkaloids. Chem. Comm. **1966**, 347.

219. HALL, E. S., F. McCAPRA, T. MONEY, K. FUKUMOTO, J. R. HANSON, B. S. MOOTOO, G. T. PHILLIPS, and A. I. SCOTT: Concerning the Terpenoid Origin of Indole Alkaloids: Biosynthetic Mapping by Direct Mass Spectrometry. Chem. Comm. **1966**, 348.

220. SMITH, G. N.: Strictosidine: a Key Intermediate in the Biogenesis of Indole Alkaloids. Chem. Comm. **1968**, 912.

221. STÖCKIGT, J., and M. H. ZENK: Strictosidine (Isovincoside): the Key Intermediate in the Biosynthesis of Monoterpenoid Indole Alkaloids. Chem. Comm. **1977**, 646.

222. SCOTT, A. I., S. L. LEE, P. DE CAPITE, M. G. CULVER, and C. R. HUTCHINSON: The Role of Isovincoside (Strictosidine) in the Biosynthesis of the Indole Alkaloids. Heterocycles **7**, 979 (1977).

223. RUEFFER, M., N. NAGAKURA, and M. H. ZENK: Strictosidine, the Common Precursor for Monoterpenoid Indole Alkaloids with 3α and 3β Configuration. Tetrahedron Lett. **1978**, 1593.

224. VAN TAMELEN, E. E., V. B. HAARSTAD, and R. L. ORVIS: Hypohalite-Induced Oxidative Decarboxylation of α-Amino Acids. Tetrahedron **24**, 687 (1968).

225. DESILVA, K. T. D., D. KING, and G. N. SMITH: 5α-Carboxystrictosidine. Chem. Comm. **1971**, 908.

226. VAN TAMELEN, E. E., and C. DORSCHEL: Synthesis of Methyl Adirubine. Chem. Comm. **1976**, 529.

227. VAN TAMELEN, E. E., and L. K. OLIVER: The Biogenetic-Type Total Synthesis of Ajmaline. Bioorg. Chem. **5**, 309 (1976).

228. — — The Biogenetic-Type Total Synthesis of Ajmaline. J. Am. Chem. Soc. **92**, 2136 (1970).

229. STÖCKIGT, J., M. RUEFFER, M. H. ZENK, and G.-A. HOYER: Indirect Identification of 4,21-Dehydrocorynantheine Aldehyde as an Intermediate in the Biosynthesis of Ajmalicine and Related Alkaloids. Planta Med. **33**, 188 (1978).

230. TIMMINS, P., and W. E. COURT: Stem Alkaloids of *Rauwolfia obscura*. Phytochem. **15**, 733 (1976).

231. MCFARLANE, J., K. M. MADYASTHA, and C. J. COSCIA: Regulation of Secondary Metabolism in Higher Plants. Effect of Alkaloids on a Cytochrome P-450 Dependent Monooxygenase. Biochem. Biophys. Res. Comm. **66**, 1263 (1975).

232. BARTON, D. H. R., G. W. KIRBY, R. H. PRAGER, and E. M. WILSON: On the Origin of the C-1 Fragment in Indole Alkaloids. J. Chem. Soc. **1965**, 3990.

233. STUART, K. L., J. P. KUTNEY, T. HONDA, N. G. LEWIS, and B. R. WORTH: The Biosynthesis of Vindoline Using Cell Free Extracts from Mature *Catharanthus roseus* Plants. Heterocycles **9**, 647 (1978).

234. MADYASTHA, K. M., R. GUARNACCIA, C. BAXTER, and C. J. COSCIA: S-Adenosyl-L-methionine: Loganic Acid Methyltransferase. A Carboxyl-Alkylating Enzyme from *Vinca rosea*. J. Biol. Chem. **248**, 2497 (1973).

235. GORROD, J. W.: Biological Oxidation of Aromatic Heterocyclic Amines. Xenobiotica **1**, 349 (1971).

236. WILSON, B. J., E. RAMSTAD, I. JANSSON, and S. ORRENIUS: Conversion of Agroclavine by Mammalian Cytochrome P-450. Biochim. Biophys. Acta **252**, 348 (1971).

237. THAL, C., M. DUFOUR, P. POTIER, M. JAOUEN, and D. MANSUY: Rearrangement of Vobasine to Ervatamine-Type Alkaloids Catalyzed by Liver Microsomes. J. Am. Chem. Soc. **103**, 4956 (1981).

238. STUART, K. L., J. P. KUTNEY, T. HONDA, and B. R. WORTH: Intermediacy of 3',4'-Dehydrovinblastine in the Biosynthesis of Vinblastine-Type Alkaloids. Heterocycles **9**, 1419 (1978).

239. POTIER, P.: La Reaccion de Polonovski Modificada. Rev. Latinoamer. Quím. **9**, 47 (1978).

240. CAVÉ, A., C. KAN-FAN, P. POTIER, and J. LE MEN: Modification de la réaction de Polonovski. Action de l'anhydride trifluoroacétique sur un aminoxyde. Tetrahedron **23**, 4681 (1967).

241. POTIER, P.: Synthesis of Bio-Active Substances: Recent Examples. In: Stereoselective Synthesis of Natural Products. Proc. 7th Workshop Conference Hoechst, Schloß Reisenburg (BARTMANN, W., and E. WINTERFELDT, eds.), **1978**, p. 19.

242. — Is the Modified Polonovski Reaction Biomimetic? In: Indole and Biogenetically Related Alkaloids (PHILLIPSON, J. D., and M. H. ZENK, eds.). London: Academic Press **1980**, p. 159.

243. STUART, K. L., J. P. KUTNEY, T. HONDA, and B. R. WORTH: Studies on the Biogenesis of Bisindole Alkaloids. The Final Stages in Biosynthesis of Vinblastine, Leurosine and Catharine. Heterocycles **9**, 1391 (1978).

244. STUART, K. L., J. P. KUTNEY, and B. R. WORTH: Studies on the Synthesis of Bisindole Alkaloids. XIV. Enzyme Catalysed Formation of Leurosine. Heterocycles 9, 1015 (1978).

245. LOUNASMAA, M., and A. KOSKINEN: Novel Applications of the Modified Polonovski Reaction. A Biomimetic Synthesis of Quinuclidines. Tetrahedron Lett. 23, 349 (1982).

246. WENKERT, E., and B. WICKBERG: General Methods of Synthesis of Indole Alkaloids. IV. A Synthesis of dl-Eburnamonine. J. Am. Chem. Soc. 87, 1580 (1965).

247. KOSKINEN, A., and M. LOUNASMAA: On the Structure of Ajmalinol. Heterocycles 19, 851 (1982).

248. ESMOND, R. W., and P. W. LE QUESNE: Biomimetic Synthesis of Macroline. J. Am. Chem. Soc. 102, 7116 (1980).

249. GARNICK, R. L., and P. W. LE QUESNE: Biomimetic Transformations among Monomeric Macroline-Related Indole Alkaloids. J. Am. Chem. Soc. 100, 4213 (1978).

250. HERLEM, D., Y. HUBERT-BRIERRE, F. KHUONG-HUU, and R. GOUTAREL: Réactions photochimiques d'amines tertiaires et d'alcaloïdes − III. Tetrahedron 29, 2195 (1973).

251. HUBERT-BRIERRE, Y., D. HERLEM, and F. KHUONG-HUU: Oxydation photochimique d'amines tertiaires et d'alcaloïdes − VI. Oxydation photosensibilisée d'alcaloïdes comportant un hétérocycle N-méthylé (nicotine, N-méthylanabasine, ajmaline). Tetrahedron 31, 3049 (1975).

252. MASAMUNE, S., S. K. ANG, C. EGLI, N. NAKATSUKA, S. K. SARKAR, and Y. YASUNARI: The Synthesis of Ajmaline. J. Am. Chem. Soc. 89, 2506 (1967).

253. MASHIMO, K., and Y. SATO: Synthesis of Isoajmaline. Tetrahedron Lett. 1969, 901.

254. — — Synthesis of Isoajmaline. Tetrahedron 26, 803 (1970).

255. — — On the Synthesis of Ajmaline. Tetrahedron Lett. 1969, 905.

256. KUTNEY, J. P., G. K. EIGENDORF, H. MATSUE, A. MURAI, K. TANAKA, W. L. SUNG, K. WADA, and B. R. WORTH: Total Synthesis of Dregamine and Epidregamine. A General Route to 2-Acylindole Alkaloids. J. Am. Chem. Soc. 100, 938 (1978).

257. BURKE, D. E., J. M. COOK, and P. W. LE QUESNE: Biomimetic Synthesis and Structure of the Bisindole Alkaloid Alstonisidine. Chem. Comm. 1972, 697.

258. — — — Biomimetic Synthesis of the Bisindole Alkaloids Villalstonine and Alstonisidine. J. Am. Chem. Soc. 95, 546 (1973).

259. LOUNASMAA, M., and A. NEMES: The Synthesis of Bis-Indole Alkaloids and Their Derivatives. Tetrahedron 38, 223 (1982).

260. WENKERT, E., D. W. COCHRAN, E. W. HAGAMAN, F. M. SCHELL, N. NEUSS, A. S. KATNER, P. POTIER, C. KAN, M. PLAT, M. KOCH, H. MEHRI, J. POISSON, N. KUNESCH, and Y. ROLLAND: Carbon-13 Nuclear Magnetic Resonance Spectroscopy of Naturally Occurring Substances. XIX. Aspidosperma Alkaloids. J. Am. Chem. Soc. 95, 4990 (1973).

261. JEWERS, K., D. F. G. PUSEY, S. R. SHARMA, and Y. AHMAD: ^{13}C-NMR Spectral Analysis of Rhazine, Quebrachamine and Rhazinilam. Planta Med. 38, 359 (1980).

262. DANIELI, B., G. PALMISANO, and G. S. RICCA: ^{13}C-NMR of Some Ajmalane Alkaloids. Tetrahedron Lett. 22, 4007 (1981).

263. HESSE, M.: Indolalkaloide, Teil 1. in Progress in Mass Spectrometry, vol. 1 (H. BUDZIKIEWICZ, ed.). Weinheim: Verlag Chemie. 1974.

264. BIEMANN, K.: Application of Mass Spectrometry to Structure Problems. IV. The Carbon Skeleton of Sarpagine. J. Am. Chem. Soc. 83, 4801 (1961).

265. — Mass Spectrometry. Organic Chemical Applications. p. 309. New York: McGraw-Hill. 1962.

266. BIEMANN, K., P. BOMMER, A. L. BURLINGAME, and W. J. MCMURRAY: High-Resolution Mass Spectra of Ajmaline and Related Alkaloids. J. Am. Chem. Soc. 86, 4624 (1964).

267. — — — — The High-Resolution Mass Spectra of Ajmalidine and Related Substances. Tetrahedron Lett. 1963, 1969.

346   A. KOSKINEN and M. LOUNASMAA: The Sarpagine-Ajmaline Group of Indole Alkaloids

268. GORMAN, M., A. L. BURLINGAME, and K. BIEMANN: Application of Mass Spectrometry to Structure Problems. The Structure of Quebrachidine. Tetrahedron Lett. **1963**, 39.

269. KLEINSORGE, H.: Klinische Untersuchungen über die Wirkungsweise des Rauwolfia-Alkaloids Ajmalin bei Herzrhythmusstörungen, insbesondere bei Extrasystolie. Med. Klin. **54**, 409 (1959).

270. PETTER, A., and K. ENGELMANN: Zur antiarrhythmischen Herzwirkung von Ajmalin. Arzneim.-Forsch. **24**, 876 (1974).

271. CASTELLUCCI, A.: Metabolic Effects of Bunaftine, a New Antiarrhythmic Agent: Comparison with Quinidine, Ajmaline, Procainamide, Xylocaine and Propranolol. Arzneim.-Forsch. **26**, 241 (1976).

272. HEEG, E.: Die Wirkung der *Rauwolfia*-Alkaloide Ajmalin, Rescinnamin und Reserpin auf den Katecholamingehalt des Herzens. Arzneim.-Forsch. **27**, 114 (1977).

273. KOCH, R.: Klinische und experimentelle Untersuchungen über Wirkung und Wirkdauer von N-n-Propylajmalinium-hydrogentartrat bei Herzrhythmusstörungen. Arzneim.-Forsch. **22**, 2079 (1972).

274. FEMMER, K., and K. GRÜNHEID: Vergleich der antiarrhythmischen Wirkung mit den hämodynamischen und kardialen Nebenwirkungen von Ajmalin, N-n-Propyl-ajmalin-hydrogentartrat und 4-[3'-Diäthylamino-2'-hydroxypropyl]-ajmalinhydrogentartrat (Tachmalcor®). Pharmazie **31**, 33 (1976).

275. HAVA, M.: The Pharmacology of *Vinca* Species and Their Alkaloids. In: The *Vinca* Alkaloids (TAYLOR, W. I., and N. R. FARNSWORTH, eds.), pp. 305 – 338. New York: Marcel Dekker, Inc. 1973.

276. HARADA, M., Y. OZAKI, and H. OHNO: Effects of Indole Alkaloids from *Gardneria nutans* Sieb. *et* Zucc. and *Uncaria rhyncophylla* Miq. on a Guinea Pig Urinary Bladder Preparation *in situ*. Chem. Pharm. Bull. **27**, 1069 (1979).

277. HARADA, M., and Y. OZAKI: Effect of *Gardneria* Alkaloids on Ganglionic Transmission in the Rabbit and Rat Superior Cervical Ganglia *in situ*. Chem. Pharm. Bull. **26**, 48 (1978).

278. — — Effect of Indole Alkaloids from *Gardneria* Genus and *Uncaria* Genus on Neuromuscular Transmission in the Rat Limb *in situ*. Chem. Pharm. Bull. **24**, 211 (1976).

279. SVOBODA, H., and D. A. BLAKE: The Phytochemistry and Pharmacology of *Catharanthus roseus* (L. G. DON). In: The *Catharanthus* Alkaloids (TAYLOR, W. I., and N. R. FARNSWORTH, eds.), pp. 45 – 83. New York: Marcel Dekker, Inc. 1974.

280. BARTLETT, M. F., B. F. LAMBERT, H. M. WERBLOOD, and W. I. TAYLOR: *Rauwolfia* Alkaloids. XLIII. A Facile Ring Closure of Deoxyajmalal-A to Deoxyajmaline. J. Am. Chem. Soc. **85**, 475 (1963).

281. HOBSON, J. D., and J. G. MCCLUSKEY: Cleavage of Tertiary Bases with Phenyl Chloroformate: The Reconversion of 21-Deoxyajmaline into Ajmaline. J. Chem. Soc. (C) **1967**, 2015.

282. HOLLAND, H. L.: Microbial and *in vitro* Enzymic Transformation of Alkaloids. In: The Alkaloids (MANSKE, R. H. F., and R. G. A. RODRIGO, eds.), vol. XVIII, 323. New York: Academic Press. 1981.

283. PINCUS, M. R., and H. A. SCHERAGA: Theoretical Calculations on Enzyme-Substrate Complexes: The Basis of Molecular Recognition and Catalysis. Acc. Chem. Res. **14**, 299 (1981).

284. KOSKINEN, A., and M. LOUNASMAA: Biogenesis of the Ajmaline Type Alkaloids. Planta Med. **45**, 248 (1982).

(Received June 23, 1982)

Author Index

Page numbers printed in *italics* refer to References

348 Author Index

Subject Index

By

A. SIEGEL, Wien

Fortschritte der Chemie organischer Naturstoffe

Progress in the Chemistry of Organic Natural Products

Volume 42:

1982. VII, 323 pages.
Cloth DM 164,—; ISBN 3-211-81706-9

Contents: Y. ASAKAWA: Chemical Constituents of the Hepaticae. — M. HEIDELBERGER: Cross-Reactions of Plant Polysaccharides in Antipneumococcal and Other Antisera, an Update.

Volume 41:

1982. 37 figures. VIII, 373 pages.
Cloth DM 196,—; ISBN 3-211-81690-9

Contents: E. HASLAM: The Metabolism of Gallic Acid and Hexahydroxydiphenic Acid in Higher Plants. — D. G. ROUX and D. FERREIRA: The Direct Biomimetic Synthesis, Structure and Absolute Configuration of Angular and Linear Condensed Tannins. — ST. J. GOULD and ST. M. WEINREB: Streptonigrin. — D. J. ROBINS: The Pyrrolizidine Alkaloids. — J. W. DALY: Alkaloids of Neotropical Poison Frogs (Dendrobatidae).

Volume 40:

1981. 21 figures. IX, 295 pages.
Cloth DM 158,—; ISBN 3-211-81624-0

Contents: P. LEFRANCIER and E. LEDERER: Chemistry of Synthetic Immunomodulant Muramyl Peptides. — SUKH DEV: The Chemistry of Longifolene and Its Derivatives. — W. HELLER and CH. TAMM: Homoisoflavanones and Biogenetically Related Compounds. — R. G. COOKE and J. M. EDWARDS: Naturally Occurring Phenalenones and Related Compounds. — C. W. JEFFORD and P. A. CADBY: Molecular Mechanisms of Enzyme-Catalyzed Dioxygenation (An Interdisciplinary Review).

Volume 39:

1980. 5 figures. XI, 316 pages.
Cloth DM 158,—; ISBN 3-211-81530-9

Contents: B. FRASER-REID and R. C. ANDERSON: Carbohydrate Derivatives in the Asymmetric Synthesis of Natural Products. — H. JONES and G. H. RASMUSSON: Recent Advances in the Biology and Chemistry of Vitamin D. — S. LIAAEN-JENSEN: Stereochemistry of Naturally Occurring Carotenoids. — T. KASAI and P. O. LARSEN: Chemistry and Biochemistry of γ-Glutamyl Derivatives From Plants Including Mushrooms (Basidiomycetes).

Volume 38:

1979. 5 figures. VII, 430 pages.
Cloth DM 195,—; ISBN 3-211-81529-5

Contents: R. W. FRANCK: The Mitomycin Antibiotics. — N. H. FISCHER, E. J. OLIVIER, and H. D. FISCHER: The Biogenesis and Chemistry of Sesquiterpene Lactones.

Volume 37:

1979. 8 figures. IX, 367 pages.
Cloth DM 178,—; ISBN 3-211-81528-7

Contents: J. M. BRAND, J. CHR. YOUNG, and R. M. SILVERSTEIN: Insect Pheromones: A Critical Review of Recent Advances in Their Chemistry, Biology, and Application. — M. McNEIL, A. G. DARVILL, and P. ALBERSHEIM: The Structural Polymers of the Primary Cell Walls of Dicots. — U. SCHMIDT, J. HÄUSLER, ELISABETH ÖHLER, and H. POISEL: Dehydroamino Acids, α-Hydroxy-α-amino Acids and α-Mercapto-α-amino Acids.

All Volumes and Cumulative Index 1—20 available
Price reduction for subscribers: 10%.

Special reduced price (20% reduction) for the complete Series Vols. 1—42 incl. the Cumulative Index to Vols. 1—20

Springer-Verlag Wien New York